Fair Weather Travel in Western Europe

Fair Weather Travel in Western Europe

by EDWARD D. POWERS

with the editorial assistance of
ROBERT D. POWERS

Mason & Lipscomb PUBLISHERS NEW YORK

Copyright © 1974 by Edward D. Powers

ISBN: 0–88405–076–9

Library of Congress Catalog Card Number:

Printed in the United States of America

First Printing

Library of Congress Cataloging in Publication Data

Powers, Edward.
   Fair weather travel in western Europe.

    1. Europe—Climate. 2. Europe—Description and travel—1945- —Guide-books. I. Title.
QC989.A1P68   551.6′9′4   74–5227
ISBN 0–88405–076–9

To Bob and Dave, may they enjoy
fair weather in all of their travels

Contents

List of Charts

Foreword

HAVE you ever tried to find a book dealing with "comfort weather"? For instance, exactly where and when can the traveler, tourist, or retiree enjoy the most ideal vacation weather in every nook and cranny of Europe?

That's what makes Fair Weather Travel a unique phenomenon—a one-of-a-kind article. Where else can you learn:

1. Which parts of Europe get 50 per cent more sunshine during the off-season months of May and June than in the busy July-August period?

2. Which canton of Switzerland enjoys a subtropical, palm tree climate?

3. How to choose the best European weather in any season?

4. Which side of Ireland gets the most sunshine and least rain—and why?

5. What are the swimming water temperatures in Europe?

6. Where is the best off-season weather in France?

7. How many hours of sunshine and daylight there are in various parts of Europe?

The potential visitor may ask dozens of questions, but the one most often heading the list is: "Where and when can I find the most pleasant weather?" with an increasing interest in the lower cost, less crowded, off-season periods. Skim through this volume and you may conclude that those answers and many more can be found between these covers.

Introduction

To quote some unknown wag, "America's two major problems are—how to control the waistline and where to park your car." By adding a third, "How to avoid vacation crowds," we get a picture of the developing situation in Europe and the need for this book. Just as in North America, the months of July and August have traditionally been the favored vacation season abroad.

But vacation patterns are rapidly changing. The greatly liberalized working schedules are making it possible for many more people to travel farther from home. That has resulted in ever-increasing crowds and mounting costs in resort areas during the most popular periods. Hurried service, more limited choice of accommodations, and the necessity of booking far, far in advance are only a few of the many gripes that are causing the search for less busy times and places.

Most people have at least some vague idea of what the weather may be in the various vacation areas during the normal popular seasons but, generally, know little of conditions throughout the remainder of the year. Off-season vacations are a real gamble and often a sad disappointment unless there has been a little pre-trip investigation of the weather prospects. On the other hand, there can be some very pleasant rewards in store for the prudent fellow who has done his homework.

As an example, visitors from North America constantly complain that Western Europe is often gray and rainy in autumn when they know it is the dependably clear, sunny season at home.

Now let's see what even a casual observation will disclose. The City Weather Table of Dublin shows that April and May are the driest and sunniest months of the year while August and September are usually the wettest with almost twice as much rainfall. Amsterdam, Paris, Copenhagen, and much of the west coast of Europe follow this pattern. Score one for a little research and off-season travel. Reading this book certainly will not guarantee perfect vacation weather, but it most assuredly will help establish more favorable odds.

It has been arranged so that the reader can either make a detailed weather analysis of a whole area or use it as a quick, ready reference handbook for finding specific information.

We strongly urge that you compare the information in the text and the City Weather Tables arranged in alphabetical order in the back of the book with that of your home area or a place with which you are familiar. That method will make it much easier to visualize the weather conditions in a foreign city or country.

Bare statistics are often a bit difficult to absorb. It is usually more informative, for instance, to know that the weather in Lisbon is akin to San Francisco's or that of Madeira matches the Southern Coast of California.

In order to illustrate the overall weather patterns of Europe, the Continent is first described by its four major climatic zones. The characteristics of each are explained in simple untechnical language with the reasons for the various conditions. Each individual country is then covered very completely with text, city weather tables, and pictorial maps which can be understood at a glance. These vacation weather maps will be of particular interest. They are really pictures which show the best vacation areas during the spring, summer, autumn, and winter seasons in each country.

We suggest that this volume of weather stories be used in conjunction with a good travel guide and the experienced advice of your travel agent. If it helps to make your next vacation or travels more enjoyable and a bit less of a weather gamble, it will have served a good purpose.

Overall Climatic Patterns
and What Makes Weather

MANY readers will turn first to the detailed weather stories of the individual countries. That's one good way to use this book. It is important, however, to realize that there are great differences between the climates of Europe and North America. These dissimilarities can be more readily understood by becoming familiar with the overall weather patterns of the whole European Continent as described in this chapter.

Comparative Latitudes of Europe and America

It is well to start by comparing the relative locations of these two large land masses. If one were to consider only the important item of latitude, a logical conclusion might be that Europe's climate must range from a chilly temperate to perhaps a sharp frigid. A glance at the globe shows that Europe is a lot farther north than most Americans realize. Actually, large portions of Norway, Sweden, and Finland extend well above the Arctic Circle.

The figures 35° to 75° N, Europe's south to north geodetic boundaries, probably mean little more to most of us than those mysterious symbols that pop out of an IBM computer. By following across the horizontal lines, however, we find that sunny Naples is a bit farther north of the Equator than New York City. London is on the parallel of latitude that runs through the middle of Newfoundland. We also learn with a shiver that the fur-clad Eskimo in Ketchikan, Alaska, is only one half a degree closer to the North Pole than the kilted Scot in Edinburgh. Chart 1 shows the relative locations of the two continents.

If armed only with that scant information, the American traveler might prepare for an overseas jaunt much as if he were heading for the Yukon Territory or Labrador. Both, incidentally, are within the same zones of latitude as northern Europe. The similarity, however, ends there. As usual, too little knowledge can be misleading. To develop a more complete and useful weather picture, it is necessary to consider many angles. While distance from the Equator is generally the most important single factor related to climate, there are many local situations where other elements may have an even greater influence on the climatic conditions.

The topography of a country, especially the altitude, has an important bearing on whether the days will be hot or cold, wet or dry, sunny or gray. Winds and ocean currents also help dictate conditions. Although the west coasts of Europe and North America are subjected to almost identical warm ocean currents and westerly breezes, the results are quite different. That is because there is little similarity in the conformation of the lands nearest the coasts.

Charts 2 and 3 show that the tremendous circulating systems in the Atlantic and Pacific oceans are almost carbon copies of one another.

The Atlantic and Pacific Ocean Currents

The Kuroshio (commonly called the Japan) Current becomes the North Pacific Drift as it flows eastward toward America. On approaching our shores, it divides. One branch heads northward, tempering the Alaskan climate and enabling our forty-ninth state to produce the largest vegetables of any grown in the United States. The other arm turns southward and, midway down the United States coast line, is temporarily converted from a warm to a cool ocean stream. The change lasts from about March through July and is responsible for the very equable climate along the Southern Coast of California. A person familiar with the American Pacific Coast can readily equate it with that of Western Europe. Residents of the British Columbia, Washington State, or Oregon coasts would feel quite at home weatherwise anywhere along the southwestern shores of England and Ireland. The upper half of the California coast gets much the same summer sunshine and gray, dripping colder seasons as The Netherlands and Belgium. The short piece of Portugal coast facing southward is quite similar to the lower coast of California.

The Gulf Stream

This mighty thermal stream, which Europeans refer to as "America's finest export," dominates the climate of many countries included in this volume. Any single factor of such importance deserves a little scrutiny.

Shortly after leaving the American shores, the Gulf Stream loses its identity and becomes the North Atlantic Drift. At about the Azores, it divides and the lesser branch, called the Portuguese Current, flows southward. A still smaller offshoot, known as the Rennell Current, bends into the Bay of Biscay and circulates upward along the coasts of Spain, France, England, and Ireland.

ITS EFFECT ON EUROPE

A minor portion of these still-warm waters enters the Mediterranean Sea through the Strait of Gibraltar. It is easy to understand why the

LATITUDES: COMPARISON OF EUROPE AND AMERICA

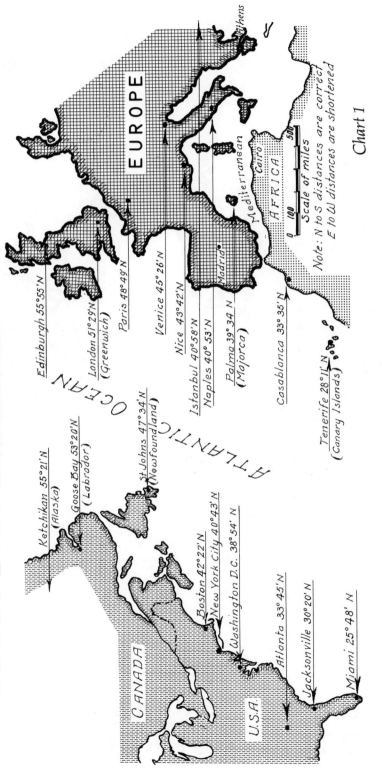

EUROPE

AFRICA

Mediterranean

Cairo

Athens

Scale of miles

0   100   500

Note: N to S distances are correct
E to W distances are shortened

Edinburgh 55°55'N

London 51°29'N
(Greenwich)

Paris 48°49'N

Venice 45°26'N

Nice 43°42'N

Istanbul 40°58'N

Naples 40°53'N

Palma 39°34 N
(Majorca)

Madrid

Casablanca 33°35'N

Tenerife 28°11'N
(Canary Islands)

ATLANTIC OCEAN

CANADA

U.S.A.

Ketchikan 55°21'N
(Alaska)

Goose Bay 53°20'N
(Labrador)

St Johns 47°34'N
(Newfoundland)

Boston 42°22'N

New York City 40°43'N

Washington D.C. 38°54' N

Atlanta 33°45'N

Jacksonville 30°20'N

Miami 25°48' N

Chart 1

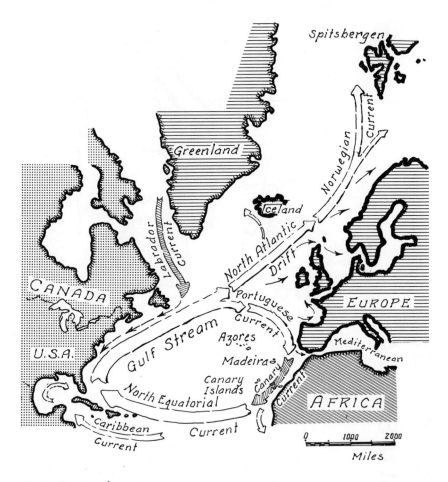

Refer to Chart 3
Note remarkable similarity of the
Atlantic & Pacific Ocean Currents,
which coupled with equally similar
wind patterns, produce almost
duplicate climatic conditions on
the west coasts of Europe and
North America.
The above is a simplified diagram.
Warm ocean current
Cool ocean current

NORTH ATLANTIC
OCEAN CURRENTS

Chart 2

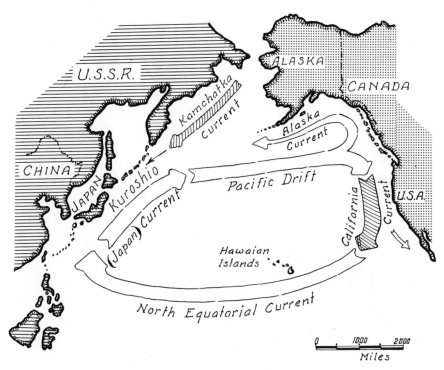

Miles

See Chart 2
Note This huge circulating
system almost duplicates that
of the Atlantic. The result is
that the west coasts of Europe
and North America have
almost exactly the same types
of climate. Being familiar with
one makes it easy to under
stand the other.

Cool current
Warm current

The above is a simplified
diagram.

NORTH PACIFIC
OCEAN CURRENTS

Chart 3

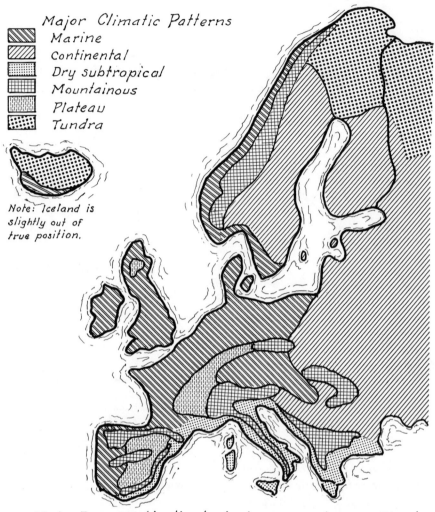

Major Climatic Patterns

- Marine
- Continental
- Dry subtropical
- Mountainous
- Plateau
- Tundra

Note: Iceland is slightly out of true position.

Marine Zone: Equable climate due to sea & sea breezes. Blends gradually into Continental Zone. Summers cool, winters not severe. Maximum sunshine in late spring & early summer. Heaviest precipitation during autumn & early winter.

Continental Zone: Subject to rapid & sometimes extreme daily & annual weather changes. Uniform precipitation but usually a little drier in spring & early summer.

Dry Subtropical Zone: Sunny. Hot, dry summers. Mild winters with moderate rainfall. Spring especially delightful

Mountainous Zone: Climate varies with altitude, exposure to southern sunshine & protection from winter winds & storms.

Plateau Zone: Tends to wide temperature extremes.

Tundra Zone: Winters long & severe. Summers short, warm in daytime, cool or cold at night.

CLIMATIC ZONES

Chart 4

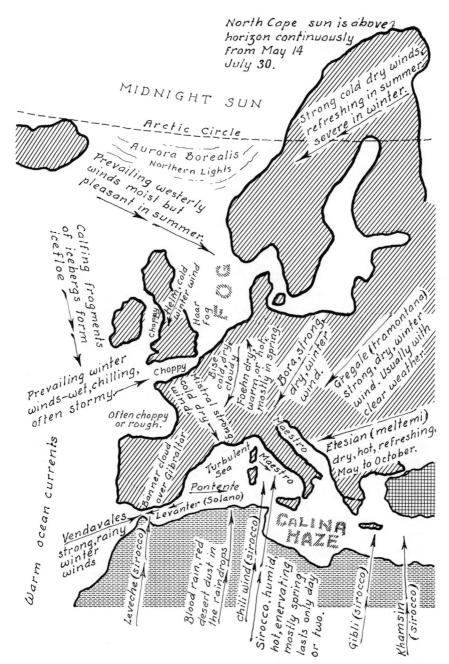

North Cape sun is above horizon continuously from May 14 July 30.

MIDNIGHT SUN

Arctic Circle

Aurora Borealis
Northern Lights

Prevailing westerly winds moist but pleasant in summer.

Strong cold dry winds, refreshing in summer, severe in winter.

Calfing fragments of icebergs form icefloe

Prevailing winter winds–wet, chilling, often stormy.

Helm, cold winter wind

Haar cold Fog

Choppy

Choppy

Warm ocean currents

Often choppy or rough.

Bise cold, dry cloud

Mistral strong cold, dry wind.

Foehn dry, warm or hot, mostly in spring.

Bora, strong, dry, winter wind.

Gregale (tramontana) strong, dry, winter wind. Usually with clear weather.

Banner cloud over Gibraltar

Turbulent Sea

Maestro

Maestro

Etesian (meltemi) dry, hot, refreshing, May to October.

Pontente

Levanter (Solano)

Vendavales strong, rainy winter winds

Leveche (sirocco)

Blood rain, red desert dust in the raindrops

Chili wind (sirocco)

Sirocco, humid, hot, enervating mostly spring lasts only a day or two.

CALINA HAZE

Gibli (sirocco)

Khamsin (sirocco)

Local air currents are described in the text. In some areas, land-sea & mountain-valley (both diurnal breezes) are actuated by heat of the sun. These air movements can produce very pleasant summer conditions.

WINDS

Chart 5

south coast of Portugal enjoys almost a subtropical climate even though it is almost as far north as Philadelphia.

The Portuguese Current becomes the Canary Current just before it reaches the Canary Islands. At this point, it changes from a warm to cool ocean stream just as happens along the California coast. So similar are the situations that Las Palmas in the Canary Islands and San Diego, California, could almost use the same weather reports.

This thermal phenomenon, Mother Nature's own gigantic air-conditioning system, is caused by a dual air-ocean activity called "upwelling." It happens when a particular combination of winds and the earth's rotation force the surface water along a western coast to move offshore. The void created by that action is filled with cold water from below. This, fortunately, occurs from March through July.

The North Atlantic Drift and Norwegian Current

The major branch of the North Atlantic Drift, veering north, surges around Ireland and Great Britain, washing the shores of almost all northwestern Europe. As a result of its soothing influence, palm trees and cacti flourish outdoors along the southwest coasts of England, Ireland, and would you believe—even a few spots up in Scotland. This amazing ocean stream then becomes the Norwegian Current. It may seem a bit incongruous that there should be a warm Norwegian Current, but it is helpful to know about it. This benevolent stream gives Norway and even Iceland open-water ports throughout the winter. While the warm ocean currents and the prevailing westerly winds would each individually influence the land climates, the synergistic effect of these combined forces multiplies the results many times.

Affect of Ocean Currents on Climate

In North America, the sea-air currents produce a marine-type climate along only a narrow strip of Pacific coastal land. The almost continuous chains of mountains extending from Alaska to Mexico form a barricade preventing this mild humid air from traveling inland. As a result, there is a sharp separation between two distinctly different types of climate.

In Europe, on the other hand, the moist ocean breezes penetrate far into the interior of the Continent almost without obstruction. This moderating influence is felt with decreasing effect as the air travels inland from the ocean. Conversely, in the absence of mountain barriers, the cold, sharp northeasterly blasts occasionally surge in from Siberia, particularly during winter.

The American tourist should remember that there is much more likelihood of experiencing an exceptionally hot or cold year in Europe than the United States. That is why it is not necessary to classify wines by vintage over here.

The Marine Climate Zone

This large segment encompasses much of Europe, which is of greatest interest to overseas visitors. Chart 4 shows that it includes Great Britain, Ireland, Belgium, the Netherlands, Denmark, much of France, Germany, Switzerland, Austia, substantial portions of mid-eastern Europe, and a bit of the Balkan peninsula. Also within this zone are the northern strip of Spain facing the Atlantic, the lower half of coastal Norway, the southwestern tip of Sweden, and even the southwestern shores of Iceland.

It is remarkable that this vast and important region should have an overall climate so very different from what its latitude says it should be. This marine weather zone of Western Europe is generally considered the most marked temperature anomaly in the world. Those portions of this wide expanse most directly exposed to the tempering sea breezes experience an unusual positive abnormity when the January temperature is 12 to 20° higher than almost all other places around the world, on the same parallel of latitude. The effect is not as pronounced during summer. Most overseas visitors will find the winter season in this zone milder than expected and the summers cooler.

The total annual precipitation throughout this whole marine zone is in the 20- to 35-inch range. Some of the western coastal portions which are most directly exposed to the moist sea breezes, however, will average 35 to 60 inches a year. This includes the northwestern corner of Portugal and Spain, the western edges of Ireland, Great Britain, Norway, and other segments of vulnerable shore line. There are a few particularly wet sections that are deluged with anywhere from 60 to 80 inches. As an example, the Bergen area on the west coast of Norway averages a whopping 79 inches as does Fort William in Scotland.

Precipitation in this marine weather zone is predominantly in the form of rain, particularly close to the shore which sees very little snow. There is much mist and drizzle which spell many gray, overcast days. Short heavy showers are not common with not more than five thunderstorms a year near the coast but usually about 20 farther inland.

Sunshine, especially near the ocean, is not very plentiful, being most shy during winter. Dublin, as an example, averages about 6.5 hours of sunshine a day in May, the brightest month, but only 1.5 in December. Throughout the western portion of this zone, the hours of sunshine generally decrease progressively from spring, which enjoys the most, to winter, which gets the least. The heaviest precipitation in the coastal areas is from late summer or autumn to the latter part of winter.

INLAND SEGMENT OF THE MARINE ZONE

As we move inland, the weather pattern gradually changes. While the spring season continues to be the driest part of the year and the fall rather wetter, the precipitation is usually more or less uniformly distributed throughout the year. Still farther inland, summers and early fall are often the wettest periods, which is quite the opposite of the wet winter, drier summer weather near the coast.

Lyon, in eastern France, averages 2.8 to 3 inches of precipitation per month from May to October but only 1.5 to 2.5 inches per month the remainder of the year. The total annual precipitation in this French city is under 29 inches. (Boston gets 41 inches, Houston 46, Jacksonville 52, and New Orleans 57.) Since Lyon is in the midst of France's famous Burgundy wine region, you can be assured of enough late summer and autumn sunshine to produce the sugar content required in the ripening of grapes for good wine. That point is well to remember. Sometimes the type of agriculture or even the farm animals can tell one almost as much as the weatherman—or this book.

Weather Differences Between Coast and Interior

We see how the climate gradually changes from the wet, cold but not usually extreme winters and cool, fairly sunny summers along the coast to the more temperate interior parts where the temperature ranges are much wider.

Munich, well to the east and at 1,807 feet above sea level, gathers a 37-inch annual precipitation, which is heaviest in summer. June, July, and August get 4.8, 6.0, and 3.8 inches, respectively, but there is a steady decrease to 1.4 in February. In spite of the rather high precipitation, there are 228 days a year when there is less than 0.04 inch of precipitation. That does not mean that they are all bright, glittering days, but sunshine is more plentiful here than toward the Atlantic.

Vienna (elevation 666 feet) continues the inland pattern. Only 1.5 inches of precipitation falls in January as against 3 inches in July. Since it lacks Munich's altitude and is somewhat in the lee of shielding mountains, Vienna averages only 26 inches. This coupled with 265 almost rainless days per year, helps establish the Austrian capital as one of the gay, bright cities of the world.

SNOW IN THE MARINE CLIMATE ZONE

As might be expected, the closer to the coast, the less snow and, of course, the quicker it melts. Most of the precipitation is in the form of mist and drizzle from dull, leaden winter skies. In such spots as Bordeaux, Plymouth and most of the south and west English coasts, and even up to Donegal in Ireland, there usually will not be more than three or four wet snowfalls a year—that quickly turn to slush and disappear. London, Dublin, and Amsterdam get 12 to 18 falls a winter, but it does not usually remain on the ground more than four or five days. Paris, which as the ballad correctly notes, "drizzles in the winter," does not get more than a dozen days a year with ground snow cover. Berlin averages about 35 such days and Vienna 44. Oslo, Stockholm, Moskow, and Leningrad, way up in the frigid penthouse of winter Europe, get from 50 to 70 snowstorms annually. That makes the generous consumption of neat aquavit and vodka quite understandable.

Mountains and High Elevation Zone

There are many mountain chains and elevated complexes scattered across Europe. None, however, include such formidable climatic barricades as the Atlas, Andes, or Rockies. The most extensive in Europe are the mighty Alps' groups which occupy large portions of Switzerland and Austria as well as sizable segments of France, Germany, and Italy.

The high Massif Central plateau of southeastern France is another large area. There are also the Pyrenees that constitute a natural division between Spain and France. Other such lengthy chains are the Apennines, which form a backbone down the length of Italy, and several chains along the common boundary separating Norway and Sweden. There are many more plateaux and mountainous sections, all of which exert an influence upon the climate. In some cases, the effect is rather limited and may result only in a miniclimate confined to a local district. A more dramatic example is the Pyrenees range which hooks down across Portugal and divides Spain into distinct wet and dry segments.

Most of the mountain areas in northern Europe, but only the high ones in the south, are popular winter resort regions. Almost all the high sections of Europe are wonderful during summer and autumn. Considerable portions of the less elevated hill country are also very pleasant in spring time. It is well, however, to check the possibility of floods, very disagreeable muddy conditions, or even snow slides.

Sunshine is generally quite abundant in the mountains even during winter. The latter is, of course, much less true in northern Sweden, Norway, and Finland where the winter days are so very short. Because of the widely varying topographical conditions, there are no large overall climatic patterns in this high zone. In many places throughout this book, we have mentioned areas which experience particularly attractive or adverse weather conditions. That information, along with the detailed data in the City Weather Tables, will identify many of the more unusual situations in these high altitude regions.

The Continental Climate Zone

This large area is less affected by topography and air or sea currents than the two previously described zones. Its climate is more directly responsive to latitude than most other parts of Europe.

It is reasonably typical of the usual continental climate found in the interior sections of large land masses of the North Temperate Zone. The broad northcentral plains of the United States fit into this category.

There are wide diurnal and seasonal extremes of temperature, and sharp, sudden changes of weather are not unusual. The overall climate is considerably less equitable than in the Marine Climate Zone and there are much more apt to be quite hot summers and cold to very severe winters.

Precipitation is lower than toward the coast, but there can sometimes be a distinct local rainy season. Note in the City Weather Tables that, while precipitation may be heavier during a particular season, it is, in the overall, more uniformly distributed throughout the year than is the case in most parts of the United States. Sunshine is fairly plentiful, particularly during spring and early summer.

The Subtropical Climatic Zone

This includes just about all of the lands bordering on the Mediterranean Sea. In many places, such as portions of the French Riviera, the mountains run down almost to the water's edge. On the other hand, there are stretches of coast where relatively flat plains extend far inland. The balmy weather carries up into the foothills and, in many cases, is felt some distance up the valleys of rivers which empty into the Mediterranean.

This subtropical zone spans from Gibraltar to Turkey in Europe. It includes the Mediterranean coasts of Spain, France, Italy, and Greece. The short southernmost part of Portugal, facing on the Atlantic called the Algarve, is also part of this zone. Even in the Swiss canton of Ticino, on the southern slopes of the Alps, palm trees flourish outdoors, thanks to the soft Mediterranean breezes.

This is one of the major areas that attract the greatest number of overseas visitors. As a climatic benchmark for Americans, it is well to remember that the weather in this entire Mediterranean area is more akin to that of Jacksonville in northern Florida than to the more salubrious Palm Beach to Key West section.

Summers are generally dry and very sunny. Some portions experience rather frequent summer droughts while others are normally semi-arid. Winters are mild. While many parts are subject to showery winter periods, very few areas in this zone experience a really wet season.

This chapter covers only an overall look at the major climatic patterns. They form a huge patchwork across Europe without regard to political subdivisions. Becoming familiar with some of the factors that "cause weather" will make it much easier to understand the detailed weather stories of the individual countries which follow.

IRELAND

THE reader will note that we have included the six counties of Northern Ireland in this section even though they are a political segment of the United Kingdom. By taking the great liberty of considering the whole as one beautiful "Green Island," we are imitating the ways of Mother Nature who ignores man-made boundaries, which, on the whole, are far less stable than weather patterns. For some years, the tourist bureaus of the Irish Republic and Northern Ireland have cooperated in joint programs. Tourism is the greatest source of foreign currency in Ireland.

Topography

The island is approximately 300 by 180 miles, but the shore line is so irregular that few spots are more than 70 miles from some coast. The country is of limestone formation, almost continuously edged with low, nubby mountains and coastal highlands. The principal ranges are the Mourne Mountains in the northeast, the Wicklow Mountains to the south of Dublin, and the Macgillycuddy's Reeks in the Killarney section of the southwest. Much of the central plain area, which averages about 300 feet above sea level, is bog land. It also, however, contains many attractive lakes, some of which form part of the Shannon River.

Ireland is about the same size as Austria or the State of Maine. Of the 32,500 square miles, 26,600 make up the Republic of Ireland, while the remainder constitutes the six northern counties. The visitor who wants to really see and get to know the country should not be misguided by these "vital statistics." There are more than 2,000 miles of magnificent coast line and many of the most scenic and enjoyable places are far out on rugged points or hidden in small, off-the-highway coves.

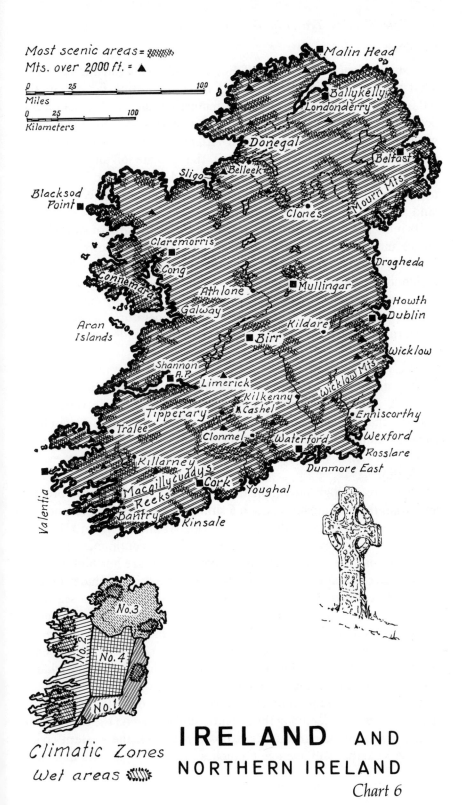

Most scenic areas =
Mts. over 2,000 ft. = ▲

0   25   100
Miles

0   25   100
Kilometers

Malin Head

Ballykelly
Londonderry

Donegal

Belfast

Sligo
Belleek

Mourn Mts

Blacksod
Point

Clones

Claremorris

Drogheda

Cong

Connemara

Athlone

Mullingar

Galway

Howth
Dublin

Aran
Islands

Kildare

Wicklow

Birr

Shannon
A.P.

Wicklow Mts

Limerick

Kilkenny

Tipperary

Cashel

Enniscorthy

Tralee

Clonmel

Waterford

Wexford

Killarney

Rosslare

Macgillycuddys
Reeks

Cork

Dunmore East

Youghal

Valentia

Bantry

Kinsale

No.3

No.2

No.4

No.1

Climatic Zones
Wet areas

IRELAND AND
NORTHERN IRELAND

Chart 6

Climate

A single glance at the map of Europe will give a fair indication of what the overall climate of Ireland must be. Sitting far out in the Atlantic, it is fully exposed to the tempering effects of a mammoth ocean current which surges in from the southwest. This benevolent water body is the warming North Atlantic Drift (a continuation of the Gulf Stream), which moderates the climate of almost all the coastal areas of the country. The year-round lush green of the Killarney lakes region shows how a warm water body can affect adjacent land areas.

Its influence is most pronounced along the southwestern shore lands where snow and freezing weather are practically unknown. Coupled with this marine action are the prevailing westerlies which pick up heat and moisture as they blow in off the ocean. In the Temperate Zone, this combination produces a marine climate that is unusually equable, with few temperature extremes. Average winter temperatures, as in much of Western Europe, are 15 to 20° higher than the latitude would indicate they should be.

After reading so much about the warm ocean currents, the visitor may be tempted by some of the lovely little sand beaches. Beware: Either a glance at a water thermometer or a toe in the water may quickly change one's mind. I've shivered many times just watching rugged Irishmen braving the surf. With a midsummer high of 60°, who other than Celts, Vikings, and polar bears, would really enjoy even a quick dip?

There are no high mountain ranges which could barricade breezes and drain them of moisture. Such a situation often causes the lands on the leeward sides to be dry or semiarid. As indicated on Chart 6, mountains and elevated lands form an almost continuous chain or frame around the outer edges of the country but are not high enough to have any climatic significance. Although this ring includes quite a few 2,000- to about 3,000-foot rounded peaks, they are rather scattered and do not prevent the moisture-laden air currents from traveling completely across the island. This allows the interior to be clothed in the same rich, green vegetation as the coastal fringes. As a consequence of the dual sea-air combination, climatic conditions are remarkably uniform throughout the entire country.

Total annual precipitation ranges from about 30 to 40 inches, being generally a bit heavier along the Atlantic shores than on the eastern side of the island. Some particularly exposed places on the western coast, such as Valentia, will average 60 inches or more a year. As indicated on the City Weather Tables, the wettest portions of the year are late summer, fall, and early winter. The months of March, April, May, and June usually receive the lightest rainfall.

It is not surprising that hours of sunshine follow much the same pattern. There is a low of 1.5 to 2 hours of sunshine per day in December and January but an overall increase to 5.2, 6.5, and 5.9 hours during April, May, and June. While these figures may not seem particularly high, Irish springtime is a delightfully bright and pleasant season to enjoy the countryside.

The tables on pages 18 and 19 show the average number of hours of sunshine per day each month in 12 locations throughout Ireland. It also lists the number of days that there will be no sunshine. We have included this rather detailed tabulation because many prospective overseas visitors think of the Emerald Isle as a gray, misty land. Most do not realize that, particularly in springtime, there may be as many hours of sunshine as in their home areas. You will note on the Sunshine Table, that while these places see the sun only 1.5 or 2 hours a day in midwinter, the months of April, May, and June will get a respectable 5 to 7.5 hours of sunshine a day. That is not the Sahara, but it is mighty pleasant weather. Incidentally, the northeastern United States averages about the same hours of sunshine during that period.

In general, temperatures in Ireland graduate downward from the southwest to the northeast. In January, the readings in Killarney will be 4 or 5° higher than Belfast, although some of the interior portions of the country less affected by sea breezes will be a few degrees colder than the northern capital. There is practically no zero weather, and the infrequent wet snowfalls usually melt rather quickly.

The summer temperatures are only a few degrees higher in the southwest than the northeast. Readings above 90° are a rare occurence; the visitor need have little more concern about heat prostration than snakebite.

Ireland enjoys a remarkably mild climate considering its location on the globe. The southernmost point in the country is north of the entire United States, except Alaska, and is, in fact, closer to the North Pole than practically all of Newfoundland, Canada. Malin Head, way up at the top of the Emerald Isle, is on almost the same parallel of latitude as Ketchikan, Alaska. In view of this, it may seem strange that palms and cacti thrive out of doors in some parts of southwestern Ireland.

The factors that "make" Irish weather, however, are few and not difficult to understand. The occasional arctic winds, blasting in from the northeast, are greatly tempered as they travel over the land masses of Europe and are much less severe by the time they reach Ireland. As a

SUNSHINE TABLE

A—Average hours of sunshine per day

B—Average number of days with no sunshine

		Dublin	Cork	Shannon	Rosslare	Kilkenny	Clones
Jan.	A	2.0	1.9	1.8	2.2	1.9	1.8
	B	11.0	12.0	12.0	11.0	13.0	12.0
Feb.	A	2.5	2.5	2.6	2.6	2.5	2.5
	B	7.0	8.0	6.0	7.0	7.0	7.0
Mar.	A	3.6	3.3	3.4	4.0	3.4	3.2
	B	6.0	7.0	7.0	6.0	8.0	8.0
Apr.	A	5.0	5.1	5.1	5.8	4.8	5.0
	B	3.0	4.0	4.0	3.0	4.0	3.0
May	A	6.2	6.3	6.6	7.5	6.3	6.1
	B	2.0	2.0	1.0	2.0	1.0	2.0
June	A	6.0	6.3	6.0	7.5	5.9	5.5
	B	3.0	2.0	2.0	2.0	2.0	3.0
July	A	4.8	5.3	4.9	6.2	4.6	4.1
	B	2.0	2.0	2.0	1.0	2.0	3.0
Aug.	A	4.9	5.1	5.1	6.0	5.0	3.8
	B	2.0	2.0	2.0	1.0	1.0	3.0
Sept.	A	3.9	4.2	3.7	5.1	4.0	3.4
	B	3.0	3.0	3.0	3.0	3.0	5.0
Oct.	A	3.2	3.0	2.8	3.8	2.9	2.5
	B	6.0	6.0	6.0	5.0	7.0	7.0
Nov.	A	2.1	2.2	2.1	2.6	2.4	1.7
	B	8.0	7.0	8.0	8.0	9.0	11.0
Dec.	A	1.6	1.6	1.4	2.2	1.9	1.2
	B	10.0	11.0	11.0	10.0	10.0	12.0
Year	A	3.8	3.9	3.8	4.6	3.8	3.4
	B	63.0	66.0	64.0	59.0	67.0	76.0

Sunshine Table

A—Average hours of sunshine per day

B—Average number of days with no sunshine

		Valentia	Birr	Clare-morris	Bel-mullet	Mullin-gar	Malin Head
Jan.	A	1.6	1.8	1.8	1.9	1.9	1.3
	B	10.0	11.0	11.0	9.0	12.0	10.0
Feb.	A	2.8	2.5	2.6	2.5	2.6	2.2
	B	8.0	7.0	8.0	7.0	7.0	7.0
Mar.	A	3.5	3.4	3.4	3.4	3.5	3.5
	B	6.0	6.0	7.0	5.0	7.0	5.0
Apr.	A	5.2	4.4	4.9	5.2	5.0	5.1
	B	5.0	3.0	3.0	3.0	3.0	3.0
May	A	6.5	6.0	6.3	7.0	6.3	6.7
	B	2.0	1.0	2.0	2.0	1.0	1.0
June	A	5.9	5.3	5.4	6.0	5.8	5.9
	B	3.0	2.0	2.0	2.0	2.0	2.0
July	A	4.7	4.1	4.1	4.6	4.5	4.6
	B	3.0	3.0	3.0	3.0	2.0	3.0
Aug.	A	4.9	4.7	4.5	5.1	5.0	4.7
	B	4.0	2.0	4.0	3.0	2.0	3.0
Sept.	A	3.8	3.9	3.3	3.9	3.7	3.8
	B	4.0	3.0	5.0	5.0	4.0	4.0
Oct.	A	2.8	2.9	2.5	3.0	2.9	2.3
	B	6.0	5.0	7.0	6.0	6.0	5.0
Nov.	A	2.0	2.3	2.0	1.9	2.1	1.7
	B	8.0	9.0	9.0	8.0	9.0	7.0
Dec.	A	1.3	1.8	1.3	1.3	1.5	1.0
	B	12.0	10.0	10.0	10.0	11.0	11.0
Year	A	3.7	3.6	3.5	3.8	3.7	3.6
	B	71.0	62.0	71.0	63.0	66.0	61.0

result, the overall climate is dominated by the warmish prevailing westerly winds.

Summers in Ireland are almost never as hot as most parts of Europe, while winters are surprisingly mild with temperature readings well above most other places in the same latitude. In describing weather, just as with many things, we refer to the normal or what statistically has happened over a long period of years. But even Ireland's record of equable weather is occasionally interrupted by an unusually hot summer or cold winter. While these may seem very harsh and work a real hardship on the residents, they are seldom as severe as the weather we are accustomed to in the northern parts of the United States.

There are many reasons for the periodic abnormal season. Sometimes, great numbers of icebergs drift unusually far south producing cold, foggy weather. If the wind shifts and blows from the northeast for an extended period, there can be a cold spell lasting many weeks. Conversely, an especially dry summer or continuous air currents from the southeast can result in hot, sunny weather. Usually, however, the visitor can reasonably expect to experience just about the conditions indicated in the City Weather Tables.

The prudent traveler, no matter in what country, always has a mental alternative plan of activities to substitute if outdoor conditions are not pleasant. And who can name a place that does not get some disagreeable weather? Fortunately, no visitor need pace the hotel room or lobby at such times since there is much of indoor interest everywhere, especially Dublin.

May we note that one of the many delights of this gracious land is its inviting pubs—easy to find in even the smallest hamlet—and a delightful haven on a rainy day. Where in America can one enjoy many hours of interesting stories and conversation over a single pint of ale? There is no law in Ireland limiting the intake, but neither will the barman or barmaid snatch away the scarcely emptied mug.

Although there are no extremely wide weather variations, meterologists most often consider Ireland in terms of four climatic zones. This is illustrated on the small map in Chart 6. Dublin is, of course, the major tourist attraction, but each area of the country has its own particular characteristics, scenery, and way of life.

Climatic Zone 1: The Southeastern Coastal Strip

This region offers a great variety of things to see and do. One feature of particular interest to visitors, is that it gets the least rainfall of any part of the whole island. The normal annual precipitation ranges from an average high of 43 inches to a low of only 28.5 inches. There is one portion of this Climatic Zone 1, however, which must be labeled as a wet spot. That is the charming Wicklow mountain country just to the south of Dublin. This southeastern coastal area enjoys somewhat more sunshine than other parts of the island. There is little temperature difference across the country but the winter winds, off the Atlantic, make that side much less comfortable than the eastern shores.

As indicated on the City Weather Tables, Dublin's 29.6 inches of total precipitation is one of the lowest in Ireland. Waterford averages 39 inches a year, and Cork 41. It is interesting to remember that Boston, New York City, and Montreal, Canada, can expect 41, 43, and 41 inches, respectively, while Chicago, and Toronto, Canada, get 32 to 33 inches a year.

Light misty showers can, and do, happen at any time. They are very frequent during the gray winter days but occur less often throughout the remainder of the year. In the springtime, they generally interrupt the sunshine only as brief sprinkles which makes that season most pleasant.

The Weather Tables of all three cities show a temperature range of less than 15° throughout the entire year. Nor are there many sudden or wide daily fluctuations, except on the infrequent occasion, which usually coincides with a shift in wind direction.

While the northwesterly coasts of Ireland are subject to constant light to brisk breezes and the occasional fierce, winter ocean storm, this eastern section is not bothered by much violent weather. There will be about four to eight short thunderstorms a year mostly in midsummer. Wind reaches what is designated as medium gale strength only four times a year in Dublin and nine in Waterford. Roche's Point, in County Cork, will record 17 annually, but the city of Cork, almost completely protected by mountains on the north and west, will be little troubled with high winds.

If the word "gale" sounds ominous, remember that air velocities between 32 and 63 m.p.h. are so classified on the Beaufort scale, and most of these wind occurrences register in the lower ranges.

EXCURSIONS

There are many one-day trips in the Dublin environs, but the prize that should not be missed is the 10-mile drive around Dublin Bay to the charming little sea village of Howth. We love the Amalfi Drive to Sorrento and Positano—especially pleasant with a little detour for a wine refresher and some gay Neapolitan music squeezed from an accordian. But even that is no more joyful than to finish off a perfect day by listening to Irish ballads at the famous Abbey Tavern in Howth.

Only a dinner of Dublin Bay prawns with Guinness stout, and crowned with an Irish coffee, could add to the pleasure.

There are many lovely if not spectacular scenes and places on the trip from Dublin to Waterford. Enniskerry, Glendalough, and Blessington are all in the Wicklow Mountains and each is ample justification for this enjoyable jaunt. And who would admit to being in Dublin and failing to see the nearby Vale of Avoca?

The greatest attraction for many visitors to the Wexford-Waterford area are the little coastal towns and fishing villages such as Rosslare Harbour, Duncannon, Knockmahon, and especially Dunmore East. The latter is a little fishing village where we stayed several times at the 160-acre Power farm which slopes down to the sea. Watching Mrs. Power making soda bread in the farmhouse kitchen, while Pat told us all about the things and places that we should see, still carries pleasant memories.

Few who get to this area can resist a peek and purchase at the world-famous Waterford glassworks, which seems much more a group of artists and craftsmen than a typical industrial enterprise. No two articles will be identical: There is so much handwork involved in their production that each is an individual little masterpiece. Even though there is usually a long list of back orders, the highest quality standards are strictly maintained and the smallest flaw causes prompt rejection.

The country between Waterford and Cork presents a serious decision. Should the visitor wander along the scenic coastal road, stop at Dungarvan, Youghal, Cobh, and the popular Kinsale, or wind inland on the narrow but excellent mountain roads, through Carrick-on-Suir, Clonmel, Cahir, Tipperary, and Fermoy? An extra dividend on the latter route is a look at Cashel Castle and the Blarney Stone. Either choice will be long remembered but, making the complete circuit and including both, will be a major highlight of any grand tour.

Climatic Zone 2: Western Coast

Although it is included in fewer itineraries than the eastern coasts because of time and distance, most of the colorful and enticing photos found in travel folders of Ireland were taken in this delightful part of the country. Bantry Bay, Killarney lakes, the Dingle Peninsula, Tralee, Adare, Galway Bay, the fabulous Connemara country, the lonely Aran Islands and much more are all here.

Remember to bring your camera, but the experienced traveler will also include Donegal tweeds and perhaps an Aran sweater purchased in this very area on a previous visit. He knows, too, that a home-knitted fisherman's bawneen of natural wool will keep one dry and comfortable in this cool and often moist country. There is a refreshing 18-mile per hour sea breeze about half of the time in summer and, while the temperature is generally between mid-50 and mid-60°, the combination will make it seem somewhat cooler.

In addition to its other distinctions, we must admit that this is also the wettest segment of a misty Ireland. The overall area averages 37 to 57 inches of precipitation a year. Valentia, Blacksod Point, and other ocean exposed spots will be in the higher ranges. From May through September, this shore fringe can expect some very light rainfall two out of every three days and one-tenth of an inch or more about every third day. There is almost always less during the earlier portion of this period.

Shannon Airport and the country a bit inland will record closer to 40 inches of precipitation. There will be a light, misty sprinkle almost every other day and one-tenth of an inch or more will fall every third or fourth day in that area. The small map on Chart 6 highlights the Killarney Lake District and the rugged Connemara peninsula as the wettest portions of this western coastal segment.

At first glance, the City Weather Tables may seem unnecessarily complete. In many parts of the world, for example, the two columns listing the number of days getting various rainfalls will show very little difference. This means that one column might be sufficient to tell the precipitation story for that particular spot.

The Weather Table of Valentia, however, may illustrate why so many columns of figures are included. Suppose we showed only the number of days with 0.1 inch of precipitation. These will range from 7 to 11 per month with an annual total of 109 such days. But if we know that there are also 16 to 25 days a month (which adds up to a whopping 245 days a year) that get 0.01 inch, the picture changes. This clearly indicates that there must be plenty of misty weather even though there may not be particularly heavy rainfall. Columns, indicating the low maximum rainfall in 24 hours and less than one thunderstorm a month, further show that this is not a land of brief torrential downpours.

Some readers may find the text and charts sufficiently informative to give them a satisfactory idea of what they may reasonably expect to encounter climatically at almost any place and time throughout Europe. Others will want a more complete and detailed weather story. They will be surprised and pleased at how much interesting and useful information can be developed from what at first glance appears to be a complex mess of hieroglyphics in the City Weather Tables.

This part of Ireland is not noted as a land of sunshine but Shannon will average 5 hours a day in April and August, 6.6 in May, 6 in June, and 4.9 in July. That may not call for very much sun lotion but the beautiful complexions of the local residents indicates that it is good outdoor weather. Even such dripping spots as the Valentia Observatory average almost 5 hours of sunshine a day through most of spring and summer, with closer to 6 in June, and a tops of 6.5 in sunny May.

This Irish weather could be much less favorable and still not deter tourism; we have yet to hear of an unenthusiastic visitor to this region. Perhaps one gauge might be the considerable number of Americans who have bought homes and live all, or most, of the year in Ireland. Acutally, as elsewhere, the weather is an uncertain thing. On one spring visit, we motored along this coast and enjoyed 3 weeks of continuously beautiful weather. Each morning, the driver (probably on the law of averages) predicted that there would be a "mite of rain." On only one afternoon did we have a 2-hour shower which was no particular inconvenience as fortunately he was much better versed in pubs than weather.

Most of the outstanding, grand old hotels of the world are gradually disappearing. Fortunately, Ashford Castle, in Cong, which rates with the top, will retain that position as long as the standard of excellence established by Noel Huggard is maintained. By Michelin standards, this fine establishment deserves a Three Star classification—"a place worthy of any length detour!" It is unknown to many, but it is locally famous, so book your reservation far, far in advance and you will carry fond memories long afterward.

If time and weather permit, a visit to the Aran Islands can be very interesting. When the sea is rough, which is fairly frequent, the usual daily trip is canceled. Ordinarily, you will see only Inishmore, the largest of the three, unless you make the tricky journey to Inishman and Inisheer in the not too stable-looking tarred canvas boats called *currachs.*

This is a stark, wind-swept, barren island with few trees and an atmosphere of frugality. The land, every foot of which is said to have been "made" from a mixture of kelp, sand, and manure, is criss-crossed with innumerable stone fences. Each tiny enclosure serves as a wind protection for cattle or meager crops. But despite the hardships, these Gaelic-speaking people are a cheerful lot whose living habits have possibly changed less than any other in all Europe over the past 100 years.

Climatic Zone 3: The Northwest Coast and Northern Ireland

This is almost a climatic extension of the west coast. It will get somewhat less precipitation, averaging 32 to 45 inches a year, and the temperature will be a few degrees lower throughout the year than the country just to the south. Here, too, there will be almost no 90° days but, toward the eastern side, in the Belfast area, 40 to 45 nights will register freezing. Few, if any, will drop to zero.

Almost all of this coastal fringe will get at least one-tenth inch of precipitation about 90 days a year, but there will be some misty weather 200 to 250 days annually and plenty of overcast skies. As is often the case, close to large water bodies, there will be few thunderstorms— perhaps five or six each summer.

This is quite breezy country and the Malin Head section can brace for winds of gale force about 35 days during the winter. Even up in these far northern reaches, there is little snow, due to the tempering effects of the North Atlantic Drift.

The Connemara shore line is rugged and wonderful, while many have described the almost continuous seascape, from Donegal around past Belfast, as one of the most spectacular in all Europe. Like almost all of Ireland's west coast, this is a fantastic combination of tiny coves, bold, rocky headlands, fishing villages, cliffs, beaches, fiordlike inlets, and the roaring sea. You will recognize the immense jumble of huge, black, basalt cubes, called the Giant's Causeway from the many travel posters.

On the landward side, you will spot glittering lakes, crystal-clear trout streams, and thatched, whitewashed cottages. Inviting roads winding upland lead to wild areas of serenity seldom now encountered in this hectic world.

Tourists visiting Belleek just across the northern border will be fascinated by the artistic skills and infinite care required to produce each eggshell-thin piece of lustrous porcelain. Few who have not seen the actual manufacture will appreciate how many individual operations are necessary to produce each delicate piece.

In Donegal, one can watch the hand looming of magnificent tweeds. These are generally quite small operations, and much of the weaving is done in little shops or even on simple looms in the homes. Very often, the individual decides upon the pattern and quantity of material to be produced. Few visitors will leave without a bundle. The honesty and reliability of these gracious people can be rated at 110%, so do not hesitate to have any purchase follow you back to America.

Hotels and other accommodations are less than plentiful on much of this lonely coast and may be fully booked in summer. One of the most delightful places that we know of in Northern Ireland is the Old Inn, 10 miles from Belfast in Crawfordsburn, County Down.

Climatic Zone 4: The Central Plains

This area is the most neglected by tourists in Ireland. Many think of it as one large peat bog, forgetting that it is coursed by the famous 230-mile-long Shannon River and dotted with glittering lakes of every size and shape. It is a wonderful boating vacationland, of greater interest to those fortunates who can linger and enjoy its beauty in leisure fashion than the high-tempoed tourist dashing from the Shannon Airport to Dublin. Its pleasures and comforts are increased by the very absence of foreign visitors.

There are also some interesting places around the outer edges of this flat country. Tipperary, with its history and horse farms, rolling grasslands and neat barns, fences and hurdles could substitute for the best in Kentucky's Blue Grass country. The limestone underlaying, just as in Lexington, is said to be the secret of the great success in producing such fine thoroughbreds. Kilkenny is a most pleasant town; it was a center for the Normans who, in spite of statute and regulation, succumbed and became more Irish than the Irish themselves. Kildare is also attractive horse country of world reknown.

But the main attractions are the fabled Shannon and the lakes that form part of its course. Killaloe, Athlone, Carrick-on-Shannon, Roosky, and Enniskillen are only some of the centers where all kinds of boating arrangements can be made. Hire your own fully equipped sleep-aboard launch or engage a boat complete with captain-guide. Either can be a most satisfyingly pleasant and lazy experience for the vacationist who wants to do nothing. Those with the time and inclination might contact the Inland Waterways Association through the Irish Tourist Board or any of the Government Tourist Information Offices.

By Irish standards, the precipitation in this area is rather modest— 35 to 40 inches per year. Temperatures are about average. The City Weather Tables of Claremorris, Birr, and Mullingar are reasonably representative of this whole broad plain. There is little snowfall and few days register over 90°. Only Birr will have as many as 50 frosty nights but, even there, little likelihood of zero temperatures. There is not much stormy weather; perhaps, one thunderstorm a month but the larger lakes can get quite rough.

Of most interest to tourists and, more particularly, the boating enthusiast, is sunshine or lack of it. The sunshine table shows that in Birr, Claremorris, Mullingar, and Clones, close to the North Ireland boundary, the average hours per day of sunshine jump from 3.5 in March to about 5 in April. May averages a little over 6 and June just under that figure. From July to the end of September, the hours per day of sunshine drop from almost 5 to a little under 4. Those figures, along with the number of days getting one-tenth of an inch of rainfall, give a good idea of the type weather that can be expected.

The Best Time to Visit Ireland

Many mistakenly think of Ireland as strictly a one-season tourist country—and a short one at that. Ireland is a crowded place in July and August, and the knowing travel agent will advise against going without confirmed reservations. Dublin is especially busy all of August, so be warned. Incidentally, the top Dublin hotels are far from inexpensive and few have space for garaging your car at any time. Americans and Canadians are most apt to arrive somewhat after July first.

If they find the weather satisfactory at that time, imagine their delighted surprise if they could experience an Irish springtime. While July and August will probably continue to be the favored season, the really choice weather normally occurs in the three months of April through June. In most cases, maximum hours of sunshine are in May, with late April and early June almost as high.

Prices during the so-called high season are more expensive and in some cases the difference can be substantial. Rates in the Donegal area may be as much as 40 per cent more in the high season than the low. Exact dates vary from place to place but in Killarney, where the differential is closer to 20 per cent, many establishments consider the months of June, July, and August as top season. A hotel in Tramore carries a prime rate in July and August; an intermediate one for May, June, and September; and a considerably lower one from October through April. Some years, the Irish Tourist Office distributes little yellow cards which grant the bearer a discount of about ten per cent in some hotels and car rental offices during the month of May. It is interesting that the discount applies to the month which enjoys the very finest weather. The airlines' low-cost period is usually August 4 to May 21 for travel to Ireland and September 29 to July 16 for westward passage. But high or low season, costs generally throughout Ireland are almost as low as any in Europe. That, coupled with excellent food, makes any stay in this delightful country a vacation bargain.

Travel patterns, like all habits, change slowly, but North Americans are discovering that there are many advantages to off-season visits in Ireland, even during midwinter. To be sure, this latter time is not the most agreeable for outdoor sightseeing and even the really robust would hardly venture to the west coast. Such places as Dublin, however, can be very rewarding at almost any time of the year.

Winters are generally gray and misty but temperatures are rarely severe. Weather conditions are certainly less uncomfortable than Chicago, Boston, or Toronto at that time. Crowds are fewer and accommodations more readily available. You will not spend 90 per cent of your waking hours with fellow tourists but can get to know and enjoy the Irish people. Food and services are also at their best. Both Dublin and London have theatrical seasons that rival Broadway, and at a fraction of the cost. For folks who enjoy music, drama, the ballet, and a chance to see uncrowded museums and art galleries, this can be a most satisfying visit.

Their first reaction could be that this might really be an indoor vacation. They forget that golf is played almost year around in Ireland. Also there are many days very suitable for jaunts to nearby places to enjoy such activities as horse racing, athletic events, or just a sightseeing tour. They may not realize how few hours are actually spent out of doors on a visit at any time of the year. Sleeping, eating, shopping, viewing exhibits, touring through the Guinness brewery and the inevitable visiting to the pub—all indoors—account for many of the 24 hours.

What soon may become popular with Canadians and Americans is a combination winter vacation. A couple of weeks of fine living in Dublin or any major European city, topped off with a week in Madeira, North Africa, the Canaries, or other convenient sun spot.

Accommodations

There are many government-rated and approved farmhouses in Ireland that accommodate guests. Many are wonderful, being better than the hotels available in some areas, while only a few are less than satisfactory. It is best, actually, to see the farmhouses rather than to make arrangements by phone. Most will accept only 1- or 2-week bookings during July and August, but overnight accommodations can usually be had at other times. The charge of about 3 pounds includes a comfortable bed and a hearty breakfast. Guests who arrive by midafternoon are often invited to afternoon tea. Besides a large pot of tea with cream or lemon, this usually includes soda bread with butter and jam and several kinds of cakes or pastries.

We should also mention that the very efficient Irish Tourist Board issues a complete and detailed booklet listing not only farmhouses but also town and country homes that receive guests. Just about all include an excellent breakfast of juice or fruit, cereal, bacon and eggs, toast, soda bread, wonderful jams and marmalades, and a pot of tea. Some also serve either high tea or dinner. The town houses can be found in most towns and cities while the country homes, which are not farms, are scattered across the country. Bedrooms usually do not have private baths but most have washstands with hot and cold water. These homes are often much more cheerful and less impersonal than hotels and give the visitor an opportunity to meet and get to know the Irish people. Rates both by the day and week are most reasonable.

Tourist Information and Transportation

Irish Air Lines and other international carriers are developing most attractive and inexpensive packages. They may be in the form of stopovers, extensions, all-inclusive tours, combinations, or other arrangements that will make off-season travel very tempting. In addition to the gracious Irish Air Lines, Shannon is well serviced by Air Canada, TWA, Pan Am., KLM, Air France, and lines from the Continent to Dublin.

Ireland does an outstanding job in tourist promotion—along with Spain and Israel, perhaps the best in the world. There are Government Information Offices in just about every town of any size—about 100 in all. The headquarters at 14 Upper O'Connell Street, in the center of Dublin, is one of the most complete information offices that we have ever visited. Some stations, in the smaller and more remote places, are only mobile trailers, but all are well stocked with complete and detailed literature and information. Equally important are the well-informed staffs, pleasant and helpful.

Even with this excellent service, it always helps to be as well informed as possible before arrival. A letter, or preferably a visit to the Irish Tourist Board Office, 590 Fifth Avenue, New York, N.Y. 10036, will be very rewarding. There are similar offices in Chicago, San Francisco, Toronto, and Montreal. The Irish Air Lines offices do much more than just sell tickets. They, too, are a fine source of information and have a surprising amount of literature.

Tourists, who plan to also visit the six counties of Northern Ireland, can get very complete and detailed information by writing to the Northern Ireland Tourist Board, Tourist Information Center, 6 Royal Avenue, Belfast 1, Northern Ireland.

Tourists return from many parts of the world with glowing praise of the scenery, the food, the culture, the architecture, or a dozen other things that may be outstanding. Ireland has a generous share of each, but anyone who has been to this soft, hospitable land almost invariably retains pleasant memories of its kindly, gracious people—and what could be more important to any guest?

GREAT BRITAIN

THE goose-pimpled natives in New York City, Montreal, and Washington, D.C. shivering in January gales may be surprised to learn that midwinter is the season when most of the cut flowers, used on Western Europe's tables, are grown out of doors in England.

No—there has not been a climatic revolution nor is Great Britain one big flower garden at that time of year. Only one little area in the southwest corner of the country or, to be more precise, the tiny Scilly Islands, about 30 miles off Land's End, is so fortunate.

Not only climate in Great Britain is varied; life styles and the physical characteristics of the country differ greatly. So whatever your tastes, Great Britain is apt to have the answers.

England, Scotland, and Wales fall within the vast Marine Climate Zone of Western Europe. Because they are a little further removed from the influence of the warm ocean currents and the westerly winds than Ireland, the climate is a trifle less equable. They also have somewhat greater exposure to the occasional harsh northeast winds and, as a result, temperatures can be a bit more extreme and uncertain. That is particularly true in the northeastern segment facing Continental Europe. Temperatures generally decrease gradually from southwestern England to the northeastern parts of Scotland.

While the whole United Kingdom averages over 40 inches of precipitation annually, England proper will record not more than 34. Precipitation is generally highest on the Atlantic shores which are exposed to

SPRING
April, May, June
Great Britain at its best. This
is driest & sunniest season.
Countryside particularly
delightful. Orchards in
blossom & flowers in full
bloom across the land.

AUTUMN
September-October
Rather uncertain weather.
Some years bright & dry,
others gray & showery.
Generally mild, cool with
limited sunshine.

SUMMER
July-August
Cool, some showers but
ample sunshine. Vacation
centers crowded but
other areas comfortable.
Some Haar fog along dry,
sunny east coast.

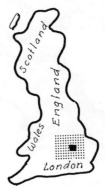

WINTER
November-March
London, a city of all
seasons -colorful, exciting
in full swing. Skies often
overcast, quite mild, with
little snow.

GREAT BRITAIN
VACATION WEATHER MAPS
Chart 7

GREAT BRITAIN
ENGLAND; WALES Chart 8

Alderney 🝰 St.Anne

France

Guernsey
Herm
Sark

Ramsey

Peel
▲ 2034

Port
Erin
■ Douglas
Ronaldsway

ISLE OF MAN

Jersey
St.Quen
Gorey
St.Helier

0   5   20 Miles
0   10   30 Kms.

CHANNEL ISLES

England
Southampton
Gosport
The Solent
Cowes
Yarmouth
Ryde
Sandown
Ventnor

Miles 5   10
Kms. 8   16

ISLE OF WIGHT

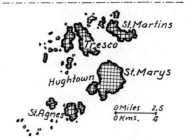

St.Martins
Tresco
Hughtown
St.Marys

St.Agnes

0 Miles   2.5
0 Kms.   4

SCILLY ISLES

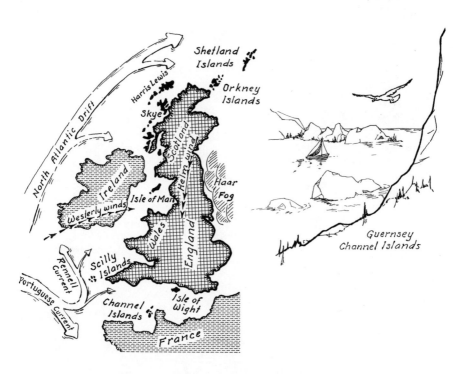

Shetland
Islands

Harris Lewis
Orkney
Islands
Skye

North Atlantic Drift

Scotland
Helm wind

Ireland
Isle of Man
Haar
Fog

Westerly winds

Wales

England

Rennell
Current

Scilly
Islands

Portuguese Current

Channel
Islands

Isle of
Wight

France

Guernsey
Channel Islands

ENGLAND'S ISLES   *Chart 9*

the moisture-laden westerly winds. As an example, Cardiff, Wales, gets 41 inches but Margate, on the eastern coast of England, averages only 21. Although distributed throughout the year, precipitation is usually heaviest during fall and winter while spring and early summer are most likely to be both driest and sunniest. Due to the prevalence of mist and fog, sunshine is not plentiful and often averages less than 1 hour a day in midwinter, but springtime sunshine will range from 5 to 8 hours in each 24.

Climate

   In spite of its relatively modest size, Great Britain enjoys a variety of subclimates. Many meteorologists divide the country into seven climatic zones—two in Scotland, one in Wales, and four in England. These segments are not generally as well defined nor do they have as widely contrasting conditions as in countries which contain high mountain ranges and very irregular terrain. Each area, however, experiences some marked climatic differences which can be of particular interest to tourists.

Castle Combe

ENGLAND and WALES

Topography

England is about 400 miles from north to south and some 300 miles across at the wide south coast. Chart 1 shows the relative latitudes of the United Kingdom and North America. Because of its comparatively mild climate, many may be surprised to learn that even the southern coast of England is north of the Canadian-U.S.A. boundary. This northerly location accounts for the delightfully long summer evenings when the sun does not disappear until 10 p.m. or later.

Although almost 40 per cent of Great Britain is farming and grazing lands, there is a wide variety of scenery. One of the glories of England proper must be the bold, rugged coasts of the north and southwest. The Cheviot Hills along the Scottish border and the Pennine mountain range, extending to the south, are fitting backdrops for the charming English Lake District in the upper western part of the country.

The wide flattish Midlands stretching almost to London contain the great agricultural districts. The western portion of this region is hilly and, while not scenically spectacular, makes a pleasant route between Scotland and the south. The large industrial centers of Manchester, Birmingham, Liverpool, and Coventry will be of little interest to most tourists, but the fascinating walled cities of Chester and York attract many visitors.

Just to the south is prime tourist country. Stratford-on-Avon, the Cotswolds with its charming thatched cottages, the stately town of Bath, Oxford, and Cambridge, Stonehenge, whose mysteries may have been recently solved, and a whole travel guide of other interesting places. One very attractive British practice, which only Ottawa, Canada, seems to have imitated to date, is the encircling of cities with a broad band of trees and landscaping called a "Green Belt."

Then there are the sunnier and drier southeastern vacation coast lands, which some call the British Riviera. This includes Ipswich, Margate, Dungeness, Brighton, and dozens of smaller and even more attractive resort spots. For many, the prize of all England is the southwest country of Somerset, Devon, Dorset, and more especially, the marvelous Cornwall coast. But beware of the July-August vacation crowds that surge like lemmings to the sea. The April, May, June, and September weather in this area is delightful, too, and few visitors are disappointed at those times.

Climatic Zone 1: Northwestern Coast of England

Average annual temperatures in this area will normally fall within the narrow range of 44 to 56° while the total precipitation varies from about 30 to 42 inches a year.

The three City Weather Tables, of Douglas (on the Isle of Man), Keswick (in the Lake Region), and Liverpool (a coastal port), will give a reasonably accurate picture of the weather one can expect in this northwestern segment of England. There is little difference among the three tables, but we suggest that temperatures be considered in conjunction with precipitation and, more particularly, the wind velocity and frequency. As indicated on the Wind Chill Table, a 35° temperature with a calm atmosphere will feel exactly like 35°. The same temperature with a 17-m.p.h. wind will seem about 18° and a 30-m.p.h. wind will feel like a shivery 8°.

Douglas, on the Isle of Man out in the Irish Sea, is a grayish, damp spot with such a mild climate that several types of palms flourish outdoors. The high 45 inches of rainfall, distributed through the year, is about twice as heavy in January as April, May, and June, which are normally the driest months. There will be some rainfall at least 200 days a year, which means many overcast skies and relatively few completely sunny days.

There is not much lasting fog, since an 18-m.p.h. breeze blows about one-third of the time. Consequently, the lower atmosphere is quite clear and bracing. Tourism is not a recent happening in these parts as the first steamer travel-tour that Thomas Cook conducted was to the Isle of Man in 1845. Ramsey on the northeast coast is the sunniest corner of this vacation island which has rated about fifteenth in the sunshine polls of several hundred resorts in Great Britain.

The visitor's choice is wide; he can select either the most raucous beach life or a quiet, palm-fringed little cove where only the crickets or frogs will be heard after dark. The quiet sections include some lovely sweeping countryside with many little glens and beautiful coastal scenery. There is golf, fishing, and the opportunity of buying a tailess Manx cat which took its name from the island.

Keswick is in the lovely Lake District. There are 16 major lakes (the largest being Windermere) and numerous smaller ponds and pools, many well stocked with hungry fish. It is a boating, sailing, and general sports center.

With almost 60 inches of rain sprinkled over more than 200 days a year, there are not many completely sunny days. While this is not exactly parasol country, the sun breaks through surprisingly often during spring and summer. The Lake Region was much more popular years ago but it is still favored by many who enjoy a calm, peaceful vacation in a pleasant setting.

The large commercial and shipping center of Liverpool is of interest to the tourist primarily as a port of entry or a starting point on a visit to North Wales and the very attractive city of Chester. Incidentally, the City Weather Table for Liverpool will apply in Chester, excepting only that the latter gets but 28 inches of rainfall as against Liverpool's 35.

Climatic Zone 2: The Midlands

The northern portion of this zone, while most important to the nation's economy, is of little interest to visitors. It includes Leeds, Manchester, Birmingham, and other mammoth industrial centers. The climate is not overly pleasant, averaging about 180 rainy days a year. It can be almost unbearable at times—when the fumes and dust pollution from the factories are heavy. This area will be a prime target in the sweeping ecological improvement program being inaugurated throughout Britain.

The southern segment of this zone, however, is quite a different tourist story. Almost all visitors staying even a week in England will get to Stratford-on-Avon. People who like to avoid the midsummer tourist rush should remember that the Royal Shakespeare Theatre has performances from April through November. Many visit either Oxford or Cambridge, but those who bypass the Georgian town of Bath with its famous spa where first the Romans and later the dandy, Beau Nash, frolicked, will miss a most pleasant experience. The attractive little towns of Broadway and Stow-on-the-Wold and many of England's most colorful inns and taverns are also in this delightful region.

Again, the temperatures show little change from the overall pattern but you will notice in the Birmingham, York, and Cambridge City Weather Tables that the precipitation is considerably lower than along the Atlantic. Even with this low rainfall, it is not overly sunny since the interior portions of England, by and large, enjoy somewhat less sunshine than the coastal areas. Occasionally, during late winter and early spring, a strong, cold blast, called the *Helm Wind,* blows in from the northeast, chilling the northern portion of this area.

Climatic Zone 3: East Coast of England

This segment is probably the least publicized in travel folders and visited by fewer foreigners than any other portion of England. It is made up of two quite distinct types of country and ways of life. The inland areas are rolling lands with some low hilly patches and marshes. It is generally quiet pastoral countryside and little changed since the beginning of this century. Rainfall is remarkably low in much of this section; many parts average only 18 to 20 inches a year. Occasionally, there is an unusually dry season or even a summer drought which can result in serious agricultural hardships but delights sun-starved vacationists.

Many English people see this area only as they hurry to the east coast resorts for summer vacation. The City Weather Tables of Tynemouth and Great Yarmouth, in the northern and southern segments of this coast, show the kind of weather these vacationists will have. Note that rainfall is down in the 24- to 27-inch range, but there is a little more misty weather than inland. Spring and summer sunshine, while not constant, often breaks through for good periods several times each day. Sometimes during the summer, a heavy, wet, sea fog called *Haar* rolls in over these coastal lands. This happens only when the wind has shifted to the east. Note on the Tynemouth City Weather Table that visibility in this general area is reduced to less than ½ mile about 40 days of the year. These coastal lands are, of course, the most exposed to the strong northeastern winds which, in midsummer, are generally brisk but invigorating and very pleasant. During much of the year, however, the weather is blustery and often uncomfortable. Great Yarmouth and Tynemouth have experienced extreme temperatures of 13 and 11°, respectively. While such low readings are unusual, even the more normal mid-30 figures seem much colder because of the sharp biting winds. Winter in this land is only for those brave souls who thrive on such rugged conditions.

Climatic Zone 4: The South Coast of England

Aside from London itself, this must be the most attractive part of all England for foreign visitors. From Margate on the east to Land's End at the western tip, it is certainly a very colorful and exciting 300 miles.

Temperatures are generally mild and very equable but somewhat lower with greater fluctuations at the eastern end than in the Scilly Islands. Precipitation follows a similar pattern. Margate averages an extremely low 21 inches a year as against the 33 inches in the Scillies. Winter sunshine is rather meager but spring and summer average 5 to 8 hours per day.

The coast from Margate around to Southampton is a continuous string of attractive towns and villages. This seaside country appeals to many kinds of people. Some portions are primarily resorts while others are quiet residential communities (the pleasant Gosport section is known as a retirement spot for retired naval personnel). It has become increasingly popular for businessmen to commute between London and the south coast.

There are all kinds of accommodations for tourists and here one is most apt to be favored with the driest and sunniest weather in all Great Britain. The Margate City Weather Table is an indication of what the visitor can look forward to. Less than 1½ inches of rainfall each month from February through June and only 21 inches for the whole year attest to pleasant conditions.

Often called the British Riviera, it also offers fine swimming, which is most popular with people from northern climes. Those accustomed to the Mediterranean or Florida beaches, however, find this water rather chilly. Because the ocean heats more slowly than land masses, these waters (including the Channel Islands) do not reach the maximum of 67° until mid-August. The Southampton Weather Table follows the coastal pattern but is a mite damper and milder. As with this whole southern strip, there are rather pleasant sea breezes with few brisk blows except in winter when it can get very snappy.

The Isle of Wight (Chart 9) is the most convenient of England's isles to visit. Sitting just off the coast, south of Southampton, it is separated from England proper by the Solent and Spithead channels. Its 13 by 23 miles is sufficient to include considerable variety. There are a dozen small villages—those toward the east like Ryde, with its 5-mile sand beach, attract vacationers who prefer well-organized holiday resorts. Most of the larger beaches are on this side of the island. The western half has more unspoiled rural villages and smaller, quiet beaches, but the entire island is well populated with visitors during July and August when accommodations are at a premium. One of the sunniest parts of Great Britain, the Isle of Wight was once favored as a health center.

The Isle of Wight is convenient for some who like to break an off-season visit in London with a few days of quiet relaxation and more probable sunshine. Temperatures on the island are substantially the same as in Southampton but the sea air is fresher and brighter.

Along the coast toward Plymouth, there is about a 20 percent increase in rainfall to 38 inches annually. The mercury seldom drops to 32° nor climbs above 90.

In the extreme southwestern corner of England are the famed Cornwall, Somerset, Devon, and Dorset counties, which are far removed from the major centers of commerce, industry, and large municipalities. There are such glamorous names as Penzance, Land's End, St. Ives, Torquay, Dartmoor, and Mousehole. Most visit St. Michael's Mount, often called England's Mont Saint Michel. This area also includes the 800-year-old "New Forest" region where wild ponies roam at will. There are cowboylike roundups twice a year.

This beautiful southwest country is an overbusy place during July and August when all of Britain seems headed in that direction. Do not arrive without confirmed reservations. Actually, the weather is sunnier with less rainfall in April, May, and June before the midsummer rush. September and sometimes October can be showery but are most often very pleasant.

Many call this England at its best, but that is a matter of choice. The Cotswolds, the Lake Region, and various other segments each has its advocates but this marvelous southwest country must be included on just about every experienced traveler's list of favorite places. Great praise must be accorded the intelligent and far-thinking National Trust which has acquired so many large choice tracts of unspoiled land for public use. Other than in Oregon, there are few places in the United States where one can enjoy such delightful stretches of seashore rambling as along the rugged Cornwall coast.

The Scilly Islands, less than 30 miles off Land's End, present a weather anomaly. Because of the many palms, feathery bamboo, and great profusion of colorful, year-round flowers, the climate of these islands is often labeled subtropical. Perhaps, "almost freeze-free" might be a more apt description. The Weather Table will verify the very mild winters and remarkably equable temperatures. Normally, the lowest winter reading will be only 43° and the highest August figure a modest 66. Zero and plus 90° may never have appeared in the island weather reports.

Sitting so far out in the Atlantic might suggest much foggy weather and perhaps gales. There are both, but in most moderate portions. Of the 27 foggy days a year, about half occur from July through October but few usually last the full day. The 24 winter gale force winds may also sound a bit wild, but remember that any air velocity above 32 m.p.h. (to maximum of 63 m.p.h.) is classified as a gale.

Rainfall is distributed throughout the year, but the 4 months, October through January, account for almost 50 percent of the 33-inch precipitation. It should be noted, however, that there is much misty weather mixed with the sunshine. More than 200 days a year will get at least a 0.01-inch sprinkle, but sunshine in spring and summer can be prolonged and brilliant.

Only five of the 145 tiny Scilly Islands are inhabited and most tourists will probably see only St. Mary's, Tresco, and St. Martins. The chief source of income for the islanders is the 100 million blossoms shipped to Western Europe from Christmastime until the Dutch takeover at about Eastertide. The Scillonians also produce beautiful spring vegetables, especially asparagus.

But the visitor wants more than fine vegetables and flowers. There is excellent fishing, bird and often seal watching, picnics, walking, and sun bathing. The absence of night life is hardly missed, and, in such a setting, who needs more than an evening pint and some good conversation? As in all this southwest country, the residents are apt to drink the locally made cider.

Although April through July are the driest months of the year, March is by far the most colorful and the time when the fields of flowers are at peak. Visitors should remember that the thermometer will show about 42 to 52° at that time. Also, while there will be abundant sunshine, they can expect a light sprinkle almost every second day.

While in this climatic zone, we must not forget those unique islands which Victor Hugo called "Fragments of Europe, dropped by France and picked up by England." The Channel Islands do indeed have a Gallic flavor; on Jersey, in particular, most of the population is bilingual. In some rural districts, many speak only a French *patois*. This island group is 100 miles from England, but Alderney is less than 10 miles off the Normandy coast.

Being closer to the Continent and subject to some crisp southeastern winds, the Channel Islands do not enjoy quite as mild or equable a climate as the Scillies, although the Weather Tables are generally similar.

Jersey, the largest island (45 square miles), has won the British sunshine poll seven times in a 10-year span (about 2,000 hours of sunshine a year). Its topography has been likened to a wedge of cheese tilted to face the southern sunshine. As one might suppose, the north coast is high, rocky, and attractive. This elevated edge also serves to shelter the lower southern part of the island from the cold northeastern winter winds which, coupled with the effects of the warm ocean currents, explains the pleasant, even climate that Jersey enjoys. Most of the 22 beaches are on the flatter, sandy south side.

Jersey is the most French of all the islands. Its capital, St. Helier, the many French names, and excellent food are hints of its previous allegiance.

The Channel Islands became a part of England and its first overseas possession when the Norman, William the Conqueror, became William the First of England in 1066. These islands are unique in that they are subject directly to the Queen of England in her capacity as Duke of Normandy.

They enjoy a considerable range of autonomy, and laws enacted by the British Parliament do not apply to them unless they are specifically mentioned. They also enjoy a special customs and tax situation and, as a result, today's tourists can purchase perfumes and other merchandise at very favorable prices.

There is a wide range of accommodations but many places accept only week-long, all-inclusive bookings, particularly during the busy July-August season. Also, be warned that the influx of April honeymooners can fill almost every bed on the island during that most pleasant of all months.

Guernsey, the island of second importance (25 square miles), is a quieter pastoral place. The coast is even more rugged with deep inlets and bold rocky headlands. St. Peter Port, the capital, is a quaint, attractive town of stone buildings and narrow lanes—a pleasant town in which to stroll since vehicles are not permitted in the central area after 10 a.m. A spot enjoyed by many is the 150-year-old French Halles, or covered marketplace, a little reminiscent of its now-departed Parisien namesake. The lobsters, giant crabs, fine fruits, and cheeses tempt many tourists to make up a picnic basket to enjoy at one of the 20 little beaches or other scenic spots. Most of the beaches are pocketed between high rocky headlands which protect them from the elements unless they open directly into the path of prevailing cold winds. This is a common characteristic of small beaches which indent rugged coast lines and an item worth remembering. In many cases, these little enclaves enjoy an individual miniclimate much more pleasant than the adjacent areas. Golf, fishing, sun bathing, or exploring can be enjoyed on both Jersey and Guernsey.

These two islands have one more thing in common—each has its famous cattle. The golden Guernsey is said to have been introduced onto that island from France by Breton monks over 1,000 years ago. Only the one namesake breed of cattle (Jersey or Guernsey) is permitted on each of the islands. Some parts of Guernsey look like one immense greenhouse where hundreds of acres of tomatoes are grown under glass.

Alderney, much less visited by tourists, is a peaceful agricultural island with limited accommodations. There are giant cliffs but also well-protected sandy beaches and quiet coves. St. Anne is a village of cobblestone streets and color-washed houses.

Tiny Sark is most noted as possibly the last feudal state in existence. It still boasts a *Seigneur,* and while most of the prerogatives are still on the statute books, few, such as "Right of the First Night," are exercised. One-thirteenth of the value of each land sale does, however, go to the Seigneur who also enjoys other unusual privileges.

Herm and Jethou are hardly more than land specks in the ocean.

The weather throughout this little archipelago is quite uniform, and the Weather Table of Jersey is reasonably representative. Eighty-eight and 26° represent the high and low temperature extremes experienced on Guernsey. The range is a little wider on Jersey with a high of 96° and a low of 12.

The Channel Islands are an unusual but extremely pleasant place to spend a rather quiet, but interesting, vacation.

LONDON

In some respects, San Francisco and London suffer from a common misconception. Perhaps, writers, poets, and songsters are at fault— neither place is always wet and foggy. As we know, the California city enjoys delightful summer weather with abundant sunshine and almost no rain for a period of several months.

London is not that fortunate but, as the City Weather Table shows, there is less than 2 inches during each of 6 months and only October approaches a 3-inch total. London's 24-inch annual precipitation is considerably lower than Montreal which gets 41 inches, Chicago 33, Boston 41, and New York City 43. So we see that London is not really wet—but damp. The answer is that while there are only 70 days a year that get one-tenth of an inch or more rainfall, 164 get 0.01 inch. This adds up to lots of misty weather which, fortunately, is so light that most visitors soon learn to disregard it.

There will be a little wet snow about a dozen times each winter but it will remain on the ground only 3 or 4 days of the year. While the London thermometer has dropped to 10° on occasion, the winter range is usually from the mid-30's to the high 40's. There are infrequent hot spells of a few days' duration, but highs are more apt to be in the low 70's than the 90's.

The biggest "happening" in London was the "change of air." After the tragic 5-day stifling smog in 1952, that resulted in 4,000 deaths, the

Clean Air Act became a law. It was the first legislation of its kind in the world.

The results show what can happen when laws are properly enforced. London now has one of the lowest air pollution ratings of any major city, and Londoners had to get accustomed to breathing a plain mixture of oxygen, nitrogen, and carbon dioxide. Visible air pollution has been reduced by 75 percent in the city and fifty percent countrywide. It is claimed, perhaps by overenthusiastic ecologists, that hours of sunshine have been increased almost 50 percent; however, the atmosphere is noticeably cleaner and brighter to visitors familiar with previous conditions. One pleasant result is that hawks, wild ducks, and other birds are once again nesting in London's many beautiful parks.

The current fear is that the increased use of automobiles will eventually offset the gains of the present antipollution program.

Hopefully, the recently developed scheme for cleaning up the rivers and waterways will be as successful. Three billion, eight hundred thousand dollars have been allocated just for England and Wales, while a similar plan is being formulated for Scotland.

Incidentally, Londoners desert the city in droves over the usual 5-day Easter weekend, making it a good time for visits, but confirmed reservations are advisable in nearby rural holiday spots at that time.

The following is the 30-year record of average hours of sunshine in London. Perhaps figures for future years may be higher.

HOURS OF SUNSHINE PER DAY

J	F	M	A	M	J	J	A	S	O	N	D
1.2	1.4	2.2	3.5	4.9	6.4	6.8	6.3	5.9	4.9	3.1	1.8

If I were limited to one general area on a visit to England, I suspect that the south coast (but including London and the Channel Islands) would be the choice. The selection of time is far easier—"What is so delightful as springtime in England!"

Climatic Zone 5: Wales

The lower two-thirds of Wales will probably never attract great numbers of overseas tourists, but a traveler who has never seen the northern Snowdon region has surely missed a choice piece of scenic real estate.

Much of Wales is mountainous and bleak, being generally suited only for pasture and grazing. There are also large, unsightly industrial and mining areas. The Cambrian Mountains occupy a major portion of the country, but there are narrow plains along the south and west—also in the Dee River valley and in the north.

Northern Wales is the lovely Snowdon country, the pride of all Welsh scenery. The imposing Caernarvon Castle is of more current interest because of the 1969 investiture of the present Prince Charles as the Prince of Wales.

Many Britains but few foreigners vacation on the pleasant, quiet island of Anglesey off the northwestern coast of Wales. This, like much of the north country, is very colorful throughout the summer with the gay spring blossoms followed by purple heather and brilliant yellow gorse.

There is a magnificent view from the 3,571-foot Mt. Snowdon, the highest peak south of the Scottish border; visibility is seldom perfect and it is well to check before making the trip to the top. The scenic north shore, from Bangor to Rhyl, is a popular Welsh resort area, Wales' answer to the Riviera. In many places, the mountains slope almost to the sea while, at other spots, there are pleasant little coves and sand beaches. The beautiful sea looks most inviting but one toe dip suggests that sunbathing might be more prudent.

Conway, Llangollen, and Bet w-s-y-Coed ("Bettoosy coed"—probably mispronounced as often as Llanfairpwllgwyngyllgogerychwyrndrobwllllantysiliogogogoch, which is now called Llanfair, P.G.) are places to see and enjoy. An interesting little tour is a circular trip along the north Irish Sea coast and a return through the hilly and mountainous country which will include these and many other interesting places.

The pleasures of this delightful land can be greatly magnified by staying overnight at some of the homes and farmhouses that provide bed and breakfast. (These are referred to as "B and B" in hotel and travel guides.) We have yet to find one that we did not enjoy. It is an excellent way to get to know and more fully appreciate the kindly Welsh people. Northern Wales, particularly in springtime, rates high in our books as a pleasant, quiet vacation area.

SCOTLAND

is made up of two climatic zones and three topographical areas. The Atlantic side of the country is generally wetter than the eastern portion but the slightly lower temperatures, coupled with strong persistent winds, make the Atlantic side feel much colder in winter.

Topography

The chain of Cheviot Hills coincides with the boundary between England and Scotland. The large, high area extending north to an imaginary line between Edinburgh and Glasgow is the Scottish or Southern Uplands. The southwestern Galloway portion of this area is known as Bobby Burns country. It includes a number of high sections such as the Tweedsmuir Hills in the center and the Pentland, Moorfoot, and Lammermuir hills just to the south of Edinburgh. It is pleasant, hilly country spotted with peaks under 3,000 feet high. There are many streams and rivers, the Tweed River being the most important.

An interesting drive is along the Atlantic coast line of Galloway from Dunfries to Prestwick. At the outer tip of the peninsula, visitors may be a bit startled to see palm trees and various subtropical plants growing outdoors in the area of Stranraer. Perhaps even more so when they remember that they are on a line with Goose Bay, Labrador. This southland atmosphere prevails because of the ever-welcome warm ocean currents and its location in the lee of the Highlands to the northeast.

Immediately above this section is the industrial and commercial belt

Shetland
Islands

Lerwick

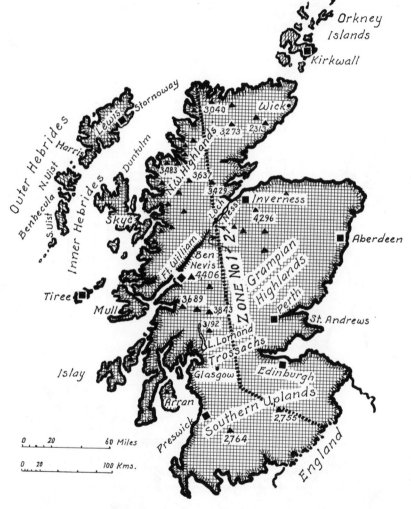

Orkney
Islands

Kirkwall

Wick

3,040
3,273   2,313

Outer Hebrides
Lewis   Stornoway
Harris
N.Uist
Benbecula
S.Uist
Inner Hebrides
Duntulm
Skye

N.W.Highlands
3,483
3,637
3,429
Loch Ness
Inverness
4,296

Aberdeen

Ft.William
Ben
Nevis
4,406

ZONE No 1 2

Grampian
Highlands

Perth

Tiree

Mull
3,689
3,843
3,192

St.Andrews

L.Lomond
Trossachs

Islay

Glasgow   Edinburgh

Arran

Southern Uplands

Preswick
2,764   2,755

England

0   20   60 Miles

0   20   100 Kms.

ıııı Boundary line between
    Weather Zones No.1
    and No. 2.

SCOTLAND

Chart 10

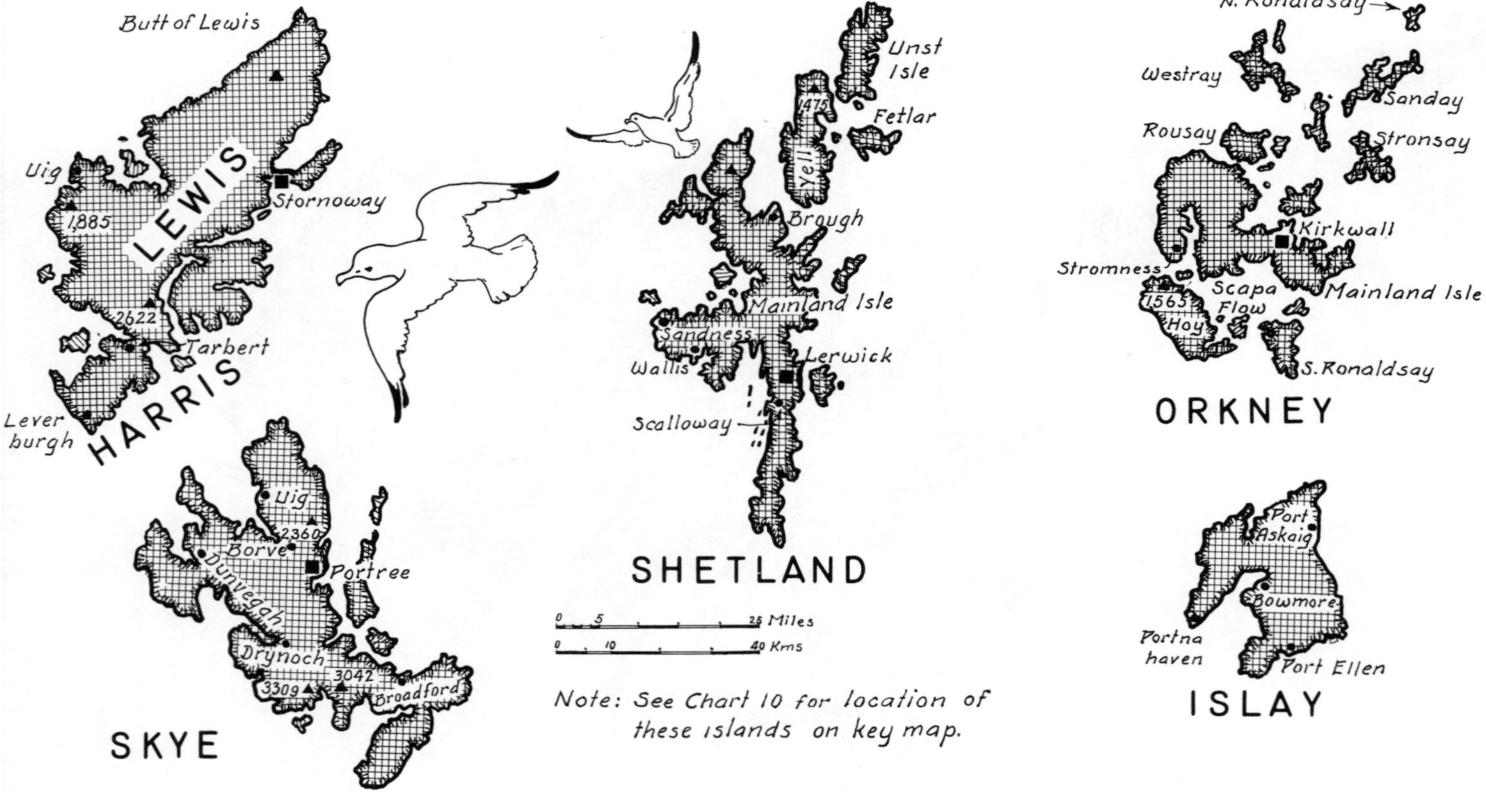

SCOTLAND'S ISLES

Chart 11

ORKNEY

N. Ronaldsay →
Sanday
Stronsay
Westray
Rousay
Kirkwall
Mainland Isle
Scapa Flow
S. Ronaldsay
Stromness
(1563)
Hoy

ISLAY

Port Askaig
Bowmore
Port Ellen
Portna haven

SHETLAND

Unst Isle
Fetlar
Yell
Brough
Mainland Isle
Lerwick
Sandness
Walls
Scalloway

25 Miles
40 Kms

0 5 10

Note: See Chart 10 for location of these islands on key map.

LEWIS

Butt of Lewis
Stornoway
1885
Uig
622
Tarbert
HARRIS
Lever burgh

SKYE

Uig
2360
Portree
Borve
3042
Broadford
Dunvegan
3309
Drynoch

called the Central Midlands. This spans the country between the Firths of Clyde and Forth. It includes the Trossachs (the "bristling country") that one travels over between Edinburgh and Glasgow. There are also several lakes in this region, the most famous being Loch Lomond and Loch Katrine. Most of Scotland's financial and commercial activity takes place in this section.

The renowned Scottish Highlands includes all of the country north of the Central Lowlands. This is the largest topographical segment and, along with delightful Edinburgh and the world-famous golf courses of St. Andrews and Gleneagles, constitutes one of the greatest tourist attractions of Great Britain. This spectacular country of majestic mountain peaks, beautiful lakes, and bold, rocky coasts, needs take second place to few spots in the world.

Much of the central portion of this great region is covered by the massive Grampian mountain complex. In the large peninsula to the north of Fort William and Inverness are the widespread North West Highlands. The spaces between are filled with glens, moors, rivers, and lakes. Little wonder that prospective visitors question if the eye-catching travel folders "are for real."

To the west and north of the Scottish mainland are some very interesting and scenic islands that all too few tourists get to see. There are the Inner and Outer Hebrides chains which include the Isle of Skye and the Lewis and Harris twins. Also, far to the north are the Orkney and Shetland island groups (*see* Charts 10 and 11).

Climatic Zone 1: Western Scotland

Many say that this, particularly the northwestern portion, is Scotland at its best. Others call it wet and wild country, but few visitors fail to express a desire for a return visit. The City Weather Table of Prestwick shows a 35-inch annual rainfall, which does not sound torrential but there will be at least one-tenth of an inch every three or four days. Precipitation is distributed throughout the year, but rainfall is lightest from February to July. There is little, if any, snow.

In spite of the tempering effects of the warm sea currents, the number of frosty and freezing nights increases with latitude and Prestwick will usually get about 50 each winter. There will be a few 90° days in July, but heat prostration has never been a serious problem in these parts. The Prestwick figures are quite representative of Galloway and all the southwestern part of the country.

Glasgow is a brash, bustling seaport that tourists can pass through quickly. The City Weather Table of nearby Renfrew and this general area shows an increase in rainfall to 41 inches a year. Just to the south, in the Troon area, however, there are many fine golf courses which remain a lush green the year round; you can be sure that the hardy Scot misses few days at his favorite game.

Fort William's 79 inches of rain water may explain that the invention of the "mackintosh" by a Scot was no mere accident. No month gets less than 3 inches and December may have about 10.

If that seems like a great amount of moisture in a short period, remember that New Orleans and sunny Miami (which averages 60 inches a year) can have a 15-inch deluge in a few hours. The explanation, of course, is the type of rainfall in each place.

Even though Fort William will get only half as much sunshine as Miami and the rainfall is spread over 240 misty days a year, it manages to get enough good weather to be an important resort center.

AVERAGE HOURS OF SUNSHINE—FORT WILLIAM

	May	*June*	*July*	*Aug.*	*Sept.*	*Oct.*
Hours Per Day	5.6	5.1	3.6	3.7	2.9	1.8

The reason for Fort Williams' unusually high rainfall is illustrated on Chart 10. Loch Linnhe forms a funnel, and the wet sea breezes accumulate in the pocket formed by the high mountains which almost surround the city. Unusual topographical features often cause such local miniclimates. Most of this northwestern area will average closer to 40 or 50 inches than the 79 inches which Fort William gets.

There is an exciting array of islands off the west coast which will some day become much better known. Even now, the names, but little else, are quite familiar to many overseas people.

Isle of Skye, the largest and most strikingly scenic of the Inner Hebrides Islands, is very popular with British vacationists. Here, in a small space, one can enjoy a bit of just about everything that Scotland has to offer. Fishing villages and tiny inland hamlets dot the soft, rolling pastoral lands. There are bold coastal headlands and quiet, sandy coves. There is startling mountain scenery in the 3,000-foot Black Coolins range whose sharp, rugged peaks provide the best rock climbing in Britain. Skye is known for its school of bagpiping. This is also the home of the perky little Skye terrier, and Scotch-based Drambuie has been produced here since 1745.

Mull, Islay, and tiny Tiree are other islands farther to the south of this group. Rugged Mull has a deeply indented coast line which measures ten times the island's 30-mile longest dimension. The tiny isle of Iona, burial place of 50 ancient Scottish Kings, is also famous for its association with St. Columba who founded a monastery there in 563.

Islay is a softer, more peaceful, island with heathery moors and hills, spotted with trout-filled lochs. The coast offers a variety of rugged cliffs and beautiful beaches, the Big Strand being 7 miles long. As in much of this northwestern region, you will see the famed Scottish long-horned cattle, the steaks of which rival the renowned Kobe beef of Japan. Equally important to some of us are the outstanding island distilleries which produce smallish quantities of premium Scotch whisky. (The Americans and Irish use the "e" in its spelling, the Scots, English, and Canadians do not.)

The Weather Table of the Tiree weather station at the strategic southwestern entrance into the island group is representative of the whole northwestern coastal region. Aside from the 45-inch annual rainfall and the 239 misty days a year, there is not much of outstanding interest in the tabulation.

HOURS OF SUNSHINE

	May	*June*	*July*	*Aug.*	*Sept.*	*Oct.*
Duntulm, *Per Day*	6.5	5.7	4	4.3	3.4	2.3
Tiree, *Per Day*	7.6	6.8	5.2	5.3	4.3	2.6

Note that sunshine is more abundant on the islands than such inland spots as Fort William and Inverness.

To the knowing traveler, the wild, haunting, wind-swept Outer Hebrides Islands are the home of the magnificent Harris tweeds and wonderful fishing. The Siamese twins, Harris and Lewis, are really a single island which is less scenically startling than colorful Skye but a visit, using Stornoway as headquarters, will not be regretted.

The City Weather Table of Stornoway, showing 268 days of misty sprinkle, might suggest that this is a land of gloom, but the following table would indicate that this is not quite the case.

HOURS OF SUNSHINE

		May	*June*	*July*	*Aug.*	*Sept.*	*Oct.*
Stornoway	*Hours Per Month*	195.0	173.0	128.0	133.0	111.0	76.0
	Hours Per Day	6.3	5.6	4.1	4.3	3.6	2.5
Benbecula	*Hours 1967*	237.0	236.0	116.0	124.0	94.0	90.0
	Hours 1968	161.0	230.0	180.0	251.0	128.0	60.0

The figures for Benbecula are given for the years 1967 and 1968 to show the possible wide variations that the tourist should always have in mind. This can happen in almost any part of the Temperate Zone but is more likely in the areas most subject to change of wind direction. In the Marine Zone of Western Europe, the greatest annual fluctuations are more apt to be in precipitation and hours of sunshine than temperature.

Climatic Zone 2: Eastern Scotland

From the English border north to the Shetland Islands is over 400 miles. If we start at Edinburgh, as many wise travelers do, we find that we are indeed on the dry side of the country. The 197 days per year when there is a heavy mist must seem to belie the very low 25-inch annual rainfall but there are only 74 days that get one-tenth of an inch or more—and that is not bad. There are 7 months that average less than 2 inches each, but unfortunately for the city's famous festival, August is usually the wettest period of the year. Edinburgh, with its whole center an immense flower garden, is such a delightful city that most visitors forget the slight moisture and enjoy its many blessings.

So much has been written about this charming place that little need be added. By good fortune, there has probably been less change in the central area of Edinburgh than any major city in the world during the past 25 years.

On the way to Perth, the visitor is tempted to stop at those great shrines of golf, St. Andrews and Gleneagles. People who have been in this area before will also want to eat at the Cramond Inn in the small village of the same name.

Perth is a little inland and, at a 400-foot elevation, gets about 60 nights of freezing weather. The 31-inch annual precipitation is in the common range of this eastern side and the spring months are usually driest.

Who would travel this far afield and not get to nearby Loch Ness on the off-chance of seeing the now world-famous "monster"? One twinkling old Scot suggested that a wee nip of mountain dew would greatly improve one's chances of success. The first photo ever taken of Nessie was signed only "photo by a doctor," and the story is that a medico was making a nonprofessional weekend call on a fair maiden at the time. But then, there are many strange tales told in this north country.

Aberdeen, the granite city, continues the overall eastern weather pattern, but even in springtime, there will be a light sprinkle almost every other day. Hours of sunshine will range from only 1.2 in January to a high of 6 hours a day in June. April, May, and June will average 4.8, 5.5, and 4.9, respectively.

Inverness is the attractive spot that many tourists wisely choose as the center of operations for exploring the Highlands. At this latitude (on a line with Goteborg, Sweden, and Ketchikan, Alaska), one is not quite in the Land of the Midnight Sun but there is ample daylight for a round of golf after a normal day of sightseeing.

HOURS OF SUNSHINE—INVERNESS

	May	June	July	Aug.	Sept.	Oct.
Hours Per Day	5.5	5.4	4.4	4.2	3.6	2.7

We should mention Braemar, at a 1,113-foot elevation, which is at the center of the Grampian Mountains. It is being developed as one of the few winter sports resorts in Britain with ski lifts and skating rink. It has long been a mecca for summer fishermen. In this high country, it is possible to get freezing weather at any time from September through May; normally, the mercury drops that low only from November through February. With 200 misty days a year and low temperatures, there can be many frosty periods.

In the spirit of one-upmanship, some tourists trek up to John-O-Groats on the northern tip of Scotland proper. A few, more adventurous, cross the turbulent Pentland Firth to the Orkneys. Twenty of the 70 islands are inhabited but Kirkwall-on-Mainland will be headquarters.

This is wild, windy Norse country that did not become a part of Scotland until 1468. The violent winter storms have eliminated most trees but the summer weather is remarkably mild as the North Atlantic Drift sweeps across the top of Scotland and tempers the climate.

Note on the City Weather Chart of Kirkwall that there are few extremes. Although most of the topography is rather flat and pictorially not too exciting, there are some spectacular coastal spots with sheer 1,000-foot cliffs. Here also is the famed Scapa Flow where the German fleet was scuttled in World War I.

But do not think that this is a poor, barren land. The trend from fishing to cattle has made the Orkadians prosperous to the extent of owning more autos per capita than any other part of Britain. While few foreign visitors get this far north, the Orkneys are a popular summer vacationland for Britains, so it is advisable to have reservations.

Farthest north are the Shetlands which are a bit above Oslo and Stockholm (in latitude) and really in the land of long summer days. But even winters are not that cold. The Lerwick City Weather Tables shows that the January temperature range is 35 to 42°. Some might be surprised to learn that Paris, 800 miles south of Lerwick, averages 32 to

42° in January and Washington, D.C. registers 27 to 42°. In spite of the latitude, there is an almost complete absence of snow.

Summers are cool by our standards, and the 250 misty days mean a lot of clouded eyeglasses. After all this travel, what does the visitor see? There are the famous tiny ponies, Shetland pullovers, colonies of sea birds, and the remains of a Viking heritage. Generally, these islands are flat and rather bare, only the 1,200-foot high sea cliffs rate a post card picture.

The Best Seasons to Visit Great Britain

Potential visitors invariably ask one of two questions—"What's the weather like?" or "When is the best time to go?" The travel bureau says "any time of the year." As indicated above, that blanket endorsement should be qualified. While it is true that one is not apt to suffer heat stroke or freeze to death, each segment enjoys its own particular "comfort weather" season. The vacation weather maps are included to show the overall weather patterns at a glance. They should be examined in conjunction with the weather tables and the detailed descriptions in the text. The maps indicate weather conditions by seasons.

Just about every square meter within the boundaries of Great Britain can be considered a springtime vacation area. It is a most favored season from the viewpoint of scenery, crowds, costs, and most important—delightful sunny weather.

Summer is almost never overly warm but is certainly a very busy time in the popular vacation areas of the country. The British and visitors from all parts of the world seem bent on occupying the same places at the same time. The safest bet is to by-pass the larger tourist cities and places highlighted in travel folders. There are plenty of delightful small villages, stretches of coast, and quiet scenic countryside off the cross-roads of most vacation travel. This is also the prime time to visit the upper parts of the Scottish mainland, the Hebrides, and Orkney and Shetland island groups.

Autumn is about as temperamental as our springtime. It may be delightful but particularly during October can be wet and gray. At its best, it is a gorgeous period and well worth the gamble.

What's wrong with London in winter? It is, of course, misty and overcast, but almost never as cold or windy as New York and Chicago. The many inexpensive theaters, concerts, musicals, restaurants, pubs, and art exhibits are in full swing and, since there is so much to do indoors, it is easy to forget a few leaky skies.

Tourist Information and Transportation

The United Kingdom is now doing an outstanding job in promoting year-round tourism. Well-stocked tourist information offices are maintained in New York City, Chicago, Los Angeles, Toronto, and Vancouver, Canada. The staffs seem especially well informed, not only about common tourist problems but also special or unusual information.

With 15 international airlines serving London and Glasgow, the choice is wide and competition keen. This results in many most attractive and inexpensive package arrangements. BOAC, Pan Am, Irish Air Lines, TWA, and SAS are among the leaders in dreaming up some almost irresistible combination jaunts. A recent United Nations survey indicates that London costs are 80 to 95 per cent as high as those in New York City but small town and rural prices are very much lower.

Britain was host to over 8 million visitors in 1971. Of the 176,000 arrivals during January, 46,000 were from the United States. There is always plenty of room for more, but with so many choosing the July-August period, some spots can be a bit crowded. As we have pointed out, England, Scotland, Wales, and North Ireland all enjoy some mighty attractive weather at times other than those two busy months of July and August and the total bill will be about 20 per cent less than the top season costs.

FRANCE

IF one could assemble segments of just about every type of climate, scenery, and terrain along with dozens of nature's finest beauty spots, the result might be a fair imitation of La Belle France. There are very few nations so uniquely favored. At least some portion of France is included in almost every grand tour of Europe.

Here one can find great variety, from the sea-flooded Norman and Breton flats, which produce that rare lamb delicacy, *agneau de pre' salé*, to the 15,871-foot Mont Blanc, the highest spot in all the Alps. The climate ranges from glaciers and perpetually snow-capped mountain peaks to the sunny Riviera and sandy beaches of balmy Corsica. Snow skiing on the slopes of Chamonix is only a few hours away from water skiing on the blue Mediterranean.

Topography

France has no long chains of towering mountain ranges such as the Andes, the Rockies, or the Atlas. Such barricades drain the sea breezes of moisture and cause wide, dry areas on the leeward sides. Neither are there vast stretches of flat lands as in The Netherlands, which allow these same humid air currents to sweep completely across the country, producing an almost uniform climate throughout.

About half of France is designated as mountainous or high plateaus. Even those portions, however, are cut through with wide valleys, interspersed with patches of low flat meadows, rolling hills, and minor tablelands. The so-called low country is ribbed with rounded hills, small mountain groups, and elevated flat lands. Consequently, while there are some substantial-sized areas having generally uniform weather conditions, they are not quite as clearly definable as in many other countries.

56

England

North Sea

BELGIUM

English Channel

Lux.

Cherbourg

GERMANY

Le Havre

Lille

Rouen

Reims

Caen

Nancy

Paris

Strasbourg

Brest

Chartres

Rennes

Orleans

Dijon

SWITZERLAND

Nantes

Tours

Bay of Biscay

Poitiers

Clermont-
Ferrand

Lyon

ITALY

Vienne

FRANCE

Briancon

Bordeaux

Biarritz

Toulouse

Nice

Monaco

Perpignan

Marseille

Pyrenees Mts.

Gulf of Lions

Pic du Midi

SPAIN

0   100 Miles
0   100 Kms.

4282

Bastia

Calvi

5794

8891

Corte

7733

Solenzara

Cargese

2969

Ajaccio

7008

Sartene

4393

Bonifacio

CORSICA

0   50   100 Miles
0   50   150 Kms

FRANCE & MONACO   *Chart 12*

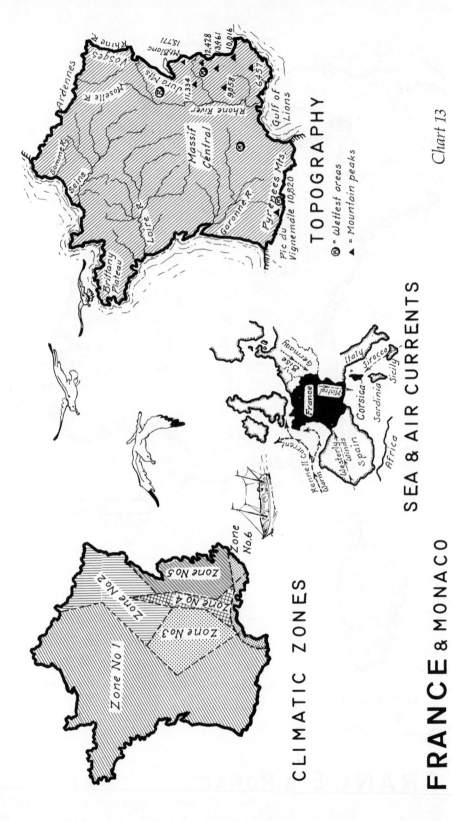

TOPOGRAPHY

Ⓡ = Wettest areas
▲ = Mountain peaks

Ardennes
Somme R.
Seine
Loire R.
Brittany Plateau
Garonne R.
Pic du Vignemale 10,820
Pyrenees Mts
Massif Central
Rhone River
Moselle R.
Vosges
Rhine R.
Jura Mts
Mt.Blanc 15,771
12,428
9,461
10,016
11,334
9,058
6,457
Gulf of Lions

CLIMATIC ZONES

Zone No.1
Zone No.2
Zone No.3
Zone No.4
Zone No.5
Zone No.6

SEA & AIR CURRENTS

Germany
Bise
France
Mistral
Italy
Sirocco
Corsica
Sardinia
Sicily
Spain
Africa
Westerly Winds
Rennell Current
Gulf Stream

FRANCE & MONACO

Chart 13

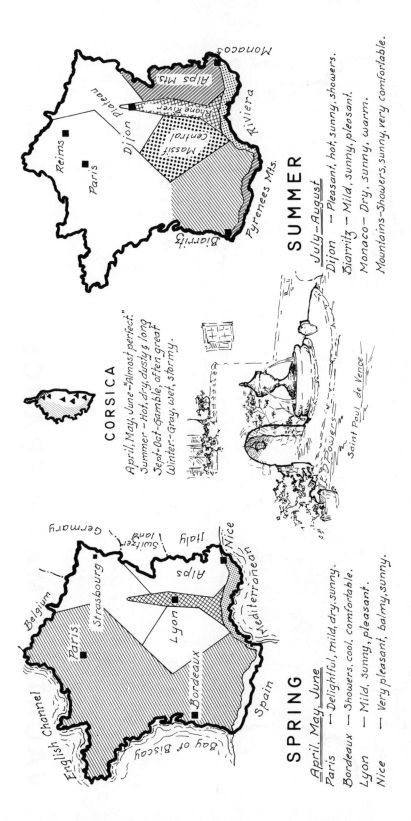

VACATION MAP of FRANCE

SPRING

April, May, June

Paris — Delightful, mild, dry, sunny.
Bordeaux — Showers, cool, comfortable.
Lyon — Mild, sunny, pleasant.
Nice — Very pleasant, balmy, sunny.

CORSICA

April, May, June — "Almost perfect."
Summer — Hot, dry, dusty & long
Sept-Oct — Gamble, often great
Winter — Gray, wet, stormy.

Saint Paul de Vence

SUMMER

July-August

Dijon — Pleasant, hot, sunny, showers.
Biarritz — Mild, sunny, pleasant.
Monaco — Dry, sunny, warm.
Mountains — Showers, sunny, very comfortable.

Chart 14

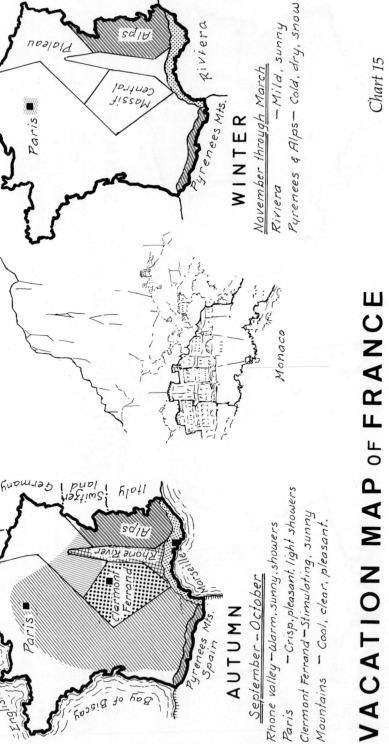

VACATION MAP OF FRANCE

Chart 15

WINTER

November through March
Riviera — Mild, sunny
Pyrenees & Alps — Cold, dry, snow

Alps
Plateau
Massif Central
Riviera
Pyrenees Mts.
Paris

Monaco

AUTUMN

September — October
Rhone valley — Warm, sunny, showers
Paris — Crisp, pleasant, light showers
Clermont Ferrand — Stimulating, sunny
Mountains — Cool, clear, pleasant.

Germany
Switzerland
Italy
Alps
Rhone River
Clermont Ferrand
Marseille
Pyrenees Mts.
Spain
Paris
Belgium
English Channel
Bay of Biscay

The tourist who enjoys the seacoast has an unusually wide choice. France has 2,000 miles of shore line, edging four large water bodies, each with its own distinctive characteristics. There is a small section bordering the chilly North Sea, a portion facing the English Channel, and a long arc around the Bay of Biscay. In the southeast is the beautiful stretch along the Mediterranean. The land boundaries are also mostly natural. The Pyrenees, spanning 270 miles from the ocean to the Mediterranean, separate France and Spain. The lofty Alps, the Jura, the lower Vosges, and the Rhine River are along the eastern side facing Italy, Switzerland, and Germany. Part of the north is hemmed by the hilly Ardennes country. Only at one point do the fields of Flanders open freely into the equally flat lands of Belgium.

The interior regions offer the same contrast. Most of the lands along the western shore lines and far up into the river valleys are relatively flat, very productive farming regions. The large Aquitaine basin in the southwest includes the fertile Garonne River valley and its widely fanned-out tributaries. This, also, is fine agricultural country and, of course, the land of the famous Bordeaux wines, truffles, and *foie gras.*

At the northern end of this strip, the Seine River system broadens into the wide, generally regular, terrain surrounding Paris. The climate is a bit more temperate than along the coast which makes this a bountiful bread basket of wheat and other grains. Since France produces some 340 distinct varieties of cheeses, it would be difficult to find even a small area not associated with a famous name. It is this Ile de France district, surrounding Paris, which produces the renowned Brie, which, during the 1815 Congress at Vienna, was designated a "Royal Cheese." Many are surprised to learn that France is also the country where oleomargarine was invented. It is seldom seen on a French table, however, as almost every district proudly produces an individual type of creamy butter. Toward the eastern edge of this section, in rather inferior agricultural country, is the equally regal champagne domain. This was the home of the good Dom Perignon who did not discover that sparkling golden fluid, but contributed much to its perfection.

Almost the entire eastern half of France is high country. The northern portion is upland plains (about 1,200 feet high) which drop down to the Rhine River, along the border of West Germany. The vast high plateau, called the Massif Central, occupies over 15 per cent of the entire country and contains peaks up to 6,000 feet.

The two other major topographical sections, both in the southeast, are the French Alps and the ever-popular Riviera.

France is divided into *départments* and *provinces.* Each has its distinctive regional foods and most produce compatible table wines. Customs vary widely, from the Bretons, who question that they are really

French, to the Basques in the Pyrenees who know that they are not. The scenery and climate may be equally diverse. The visitor who travels extensively through France has the exciting experience, in effect, of seeing many small but different countries. That is one of the many joys of a lengthy stay in France. Each of these areas is covered in a separate little booklet available at the many excellent Government Tourist Offices in New York City, Beverly Hills, and San Francisco, California, Montreal, Canada, and other large cities around the world.

Climate

France is one of the few countries that experiences three of the dominant European climates—the Marine, the Continental and the Mediterranean. In addition, there are the high plateaux and mountainous areas which include a mixture of these plus a little polar weather.

Although meteorologists often view France as being made up of six major climatic zones, there are actually countless additional local sub- or miniclimates, which vary greatly from their surroundings. Most of these anomalies are the result of topographical factors, often coupled with the effects of sea- and land-air movements. An example might be the position of a valley in relation to the prevailing winds or the angle at which the sun strikes a hillside.

The simple vacation maps (Charts 14 and 15) have been designed to show, at a glance, the most attractive vacation areas during each season of the year. To some travelers, enjoying agreeable weather is paramount, while to others, only seeing a particular place or area is important. Most often, it is possible to attain both.

Climatic Zone 1: Western France

This segment, the largest by far of the six, falls within the vast Marine Zone that envelops much of Western Europe. Its climate is mostly controlled by the prevailing westerly winds that have absorbed moisture from the warm ocean currents. These latter are all branches of the North Atlantic Drift. Their effect gradually decreases as these winds, penetrating far into the interior, encounter mountains or come up against air movements from other directions.

The climate is quite equable with little snow but abundant rainfall distributed through the 12 months. From early spring (in some parts starting in January) until midsummer is usually the driest part of the year, while April, May, and June are often the most pleasant with greatest sunshine.

Those 3 months are by far the best periods to visit the Normandy and Brittany peninsulas. These interesting areas are noted for fine foods,

Calvados, and delicious Camembert cheese, but certainly not for having Florida weather. Springtime, however, is not only the most agreeable part of the year, it happens to also be the uncrowded, low cost off-season when service is unhurried and accommodations most available.

Most North Americans are accustomed to enjoying glorious sunny autumns and more tempermental springtime seasons which may be delightful or quite often never happen at all. They are the ones who arrive in Western Europe in late summer and frequently are disappointed with the cool unpleasant weather. July and August are the particularly busy months when prebooking is advisable. Note on the City Weather Tables of places on these peninsulas that September and October also often enjoy very good weather. That, of course, is a less crowded period than midsummer. Be warned, however, that the weather during those two months is somewhat of a gamble. There is always the chance that there will be an exceptionally fine season but much more likelihood of experiencing real webfoot weather. The following local expression probably best describes this general area, "If you can't see a mile out to sea, it's raining. If you can, it's about to rain."

This whole coastal strip is much like the western shores of Ireland and Great Britain—gray and wet, particularly in midwinter when 4 to 6 inches of rainfall a month is not uncommon. As indicated on the City Weather Tables of Cherbourg, Brest, and Biarritz, the total annual precipitation ranges from 37 to 49 inches.

Winters are generally mild, but storms from the ocean can be violent with gale force winds. Some of the highest tides in the world occur along the coast of Brittany. The sea can rise and fall as much as 44 feet twice a day, and a unique installation near St. Malo generates electric power from its action.

Farther inland, precipitation decreases, and Paris, Orleans, and Reims average about 25 inches a year. In common with most of this zone, spring is long and arrives early, while unlike America, autumn is short and most often quite wet.

At this distance from the ocean, there is a tinge of the more temperate Continental climate. Temperature ranges tend to be a bit wider. There are more apt to be freezing winter nights and hot summer days when the mercury will occasionally climb to 100°. Quite rapid temperature changes are not uncommon and usually occur when there is a shift of wind direction. If it is from the east, the weather will most often be cooler but drier. The ocean breezes are mild and humid.

While Paris is almost always just about perfect in the springtime, it is a city of all seasons. There is so much to see and do, both indoors and out, that no day need ever be a loss. It is also a convenient headquarters for visiting all of the exciting places and sights in the surrounding country. Many who are on short visits take advantage of the helicopter tours to the chateau and wine country of the Loire Valley and other nearby towns and places of interest. But whether the stay is a day or a year, there will always be many more fascinating things to see and do on the next visit. The Orleans and Nantes Tables show the type of weather which can be expected in that area. Lille, in the northern industrial region, is not a place of great tourist interest.

As might be expected, the average temperatures of Bordeaux, Toulouse, Biarritz, and Perpignan, across the southern part of the country, are 5 or 6° higher than most other sections of this zone. Note that in both of the wine districts, Bordeaux and the Loire, the late summer grape ripening season is rather dry and sunny. The abundant sunshine and warmer weather in the south accounts for the sweet dessert wines such as the premium Chateau d'Yquem.

The Pyrenees, separating France and Spain, should also be mentioned briefly. The Western or Atlantic end is quite wet as indicated by the 49-inch annual rainfall at Biarritz. The very high sections, mostly toward the center, get even more. Pic du Midi, directly south of Lourdes but at 9,378 feet, averages a whopping 75 inches a year. At that elevation, of course, it falls mostly as snow. Perpignan is representative of the Mediterranean side. The 23 inches is distributed rather equally through the year.

This part of France has long been popular with English tourists and many vacation in the Pau area. It is delightful country with a great variety of scenery. A drive from the Mediterranean to the Basque country at the western end can be a most enjoyable vacation. There is considerable summer sunshine and the mountain air is fresh and invigorating.

Climatic Zone 2: Northeastern France

This rather rough upland plain country (elevation about 1,200 feet) is a little off the beaten path of tourism. Strasbourg, the German-French city on the left bank of the Rhine River is the only place apt to be included in many itineraries.

The weather in this overall area is inclined to be somewhat extreme and uncertain. Strasbourg, which is a most colorful city, has experienced both 102 and minus 10° readings. While those are not common occurrences, there will be about 100 nights a year when the mercury drops to freezing.

The reason for this rather harsh, erratic weather is twofold. Marked differences in topography cause distinct types of subclimates. This plateau varies from rocky elevations to flat peat lands, cut through with deep valleys. There are the Ardennes heights in the north and the 4,000-foot, peaked Vosges on the east. The second factor is the shifting wind. The warming breezes from the west have lost much of their effectiveness this far inland. The occasional cold, dry, northerly wind, called *bise,* brings with it heavy clouds and discomfort. Neither scenery nor climate in this region are particularly inviting.

Climatic Zone 3: Massif Central

The major headwaters of the extensive Loire and Garonne river systems originate in this high, rugged plateau which covers the whole southcentral portion of France. It has a most irregular terrain, being a complex of deep valleys, relatively flat patches at various elevations, and a scattering of mountain peaks of over 6,000 feet. The southeastern portion is the highest part of this plateau. As in most such rugged and diverse country, there is a wide range of climatic conditions, dependent on altitude and configuration of the land. Exposure to, or protection from, the wind and sun also have an important bearing on the situation.

The warm, moist breezes from the far-off Atlantic coast temper the weather on the western slopes of the Massif, giving them a modified marine climate. The opposite sides, in most cases, are open to the sharp winds from the northeast and are generally drier. Conditions are also more severe in the large interior portion which has a typical high-elevation Continental climate. Winters are long, cold, and snowy; summers are bright, dry, and hot, while autumn gets the highest precipitation. Although spring can sometimes be a showery season, most often January, February, and early spring are the driest and sunniest parts of the year. This is true in Clermont-Ferrand, Vichy, and to a lesser extent, Millau.

A crescent fringing the southeastern edge of this high country enjoys a mild Mediterranean climate and except for differences due to elevation, weather conditions are somewhat akin to those along that sea. This favorable condition is common to other similarly situated areas. The canton of Ticino, Switzerland, is also on the leeward side of the Alps and sheltered from the cold northern blasts. At the same time, it gets a maximum of exposure to the southern sunshine and the mild breezes from the warm Mediterranean.

There is considerable choice in both climate and scenery which makes this large, generally overlooked area very fine touring country.

Climatic Zone 4: The Rhone Valley

As is well known, this bountiful valley has a chain of Michelin-starred restaurants, magnificent wines, and comfortable medium-class accommodations. It is difficult to imagine anything more pleasant than a leisurely tour from one end of the valley to the other during the grape gathering and pressing season. There is an infectious gaiety at that time of year. If the budget can stand the strain, how better to top off this memorable jaunt than dinner at the Pyramide Restaurant in Vienne, which many call the finest in all France. The Confrerie des Chevaliers du Tastevin celebrate with an annual banquet at the Clos de Vougeot in late autumn. That event, of course, is a winetaster's dream.

The City Weather Tables of Dijon, Lyon, and Marseilles span the weather conditions of the Rhone Valley from north to south. The modest 22 to 29 inches of rainfall is spread throughout the year with late fall and early winter getting the heaviest precipitation. Both daily and annual temperatures fluctuate in a narrow range. The south end will average 4 or 5° higher than the northern sections. Note that there are usually many more 32° than 90° readings although 0 weather, while rather unusual, is not unknown. Occasionally, the delta mouth of the Rhone will experience the tail end of a *sirocco,* a wind which originates in northern Africa. This is hot, humid, and enervating and, adding to the discomfort, is sometimes dust-laden. Fortunately, it appears very infrequently and usually lasts only a day or 2, generally being followed by a cooler northerly wind. While it may occur at any time of the year, such summer winds are rare.

Of far greater concern is the notorious *mistral* which is featured in so many of the stories of this area. This cold, dry, northerly wind blasts in most often during the winter and usually at a velocity of 30 to 40 m.p.h. It has, however, been recorded at up to 80 m.p.h., overturning trains. Although it sweeps down across this whole section of southern France, this violent air stream is mostly funneled down the Rhone Valley until it hits the Gulf of Lions, making passage across that body of water a rough experience. This persistent wind, even when not strong, can be most annoying, particularly to residents who must live with it day after day. Visitors may notice that dense rows of tall cypress trees are often planted as windbreakers, and many houses and buildings are windowless on the windward side.

Climatic Zone 5: The French Alps

These spectacular mountains should be considered in two sections. The area to the north is much higher, colder, and wetter. Here are the lofty Mont Blanc and other towering peaks. Close by are Chamonix, Megève, and a dozen smaller skiing meccas. Snow is deep and plentiful during long periods of the year and mountain passes may not open until late spring. It was in the awesome chasms of the Annecy district that sky-flying was invented. Summers can be hot, sunny, and wonderful. At such altitudes, there are rapid and pronounced changes in temperatures particularly just after the sun sets. Rainfall is plentiful, but mainly during fall and early winter. At the higher altitudes, precipitation is in the form of snow—in some spots throughout the entire 12 months.

The southern Alps, affected by proximity to the Mediterranean, are much less cold. Except at the higher elevations, snow is not heavy. Summers are very bright and pleasant, with only moderate rainfall. The City Weather Table of Briançon, at a 4,613-foot elevation, shows what the temperatures and precipitation are in this general area.

Another phenomenon, experienced more in the Swiss Alps, is the *foehn*. Canadians and Americans, familiar with the *Chinooks* which blow down the eastern slopes of the Rockies, will recognize the foehn. These are dry, hot winds which descend the leeward side of mountains and can raise the temperature 30 or 40° in 15 minutes. They are generally beneficial but, on occasion, can cause some damage. The foehn occurs most frequently in spring when it can sweep down and melt residual snow patches in a matter of minutes. The autumn foehns are usually welcomed as the resultant rise in temperature helps ripen crops, particularly, the grapes.

Climatic Zone 6: The French Riviera

This sunny strip ranks second only to Paris itself as a tourist goal. While that is quite natural, it is also unfortunate because after a visit to these two star attractions, all too many visitors assume that they have seen France. It is a glorious strand, however, and well deserves its great popularity.

The word "Riviera" brings to mind sunshine, water sports, boating, and swimming. The following table is the sunshine record of six spots along this coast:

These are very attractive figures, particularly to people in the northwestern part of Europe which experiences so many gray and often misty days. It might be interesting to compare the above table with a few of the bright spots in America. Miami averages 2,903 hours of sunshine a year, which is in the range of these Riviera cities. It is more equally distributed throughout the year in that part of Florida which gets from 209 to 227 hours per month from November through February.

Los Angeles averages a total of 3,284 hours a year and gets from 217 to 249 hours of sunshine per month during the winter.

AVERAGE HOURS OF SUNSHINE PER DAY AND YEAR

	J	F	M	A	M	J	J	A	S	O	N	D	Year
Perpignan	5.2	6.1	6.6	8.1	8.2	9.2	10.3	9.1	7.6	6.0	5.3	4.5	2,644
Montpellier	4.6	5.9	6.6	8.3	8.9	10.6	11.9	9.9	8.3	6.0	4.8	5.9	2,742
Marseille	4.3	5.6	6.6	8.3	9.1	10.6	11.9	10.3	8.5	6.0	5.0	5.9	2,763
Toulon	4.6	6.1	6.9	8.4	9.3	11.1	12.3	10.6	9.1	6.5	5.3	4.5	2,881
St. Raphael	4.6	6.1	7.0	8.6	9.5	11.3	12.6	11.3	9.2	6.6	5.4	4.5	2,959
Nice	4.6	5.9	6.3	8.1	8.9	10.4	11.8	10.3	8.6	6.4	5.3	4.5	2,778

As the Nice Weather Table shows, there are few extremes in temperature. Except for October, November, and sometimes December, rainfall is quite modest. It is easy to understand why this coast is so popular at all times of the year. Generally, the high seasons along the Cote d'Azur are December 15 through March 15, Easter Week, and July 1 through September 15. There are also high crowd periods such as the May Grand Prix and the midJanuary Rally. In between times, many of the hotels reduce rates about 20 per cent.

The sea temperatures are not as high as might be expected, and particularly during the winter, more swimming is probably done in pools than in the surf. Only a small percentage of the French Mediterranean coast is sand beach, most being either beautiful rock formations or pebble flats. The sea does not reach a maximum temperature of 70 to 73° until mid-August. The following table is quite typical of this whole Mediterranean coast.

AVERAGE SEA WATER TEMPERATURES (°F)

	J	F	M	A	M	J	J	A	S	O	N	D
Marseille	55	52	56	59	64	69	71	71	70	67	61	57
Monaco	57	55	55	57	63	69	72	76	73	69	63	59

Overseas visitors should realize that the Riviera winter does not carry the same weather guarantee as, say, Marrakech or the Caribbean. Just as in Florida, there can be cold disagreeable spells which fortunately do not happen very often and winter weather is ordinarily very pleasant.

CORSICA

In ancient times, Corsica, as many such situated islands, was a cross-roads for both the traveler and conqueror. It is again becoming popular with the alert present-day traveler who hopes to enjoy this pleasant spot before the next invasion—this one of mass tourism.

Topography

Because of the varied topography, this 300-square-mile island boasts quite a range of little climates. A chain of mountains, containing a number of peaks up to almost 9,000 feet, extends roughly down the center from north to south. The west coast is high and rugged with deep inlets, resembling the shores of Brittany and western Ireland. The east side of the island is generally flat land and, until the marshy portions were drained, was malaria-ridden.

The west side is still by far the more popular and contains the most attractive hotels and accommodations. The many beautiful, white sand beaches, now so peaceful and undeveloped, are certain to attract great crowds in the near future. Corsica is more sparcely vegetated but a bit greener than the Island of Sardinia, just 15 miles to the south. It has retained considerable flavor of its former Italian rulers but also contains the almost wholly Greek town of Cargese, where that language is still spoken and taught in the schools.

Climate

Summers are long, hot, dry, and dusty, particularly on the eastern side which can be most uncomfortable in June, July, and August. The mountains barricade any relief from the west and breezes are apt to be hot, humid, and sometimes in the form of dusty blasts from Libya in North Africa where the Arabs call these desert winds the *gibli.* Occasionally, there are rain showers—also originating across the Mediterranean—known as blood rain because they stain every exposed surface with a fine red dust. Another weather occurrence, much more common in the eastern end of the Mediterranean, is the *calina haze.* This occurs during July and August and consists of powdery dust clouds which can turn a bright, blue sky to a leaden gray in minutes. Fortunately, all of these climatic freaks appear very infrequently and then only for short periods. Usually, there are alternate land-sea breezes which are very refreshing.

Because of the refreshing westerly winds, that side of the island, especially toward the north, is quite tolerable in warm weather. The ideal spot in midsummer is at about a 500- to 1,000-foot elevation in the western hilly interior. There it is sunny and pleasant, with almost no rain.

The following table shows a 15-year sunshine record in Ajaccio which is quite typical of the island.

AVERAGE HOURS OF SUNSHINE PER DAY AND YEAR

	J	F	M	A	M	J	J	A	S	O	N	D	Year
Hours Per Day	4	5	7	8	9	11	12	12	9	7	5	4	2,790

The out-of-season periods of May, June, late September, and early October are likely to experience the most perfect weather. However, there can be some heavy showers and perhaps a few stormy days in the fall. Spring is long, and because it arrives so early, April is usually a wonderful weather month.

Unless there has been a particularly disastrous summer, there will be good swimming from late June until the first of October. Some of the really hardy take a dip in the more protected coves even during midwinter—but that is not for everyone.

As indicated in the Weather Tables of Ajaccio, Bastia, and Solenzara, winter is the wet season. It is the time of heavy showers along the coast and plentiful snowfalls in the high mountains. While the low country can be sunny and pleasant at that time, the prudent visitor chooses a location sheltered from the wind. Several winter resorts are being developed in the higher, interior mountains.

The temperature range throughout the year along the coast is rather narrow, with few extremes. There are some 90° days and, on occasion, even 100° has been recorded. A zero reading is almost unknown and even 32° weather is rather rare at these lower elevations.

General Information

Corsica is still in the delightful stage of being uncrowded, but because there has been no large tourist hotel development, it is well to book accommodations in advance. The French Government is sponsoring a group of very sensibly priced hotels and inns which will go far in attracting visitors. Although no tourism boom has inflated prices, costs are not low as most foodstuffs must be imported. Nor will one find the haute cuisine of Paris. There is, however, wonderful fresh fish in great variety, savory dishes of beef, lamb, goat, and even wild boar. It is said that feeding on *maquis* is the cause of the peculiarly intriguing flavor of the meats. That is the shrub which gave Corsica the name "Scented Isle." Also, of course, the designation of the first French underground which could quickly disappear into its dense growth.

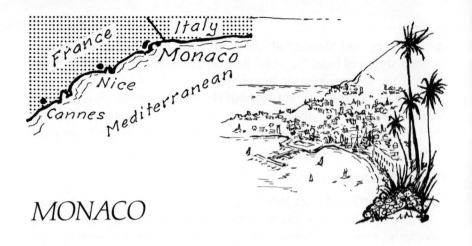

MONACO

Any country that can satisfy the demands of tourism through all seasons of the year is certainly a most fortunate one. To have all of this fitted into an area scarcely half the size of New York's Central Park is a bit unbelievable. But it is true. The place—Monaco; the time—any day of the year.

Accommodations fit every purse and the same is true of restaurants in even greater variety. There is golf, tennis, sailing, swimming, fishing, and general beach and sea life in every form. To be sure, one must travel a couple of hours to snow ski but need walk only a few hundred yards to enjoy its marine counterpart.

Climate

But—and this is a very big important question—what is the weather like? Let us start by saying that, on the average, there are only about 60 days a year when the sun does not shine. The following table will show in which short period of each month those gray days usually occur.

These figures alone almost tell the beautiful weather story of Monaco. It is obvious that there is no real rainy season and, unlike so many places, there is little likelihood of one's stay being dampened by continuous rainfall.

Although Monte Carlo gets about 30 inches of precipitation a year (New York City averages almost 50 per cent more), much of it falls as infrequent but heavy showers. The fact that there are also about 30 thunderstorms annually indicates that moisture is built up in the atmosphere and usually released as short downpours. These are often only brief interruptions in an otherwise sunny day.

Prospective visitors also will want to know whether long woolens or bikinis will be more comfortable. The surprisingly equable temperature statistics shown on the Monte Carlo City Weather Table give adequate answer. Note that the spread between the average monthly high and low readings seldom exceeds 10°. The range throughout the whole year is very narrow and even the very unusual record extremes have registered a high of only 93 and a low of 27°. A minimum of clothing will easily take care of almost any length stay as far as weather is concerned.

AVERAGE DAYS OF POOR WEATHER

Beginning of January	3 days
End of February	3 days
Mid-March	12 days
Beginning of April	10 days
Mid-May	3 days
Mid-June	2 days
End of July	2 days
Mid-August	3 days
Mid-September	3 days
End of October	3 days
Beginning of November	8 days
Mid-December	8 days
	60 days per year

Being in the sheltered lee of the inland high country, Monaco suffers little discomfort from the cold northern blasts. This coastal area does, however, enjoy almost daily sea-land breezes. On occasion, these Mediterranean shores are exposed to strong *levanters,* strong winds from the east, or *ponentes* from the west. Even less often, a hot, wet sirocco, sometimes dust-laden, will blow in from North Africa. All of this adds up to wonderful overall weather specifications that many areas around the world would love to have.

While most swimming is done in the numerous pools, the blue Mediterranean also attracts a great number. As indicated in the following table, the sea heats up more slowly than land, and it is midsummer before it becomes very tempting.

TEMPERATURES OF MEDITERRANEAN SEA (°F)

J	F	M	A	M	J	J	A	S	O	N	D
57	55	55	57	63	69	72	76	73	69	63	59

Accommodations

Although Monaco is popular the year round, there are several high seasons which are particularly busy and confirmed reservations are advisable. They are from December 15 to April 30, Easter holidays, and July 1 to September 15. Your travel agent can tell you the dates of the Grand Prix, in spring, which attracts racing enthusiasts from around the world, the January Monte Carlo Rally, and other Monegasque events that draw great crowds. Hotel and other costs may be as much as 20 percent lower during the in-between periods.

This delightful little principality is a spot that should be included in every Mediterranean area itinerary.

THE IBERIAN PENINSULA

DO NOT jump to conclusions, if your travel agent should mention bagpipes, plaids, and windmills. He could be referring to the Iberian Peninsula. There are probably more windmills in Portugal and Spain than The Netherlands. And almost every Portuguese fisherman in Nazare is clad in the type of plaid patterns that we call tartans. But he wears them as sea jacket and trousers, not kilts. The folk music of Galicia, in northwestern Spain is often played on bagpipes, and the lonely shepherd in the Balearic Islands may blow the pipes for his own pleasure. The Iberian Peninsula is fortunate in its location and relative isolation from the remainder of Europe. The stockadelike Pyrenees, stretching completely across its only land boundary, have long served as a military barrier. Their current function is mainly a divider of two distinctive climatic regions and a good area in which to ski and play. This border is also the place to change trains since the standard European rolling stock cannot operate on the wider-gauge Spanish tracks.

The 2,000 miles of Continental coast line extend along the often choppy Bay of Biscay, the Atlantic Ocean, and the 1,000-mile stretch of Mediterranean shore lands. Both Spain and Portugal were rather late in developing as tourist attractions as far as North Americans were concerned. In recent years, however, they have surged to the forefront as favored nations for overseas visitors.

Many delightful villages and tiny fishing ports, which had long been quiet little hideaway spots for knowing Europeans, are now busy resorts. Some are little Miami Beaches studded with multistoried hotels and all that goes with overrapid development.

But be of good cheer. There are still thousands of unspoiled little communities in both Portugal and Spain which seldom see an overseas tourist. Hopefully, many will remain undiscovered for years to come.

Some countries are limited to two seasons, one pleasant, the other less so. Topographic and other conditions vary so greatly throughout the Iberian Peninsula that there are many little climatic areas. As a result, there is excellent vacation weather in several parts of this large region every month of the year. You can either decide on the time and check the areas where your favorite weather is most apt to prevail or, if there is some particular place you would like to visit, it is easy to find what kind of weather to expect there at any period of the year.

SPAIN

Some think of Spain as a land of perpetual sunshine. It is generally just that—at least in most areas—but also much more. While the major portion of the country is one where olives, grapes, figs, Valencia oranges, and almonds are grown in sunshine, it also includes a few gray, misty sections and towering, snow-capped mountain areas.

It certainly does not lack in variety. The calendar of winter activities lists skiing and outdoor ice-skating. At other locations, however, during the same season, snorkeling, sun bathing, and water-skiing can also be enjoyed. In some areas, the snow-skiing season extends well into June, but it is also possible to find snow and glaciers in mid-summer.

Spain, of all Western European countries, is one of the easiest on the

Bay of Biscay

FRANCE

ANDORRA

Atlantic Ocean

La Coruña
Santander
San Sebastian
Vigo
Leon
Burgos
Bragança
Zaragoza
Barcelona
Porto
Segovia
Tortosa
Ávila
Madrid
Toledo
Valencia

BALEARIC IS.

Mallorca
Ibiza
Formentera

PORTUGAL
Coimbra
Lisbon
Évora
Nazaré

SPAIN

Alicante
Murcia
Córdoba
Seville
Granada
Lago
Faro
Jerez
Málaga
Cádiz

Mediterranean

0 20   100 Miles
0   100 160 Kms.

North Front
Gibraltar
Africa

France

Les Escaldes
Andorra
Spain

0   5 Miles
0   8 Kms.

ANDORRA

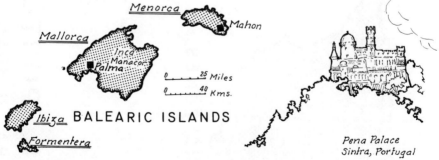

Menorca
Mahon

Mallorca
Inca
Manacor
Palma

0   25 Miles
0   40 Kms.

Ibiza

BALEARIC ISLANDS

Formentera

Pena Palace
Sintra, Portugal

SPAIN PORTUGAL
ANDORRA, GIBRALTAR (U.K.)

Chart 16

TOPOGRAPHY

CLIMATIC ZONES

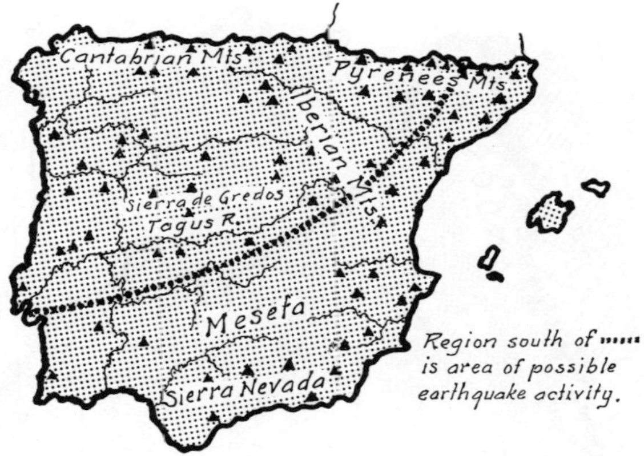

SPAIN PORTUGAL
ANDORRA, GIBRALTAR (U.K.)

Chart 17

SPRING

April, May, June
This is the sunniest and most pleasant season to visit practically all of Spain and Portugal. The southern parts are particularly delightful with blossoms and festivals.

SUMMER

July – August
Most popular tourist season. Interior: northern half pleasant, southern part too warm for some. Mediterranean, Bay of Biscay and Atlantic coasts are delightful.

AUTUMN

September-October
Generally agreeable but showery along Atlantic and Bay of Biscay. Southern interior very pleasant. Not quite as dependable as the spring season.

WINTER

Skiing in Pyrenees. South Atlantic and Mediterranean coasts among best in Europe. Madrid & Lisbon pleasant almost all of the year but about perfect in springtime.

VACATION WEATHER MAPS
SPAIN PORTUGAL
ANDORRA GIBRALTAR

Chart 18

purse. Equally important, even the most modest accommodations, are spotlessly clean. By using one central location as headquarters and radiating to other points of interest, it is possible to live very well and very inexpensively. Spain offers a wide range of accommodations and eating places. Of exceptional quality and value are the government-operated Parador, Albergue, Hosteria, and Refugio establishments which, almost without exception, are in choice locations.

Yes, this is a remarkable country of wide scenic and climatic contrasts, and well deserves its reputation as a year-round vacationland.

Topography

At first glance, there may appear to be an overall uniformity about the country. Actually, Spain is made up of five distinctive areas as shown in Chart 17.

By far the largest topographical segment is the high plateau (2,000-foot average) called the *meseta,* which occupies most of central Spain. It is a harsh, rather desolate region, with lush vegetation limited mostly to the river valleys and a few isolated spots, but these wide spaces are not without scenic interest. Criss-crossed by rivers and mountain ranges with peaks up to 8,000 feet, it resembles dry, sunny Nevada, Arizona, and New Mexico in the southwestern United States.

The fertile band in the north, stretching along the Bay of Biscay from the French border to the Atlantic, is one of the fine agricultural areas of Spain. It is separated from the meseta, to the south, by the Cantabrian mountain range. This coast is indented with many coves, inlets, and bays, but the western Atlantic end in particular is very rugged and spectacular.

The Pyrenees form a towering, natural barrier between Spain and France. Their average height is greater than that of the mighty Alps although the individual peaks are not as lofty. The highest part of this zone is the central portion where summits reach 9,000 to 10,000 feet. The terrain tapers down at both ends, the drop being abrupt toward the Mediterranean and more gradual in the Atlantic Basque section.

The 1,000-mile Mediterranean coast is, most fortunately, sandwiched between the blue sea and many chains of high, sheltering mountains. The Iberian ranges in the northeast and the Sierra Nevadas along the southern section are two of the most important.

This shore line varies considerably. There are broad flat plains which interrupt the narrower ribbon that, in some sections, is not much wider than the roadway. Strangers are delighted to find the almost continuous chain of beautiful sand beaches, the equal of any in Europe.

While the lower Atlantic coast may seem like a continuation of the Mediterranean coast, there are many differences. It is an agricultural land of the grape, fine wines, farms, and a colorful Moorish atmosphere. The fertile lowlands of the wide Guadalquivir Valley extend well into the interior.

Climate

The major portion of Spain is classified as "dry." This is shown on the small center map in Chart 17. The northwestern segment of the whole Iberian Peninsula is exposed to the humid ocean breezes which produce an overall marine climate.

As these air currents travel inland, they give up moisture. The barricading mountain ranges in both Portugal and northeastern Spain drain out almost the last traces, and as a result, the vast region to the southeast of these mountains is dry to semiarid.

The climatic zones of Spain follow much the same general patterns as the topographical conformations.

Climatic Zone 1: The Meseta

This entire zone falls within the low precipitation classification. However, there can be wide annual weather variations. As an example, over the years, the average precipitation of this whole area has ranged from a modest 25.5 inches down to a very meager 8.5 inches a year. (The parched State of Nevada averages just over 9 inches.)

The City Weather Tables of Zaragoza, Madrid, and Granada (spanning this zone from north to south) indicate that the normal average rainfall in these three districts is a very low 12, 20, and 17 inches, respectively.

The precipitation pattern throughout this region is quite uniform. While every month gets some rainfall, there is one not very pronounced moist and drier period each year. The longest and most definite dry, sunny season is summer, which sometimes begins early and may extend for several additional weeks. In the southern part of this zone, precipitation tapers off in April or May and starts to increase again about the middle of September. In the northern section, this drier time is more apt to be from May or June to October.

Spring and autumn are the most pleasant periods for visitors even though they get the highest of a very meager total rainfall.

Note that the total rainfall within a 24-hour period can be quite high but there are few rainy days per month. Couple that with the 12 to 25 thunderstorms occurring during the summer, and it is obvious that much of the precipitation is in the form of short, heavy showers. Most are only refreshing intervals in otherwise bright, sunny days.

A few may find midsummer in these parts a trifle warm. Although the normal July and August readings are usually below 90°, there can be the occasional hot spell when the mercury boils over the 100° mark. Even at the higher temperatures, most tourists seem to suffer little discomfort since the relative humidity seldom averages over an agreeable 50 percent. The high-plateau altitude insures a comfortable drop in temperature after sundown.

Conversely, winters are not severe. About 35 nights a year, the mercury will touch 32° but never sinks to 0°. The winter humidity, however, is quite high and, even during rather mild weather, the North American visitor may shiver a bit in a hotel with marginal heating. As a gauge, notice in the City Weather Tables that Zaragoza, well up to the north, gets about 24 below 32°-nights a year. Granada, in the extreme south, averages 70 days when the thermometer will register 90° or better, which is about twice as many as Zaragoza will experience. The meseta is classified as having an overall Continental climate with hot summers and rather mild to quite uncomfortable or even harsh winters. Spring is usually glorious, while autumn can be either pleasant and bright or somewhat gray and showery, most often the former.

SUNSHINE IN THE MESETA

Average hours per day and year

	J	F	M	A	M	J	J	A	S	O	N	D	Year
Zaragoza	4.2	6.0	6.4	7.3	9.0	10.5	11.6	10.5	7.5	6.3	5.3	4.0	2,719
Madrid	5.0	6.2	6.0	7.8	9.0	10.5	12.5	11.0	8.5	6.6	5.3	4.2	2,859
Granada	5.2	4.5	5.9	7.3	9.2	10.6	11.5	10.9	8.0	6.5	5.9	3.4	2,782

Particularly during July and August, a dense haze, called the calina, can quickly turn a bright, blue sky into a leaden gray. This powdery dust, picked up by the wind, is much more common in the eastern end of the Mediterranean. Fortunately, it usually persists for only a matter of hours or occasionally a day or two. This condition may sometimes be associated with the hot, dry *leveche* winds which blow in from the south. They are siroccos and are more pronounced in the southern parts of the meseta. The *leveches* may appear at any time of the year— strongest and most common in early spring and rarest during the summer. If they pick up moisture in crossing the Mediterranean, they are particularly enervating.

MADRID

The City Weather Table of this magnificent capital is quite representative of the general region. Very important, of course, are the 2,800 hours per year of sunshine. While that rates a bit below the 3,200 figure of southwestern United States, it is about the same as the overall Florida average. Temperatures extend over a wide range. Although there are usually about 40 to 50 days over 90° and the same number of nights when the mercury drops to about 32°, really uncomfortable extremes are quite rare. The low midsummer humidity of 40 to 65 per cent, coupled with the frequent light breezes, makes that season not too uncomfortable for most, even at those times when the thermometer may register up toward the 100°-mark.

The modest 20-inch total precipitation is distributed through the year but is heaviest between October and December. There are only two or three showery days in each month of June, July, August, and September. Actually only about 50 days a year get as much as one-tenth of an inch of rainfall. Some people find the much higher humidity from November through February uncomfortable even when the weather is quite mild. But make no mistake, delightful Madrid is a place of all seasons.

Madrid can no longer be classified as one of Europe's smaller capitals. It is now dotted with huge structures of all kinds. There are towering hotels, apartments, office and government buildings as well as a 120,000-spectator stadium. In spite of all this recent building, however, this beautiful city has retained the feeling of spaciousness because of the very wide boulevards, expansive étoiles, monumental fountains and statues, but more particularly because of the delightful flower-filled parks and landscaped areas. There are probably more attractive sidewalk cafes along tree-shaded streets than in Paris and Vienna combined.

Pleasant 2-day trips can be made from Madrid to Toledo, 40 miles to the south, or to the huge castle monastery of El Escorial and Valle de los Caidos (Valley of the Fallen) in the opposite direction. This latter is a colossal subterranean memorial to the dead of both sides in the Spanish Civil War. Too many miss the colorful Segovia just 30 miles farther north, where the camera bug can have a field day with the many castles, moats, quaint streets, and gorgeous ceramics that only the Spanish can produce. In the same area is the pleasant La Granja, the Versailles of Spain.

Climatic Zone 2: The Northern Coast

By studying the air and sea currents on Chart 17, one can deduce what the weather in this zone must be. The prevailing westerlies funnel into the Bay of Biscay which borders this coastal strip. Picking up heat and moisture from the Portuguese and smaller Rennell ocean currents (branches of the Gulf Stream), they produce an equable marine climate.

La Coruña at the western tip, Santander near the center, and San Sebastian at the French border all get plenty of rainfall. The annual precipitation is 38, 47, and 53 inches, respectively, with the heaviest downfall extending from October through January. Except in La Coruña, where the June, July, and August figures are less than 2 inches per month, most of this area gets substantial precipitation throughout the entire year. Note that there are about twice as many rainy days as in the meseta. There will usually be 150 to 200 days a year that get at least light, misty sprinkles.

The highest annual average precipitation ever recorded for the entire zone was an ample 52.7 inches. The lowest was a still-hefty 37.5, which is a fair indication of a generally wet climate. Weather conditions are quite similar to those found along the western borders of Ireland, Great Britain, and the Low Countries—and for the same reasons. The northern coast will average 2 or 3 hours less sunshine a day than most of Spain. Note, however, that these areas will get from 5 to 8.5 hours of sunshine a day during the entire 6 months of April through September.

AVERAGE HOURS OF SUNSHINE PER DAY AND YEAR

	J	F	M	A	M	J	J	A	S	O	N	D	Year
La Coruña	3.1	4.4	4.6	6.3	7.0	7.7	8.5	7.9	6.3	5.0	3.7	2.7	2,062
Santander	2.7	3.5	4.5	5.5	6.0	6.4	6.9	6.3	5.0	4.2	3.1	2.5	1,747
San Sebastian	3.0	3.4	5.0	5.3	6.4	6.5	7.0	7.0	5.4	4.6	3.2	2.7	1,831

This northern coastal region is relatively unknown to overseas visitors. It is a delightful summer vacation area with every kind of accommodation from the superdeluxe to the very modest. San Sebastian corresponds to the fashionable Biarritz just across the French border. Both are handsome holiday resorts in Basque country, where the proud people consider themselves apart from other Spaniards; in truth, their ethnic background has never been fully established.

This whole coast is a paradise for those who enjoy seafood. In addition to a wide variety of grills and regional fish dishes, the natives of this magnificent rugged coast have a magic with fish soups and stews. There are many tiny fishing villages in the numerous little coves and inlets—such as the San Pedro and San Juan twins—which are noted for fresh-from-the-sea lobsters, oysters, and other shellfish.

This is really the green belt of Spain and differs widely from the almost semiarid interior region just south and east of the coastal mountain ranges. Bilbao, an industrial city, is of little interest to tourists, but there are many inviting places all along the coast to the west.

The Santander section combines fine beaches, excellent salmon and trout fishing as well as the deep sea variety, golf, and lively night life for those with that much stamina. La Coruña, on the very northwestern tip of Spain, is a windy place during the cold season but most pleasant and refreshing in summer. Few Americans, but English and other Europeans, vacation here in Galicia Province. The nearby Santiago de Compostela, a national monument city like Toledo, is well worth a visit. It is a fascinating strolling city with something of interest at every turning.

Neither the average monthly nor annual temperature spread, along this north coast, normally exceeds 10 to 15°, with few 0 or 90°-days. On occasion, the mercury can climb to 100° or a bit higher but seldom sinks much below the freezing mark. Potential visitors should remember that, while winters are generally rather mild, the frequent sea winds are sometimes sharp and uncomfortably chilling. La Coruña, as an example, will experience 18-m.p.h. winds about 20 per cent of the time during winter and quite often must brace against blasts of 30 m.p.h. or more. As indicated on the Wind Chill Table, 35° in an 18-m.p.h. wind feels like 16° and in a 30-m.p.h. blow feels like 5°. This, in combination with the drenching winter rains, explains why there is no tourist boom at that time of the year.

Although spring and fall can be very pleasant, the prime periods in this coastal strip are July, August, and September. It is the season of lightest rainfall with wet weather or perhaps only a shower about 1 day in 8. This period will also average 7 or 8 hours of sunshine a day which adds up to very pleasant vacation conditions. Europeans, particularly the British, enjoy the tangy, summer sea breezes which they find pleasant and invigorating. As an extra bonus, most places along this coast are not jammed with tourists even in midsummer.

Climatic Zone 3: The Pyrenees Mountains

As in most mountainous terrain, the climate varies considerably depending upon altitude and exposure. The western segment accumulates the highest precipitation from the wet Atlantic winds and the annual rainfall, which occurs mostly in autumn, can exceed 60 inches. Rain, at the milder Mediterranean end, is heaviest during the summer months, but total rainfall is lower than toward the west.

The central section is unusually dry, being limited to about 20 inches, but since 1 inch of rainfall equals about 10 inches of snow, this adds up to plenty of skiing.

As might be expected, there are many little enclaves that have local miniclimates. Some are severe, others mild, depending upon whether exposed to sharp wintery blasts or in a protected location available to sunshine. While spring arrives earlier than on the French side of the mountains, winter, even in the valleys, can be long and sometimes quite harsh. This is both a summer and winter vacation country with skiing resorts very much on the increase. A vacation tour from the Bay of Biscay along the spectacular Pyrenees to the shores of the Mediterranean in midsummer can be a delightful experience. A detour through historic Pamplona where, for 1 day in mid-July, a herd of fighting bulls racing through the streets and scattering crowds of youths who engage them in mock combat, adds to the excitement of the trip. A visit to attractive Zaragoza and a day in the unique little mountain State of Andorra also fit well into this itinerary.

Climatic Zone 4: Mediterranean Coast

Aside from Madrid itself, this is the most popular tourist attraction and is delightful during all 12 months. It has a largely subtropical Mediterranean climate with short, mild winters and dry, sunny summers. Most of the relatively low rainfall occurs in early spring and fall, being a bit heavier during the latter period. Either October or November usually has the honor of being the most moist month in this dry country. Note on the City Weather Tables that the north and south ends of this great arc, Barcelona and Malaga, get 20 to 24 inches of precipitation, while Valencia and Alicante, nearer the center, average only 16 and 13 inches annually. Almeria, in the southeastern corner, getting a scant 8.6 inches, is one of the driest and sunniest spots in all Spain. There is hardly any rain along this coastal band from April through August, usually about one to five showers per month. It is not uncommon for them to be in the form of brief thunderstorms which may interrupt the sunshine for an hour or less. There is no rainy season in this part of sunny Spain.

Temperatures are neither high nor low; normally, the summer range is about 60 to 85° with perhaps 5 to 15 days when the mercury will top 90°. While unusual, there are occasional 100° hot spells which may last several days. Winters are even more moderate with very few 32° readings, while 0 is unknown. The relative humidity is remarkably uniform throughout all 12 months.

As indicated in Chart 17, most of this coastal band is protected from the north and western blasts by long, almost continuous, high mountain ranges that run parallel to the shore line. There is, however, complete exposure to the sea, and the leveche often blows in from North Africa.

The south coast as far west as Gibraltar is also occasionally swept by the levanter wind. These air movements can be either moderate or quite violent, sometimes generating wet, gusty storms.

Another climatic happening which surprises most overseas visitors is the occasional blood rain. This is very fine red dust which has been picked up by the winds that sweep across the African deserts. The tinted raindrops stain any exposed surface.

Most often, however, there are pleasant breezes, just right for sailing but disturbing the sunbather very little. In general, the weather picture along this beautiful coast adds up mighty close to 100 per cent perfect.

The designation, Costa del Sol, could fittingly be applied to almost this entire zone as suggested by the following sunshine table.

AVERAGE HOURS PER DAY AND YEAR

	J	F	M	A	M	J	J	A	S	O	N	D	Year
Barcelona	4.4	5.9	5.6	6.9	8.1	9.2	10.0	7.6	6.6	5.6	5.0	4.3	2,437
Valencia	5.2	6.2	6.1	7.3	8.5	9.6	10.6	9.1	7.6	6.1	5.1	4.6	2,643
Alicante	5.9	6.9	6.9	8.5	9.7	10.7	11.4	10.6	8.3	7.0	6.2	5.5	2,974
Malaga	5.9	6.3	6.2	7.8	9.6	10.9	11.3	10.6	7.9	6.5	6.3	5.6	2,901
Seville	5.3	6.5	5.9	7.8	9.6	10.9	11.5	11.0	8.9	6.3	5.6	4.7	2,862

Notice that sunshine increases toward the south. Barcelona, with cloudy skies one day out of two (190 per year), gets much more gray weather than either Alicante or Malaga. In spite of its rather generous sunshine record, it will average only about 90 days a year classified as completely sunny. Conversely, there are not more than 70 rainy days, so Barcelona's weather report will be generally clear but not as sunny as points farther south.

In addition to these places, there are many spots along both the east and south coasts which enjoy their own individual climate. That is generally the result of unusual land configurations. As an example, Calella de la Costa, some 35 miles north of Barcelona, faces southward and is protected from the northern winds by a chain of wooded mountains. This situation produces a surprisingly pleasant year-round climate.

Arenys de Mar, just to the south, has what visitors call a 12-month springtime weather. This town stands on the beach at the foot of some massive mountains that insulate it from the severe inland climatological influences.

Valencia boasts a climatic track record much to be envied. An average year will experience only about 50 rainy or cloudy days, four foggy ones, and nine that can be called stormy weather. Add to that over 2,600 hours of sunshine, a mild equable climate with almost never a freezing or 90° day, and you really have something to shout about. This citrus center is a delight in springtime when the orchards are in bloom or in the winter when the golden-red fruit ripens. But why go on—just look at the figures on the City Weather Table. With a skimpy 16 inches of rainfall for the whole year, there just have to be plenty of blue skies and sparkling days. Much of this coastal segment would' fit into the same weather specifications.

There are a number of similar locations along the glittering Costa del Sol in the south. The Sierra Nevada and other steep mountain ranges shield the coast from the chilly inland winds, giving the shore lands luxuriant subtropical conditions. Some are particularly well protected. Almeria is one such, being almost completely surrounded by towering heights. There is one 11,421-foot peak between Almeria and Granada.

Conversely, some spots on the coast are opposite gaps or openings in the mountain ranges. These places can be mighty chilly and uncomfortable as goose-pimpled winter tourists can attest.

The visitor can find just about perfect weather at some point along this golden coast almost any day of the year. The Spanish Government Tourist Offices have excellent folders of each section of this coast. They contain detailed descriptions and photographs and unusual, large-scale topographical maps, which are generally very difficult to find.

This coastal area is a unique piece of real estate. Within almost any 25-mile radius, one can enjoy practically every type of summer and winter sport and recreational activity. From the blue Mediterranean to the ski slopes is a matter of hours. Few other places can so honestly use the motto, "snow and surf."

Who would dare visit Spain and fail to see glamorous Granada? Fountains, flowers, sunshine, fantastic Moorish architectural gems, and the fabulous Alhambra are but a small part of its lures. And we should mention a most unusual treat, the delightful sixteenth-century Parador de San Francisco within the Alhambra itself. As an extra bonus, inland Granada sits comfortably in the lower mountains which assures a most pleasant summertime climate.

Climatic Zone 5: Lower Atlantic Coast

This small segment might be described as a combination of the Mediterranean and the North Atlantic coasts. It enjoys the equable weather of the northern Marine zone, coupled with the drier and very much sunnier climate of the south.

Andalusia's charming Seville is rivaled in southern Spain only by the equally fascinating Granada. It is a city of music, bull fighting, history, fine regional foods, festivals, flowers, more Moorish architectural gems, and of course—sunshine. The heat-sensitive should either enjoy the comfortable siesta custom or avoid the usually hot midsummer period. Seville is one of the most delightful spots on earth during the glittering blossom-scented springtime season. That also happens to include two short periods when the alternative to preconfirmed reservations is sleeping in one of the many beautiful parks. During Holy Week (just before Easter) and the April Spring Fair, Seville seems to be trying to accommodate the entire population of Spain plus an army of foreigners. Get the exact dates of these two events, although the days just before, after, and between are equally bright and balmy.

There are several alternatives for visitors who find themselves in the middle of a heat wave. Refreshing swimming is close by, and it is only a hop, skip, and jump to higher elevations where days can be warm but nights are almost always cool and pleasant.

This Seville area may see several midwinter nights when the mercury touches freezing but few suffer chilblains in this country. The 24-inch precipitation is about the same as Malaga.

The Balearic Islands

The two islands of most interest up to the present have been Mallorca and Menorca. With the ever-increasing number of visitors, it is probable that Ibiza and tiny Formentera will soon become equally popular. The former boasts some fine rocky coasts, indented with inviting little coves and excellent sand beaches. Formentera is quite flat and rather windy but also has the same good beaches.

As indicated on the City Weather Tables of Palma and Mahon, both of these pleasant spots enjoy fine year round climates. They are seldom overly hot but either can experience the occasional 100°-midsummer day.

The usual summer temperature range is about 65 to 85° with anywhere from 5 to 20 days that will register 90° or over. They are even less often troubled with cold weather. Note that winter figures span only from 42 to 63° but readings below 32° are not unknown.

The rather moderate 20 to 25 inches of rainfall is heaviest in late fall and early spring. Late spring through summer are the driest and sunniest periods, but, as indicated on the City Weather Tables, there is really no rainy season. Actually, there will be more than one-tenth of an inch of precipitation only about 60 times in the entire year.

Prospective visitors should realize, however, that while winters are normally mild and agreeable, there can be blustery days, heavy overcast dripping skies, or even traces of snow. The Balearics are not quite as fortunate in weather as, San Diego, California, or Palm Beach, Florida. But then, those two attractive vacation spots do not quite measure up to the winter weather in some of the Caribbean Islands or Mexico. The ace in the hole for Europeans is North Africa or the Canary Islands where pleasant winter weather is even more certain.

The single weather element of most concern to tourists is, of course, sunshine. Although these islands get a little less than the brightest portions along the Mediterranean coast in total hours per year, they fare very well indeed, as shown in the table below. The sun appears about 300 days each year in Mallorca, which does not leave too much time for bad weather.

AVERAGE HOURS PER DAY AND YEAR

	J	F	M	A	M	J	J	A	S	O	N	D	Year
Palma	5.0	6.2	6.5	7.3	9.6	10.0	11.3	10.6	7.6	6.2	5.6	4.6	2,796
Mahon	4.0	5.1	5.9	6.7	9.0	9.8	11.3	9.8	6.9	5.3	3.2	3.5	2,494

Menorca, the most easterly of the group, is more exposed to the strong northeasterly winds and, as a consequence, does not enjoy quite as mild a climate as the others. This island is an elevated limestone tableland averaging about 1,100 feet, with a higher ridge down the center. There are some low hills, an interesting rugged northern coast, and several fine sandy beaches on the lower south side.

Mallorca, the largest island, has an interesting and varied coast line. It is a succession of small coves, many beautiful, steep, rocky stretches, a great number of excellent beaches, and lots of material for photographers. The land drops down from the high northern section, containing several 4,500-foot peaks, to the 1,500-foot lower end where a lesser mountain range parallels the southeastern coast. Vegetation is generally sparse rather than lush. The higher land toward the north side of the islands provides some measure of protection to the lower southern shores.

PORTUGAL

What makes sunny little Portugal one of the most popular among tourist countries? First among its assets must be its kindly, gracious people. Few tourists, in relating the highlights of their travels, fail to mention the appealing characteristics of the Portuguese. Then there is attractive scenery in great variety. Also, in spite of its modest dimensions, Portugal is a year-round tourist country. Many parts are particularly delightful in the spring while, in others, summer and autumn are generally the driest with most sunshine. The golden Algarve coast ranks up among the top as a winter spot but is also popular at all other times. Lisbon, of course, is a place to visit whenever you are within two days' travel time. Some writers, a bit overenthusiastic, describe Portugal's climate as perpetual springtime. That designation does apply to a few spots, and many flowers do bloom twice a year.

Overall Climate and Topography

Portugal is made up of two major climatic zones—the coastal strip and the northern and eastern interior plateau country. Proximity to the warm Portuguese Current (a branch of the Gulf Stream) gives the immediate shore lands quite an equable climate, with warm, dry, sunny summers and mild, rather wet winters. The temperature range is quite narrow on both a monthly and annual basis, usually not more than about a 10°-spread in either case. Note, however, that there are sometimes abnormally high or low extremes—most often lasting only a few days. There is a wider thermal range through most of the interior regions. As the moist sea breezes impinge on the high mountains, a deluge is released, making this elevated part of Portugal one of the wettest areas in all Europe.

The City Weather Tables show the weather conditions that a visitor can expect although, occasionally, there will be a period of unusual weather. Records show that the whole coastal strip once experienced a year when the average precipitation was an amazing 57.2 inches while another only brought 15.2 inches for the 12 months. The interior during one dry 12-month stretch had only 24.8 inches of precipitation but in another exceptionally wet year was deluged under 86.9 inches. Few visitors will ever encounter such extremes.

The following table shows how precipitation decreases from Braganca down to Faro with a converse increase in hours of sunshine.

| | SUNSHINE | | PRECIPITATION |
	Hours Per Year	*Per cent of Possible Maximum*	*Total Inches Per Year*
Braganca	2,724	60	54
Porto	2,666	60	46
Coimbra	2,604	58	38
Lisbon	3,022	68	27
Evora	2,989	66	25
Faro	3,158	71	17

The City Weather Table of Braganca, at 2,356 feet, in the extreme northeastern corner, shows the typical weather conditions in the mountain country. There have been occasional temperature extremes of 10 and 103°. In this high country, a shift of wind direction can bring with it a sudden and substantial change in weather. With a northeasterly blow, the air usually becomes clearer and drier—accompanied with a noticeable drop in temperature. Such a variation in summer signals more sunshine along with dry but refreshing weather. In winter, it may mean blustery and freezing or perhaps only crisp, bright days. Even with a 54-inch average precipitation, summers are usually very dry, sunny, and pleasant. The southern third of the country is, of course, quite dry but it is the data, shown in the above Table's first and second columns, that establish Portugal as among the sunniest places in all Europe. Few countries can claim that the sun will shine an average of 60 to 70 per cent of every daylight hour. Some can in one or two segments but only Greece will match this countrywide 12-month record. Note that the 4 months of June through September account for little more than ten per cent of the annual total precipitation which makes that season almost weatherproof. And in this whole 120-day period, only about a dozen will get more than one-tenth of an inch of rainfall. While there may be a day or two when the mercury hits the century mark, the normal high and low temperature ranges will be 65 to 81° and 49 to 55°. There can be some snow during any of the other 8 months although December, January, and February will get the heaviest falls.

The City Weather Table of Evora, at 1,053 feet and much farther south, gives a good picture of the weather to be expected in that area. The average annual high and low temperatures are 69 and 51° as against 62 and 42° in the high northern Braganca. There have been 100°-heat waves during the months of June through September, but there usually will not be more than 35 days a year when temperature readings will be much above 90°. Conversely, many winters will experience fewer than five nights of freezing weather; snow usually occurs only during December, January, and February. The relatively dry season, extending from May through September, is longer than farther north. There are usually only one to three light sprinkles in each of the months of June, July, and August. Actually, only 60 to 70 days a year will see any substantial rainfall, the remainder are either sunny and bright or quite clear.

The Porto and Coimbra City Weather Tables show what to expect in the mid to northern parts of Portugal. Aside from the greater precipitation and winter fogs, the only differences are a matter of latitude. Another noticeable feature are the two dozen or so thunderstorms which signify short, heavy downpours. On the other hand, there is just as little likelihood of a 100°-hot spell here as farther south. Low temperature extremes will be equally infrequent.

Winds of up to 20 m.p.h. are common throughout most of Portugal, which accounts for the many active windmills. Very strong winds are rather infrequent and are most apt to be encountered during the off-ocean storms or when the cold northeast blasts howl through the mountains. The sometimes rather violent *vendavale* winds along the northern portion of the Atlantic coast line can generate some disagreeable weather, especially during winter and early spring.

An earthquake belt that includes all of Turkey extends westward across Europe to the Atlantic Ocean. The southern portions of both Spain and Portugal lie within this zone of possible earthquake activity. This is shown on Chart 17. It may be reassuring to the more timid to know that about 90 per cent of all the earthquakes and tremors in the world occur on the perimeter of the Pacific Ocean.

The Atlantic Coast

The whole glorious sunny coast of Portugal is one of Europe's most favored playgrounds. Wonderful climate, excellent beaches, and low costs also make it a popular retirement place for people from all parts of the world. Some prefer this to the French and Italian Rivieras, but that is a matter of taste.

Many crowned and uncrowned heads of state have zeroed in on the Estoril segment—which is very convenient to Lisbon. Here is a place where money can be spent or not as you choose. There are deluxe hotels and the lively casino for those with unlimited budgets, but accomodations of every type and price are available and all will be clean.

One of the many delights of a visit to Lisbon is the ease of getting to the great variety of nearby places of interest. There is a continuous chain of tiny fishing ports, beaches, and resorts that can be reached on the very efficient electric railroad. A bit inland is the fascinating little town of Sintra surrounded by beautiful forests and its storybook Castle of Pena.

Nazare should be on every itinerary. A unique fishing village where all wear colorful Scottish plaids, women in yards of billowing skirts and men in varied, patterned jackets and trousers. Many still haul their fishing boats up on the sandy shore, using several span of oxen for motive power. A short drive inland from Nazare is the pilgrimage town of Fatima, which, like Lourdes in the French Pyrenees, attracts people from all over the world. There is an especially important church ceremony the thirteenth day of each month.

Farther to the north is the port wine center of Porto. Both sides of the Douro River banks above the city are lined with vineyards where the entire production of wine is as carefully regulated as in France.

The lower half of the Atlantic Coast is more popular than the northern portion which does not enjoy quite as perfect a year-round climate. Lisbon toward the center and Faro on the south coast are quite representative of their respective areas.

LISBON

Even a hurried glance at the City Weather Table will show that there are only 27 inches of precipitation spread over 365 days, which does not allow for a very intensive wet season. Usually less than 70 days a year will have more than one-tenth inch of rainfall. July and August get almost none. Every month from October through February or March, however, will have rain, or more often, a light shower, about one day in three. Temperatures are remarkably equable with few high or low extremes.

Perhaps one of the better-known spots in the United States with a climate quite similar to that of Lisbon might be San Diego. While the Portuguese capital and this southern California city are not identical twins, most often they could use a single weather report.

Both are on western coasts, exposed to prevailing sea breezes which generate equable temperature conditions. The high and low annual averages of Lisbon are 67 and 55° while those of San Diego are 68 and 55°. The humidity figures are almost interchangeable, with just slightly higher readings in San Diego from December through February. Lisbon gets 27 inches of precipitation as against only 10 for San Diego, with most of the difference occurring during the winter. In spite of the higher rainfall, Lisbon glitters with 3,022 hours of sunshine a year while San Diego averages 2,959. Neither is overconcerned about extreme temperatures, but the Lisbon thermometer will register over a slightly wider range.

There is an interesting sidelight to this comparison. These two cities have remarkably similar weather conditions even though Lisbon is about 850 miles farther north than San Diego (Leningrad is about the same number of miles north of Venice and Toronto, Canada, above Savannah, Georgia). As always, there is a reason—in this case, it is quite simple. While the climate of each is dominated by the ocean currents, the warm waters of the Portuguese Current sweep down along the Iberian coast, while a cool water body called the California Current flows past San Diego. The action and characteristics of both of these ocean streams are described in detail in Chapter 1.

ALGARVE COAST

Many associate Portuguese sunshine with the Algarve coast, as well they might. Because the golden days there are such a strong tourist magnet, we give the following table in greater detail.

HOURS OF SUNSHINE PER DAY AND YEAR

J	F	M	A	M	J	J	A	S	O	N	D	Year
5.6	6.7	7	9	10	12	12	12	9	8	6	6	3,192

NUMBER OF SUNNY DAYS PER MONTH AND YEAR

J	F	M	A	M	J	J	A	S	O	N	D	Year
29	26	30	30	31	30	31	31	30	30	29	29	356

This is a coast of almost continuous sandy beaches. If the sea seems a bit chilly, remember, too, that aside from the few city outlets, the water along these shores is probably as clean and free of pollution as any similar stretch in all of Europe or America.

AVERAGE SEA-WATER TEMPERATURE (°F) ALONG THE ALGARVE
COAST

J	F	M	A	M	J	J	A	S	O	N	D	Year
59	62	62	63	67	68	71	73	72	69	62	59	66

Visitors enjoy this wonderful southern seacoast every season of the year, but when the almond trees are in blossom during late winter or early spring, it is a glorious place to be.

ANDORRA

This 191-square-mile state, known as the Valleys of Andorra, sits high in the Pyrenees Mountains in somewhat isolated splendor. The average altitude is over 6,000 feet and even the valley beds are at least 3,000 feet above sea level. The Coma Pedrosa, at 10,170 feet, is the highest peak, but there are several others almost as lofty.

Topography has dealt kindly with Andorra. The two main branches of the Gran Valira form the upper part of a large "Y"-shaped valley. They extend northeasterly and northwesterly from Andorra La Vella (Andorra city) which is in the southcenter of the country at a comfortable 3,000-foot elevation. The main river stream or trunk flows southward into Spain. These and many smaller valleys, facing southward, are almost completely shielded from the north winter winds by ranges of high mountains. A glance at a relief map shows why this little country is called "Valleys of Andorra."

As is usual in such high country, the climate varies and is determined by altitude and degree of exposure to the sun as well as protection from the wintery blasts.

Winters are generally quite severe and many of the valleys and passes are completely snowed in for several months. Entry by road from France, once possible only in summer, is now open year round. The easiest access for most tourists is via Barcelona but there is also train service from France. The mercury has been known to drop below −20° but 5° is a more normal midwinter reading. As might be expected, there is a sharp and sudden drop in temperature at sundown even during summertime.

In some places, there is a very marked difference in climate between the north and south sides of a mountain. The portions facing France are somewhat affected by the moist ocean breezes and as a result get higher precipitation. They are also colder and spring is noticeably later in arriving. Most sections of the southern slopes are somewhat drier and generally quite sunny. This makes for a nice combination. There is plenty of long-lasting snow for skiing and other winter activities while the more sedentary, and those with broken limbs, can bask in the warm sunshine.

The ample precipitation is highest during April, May, June, and October. Summers are generally dry and on the sunnier slopes the temperature may reach 90°. In shaded places, however, it can drop to as low as freezing.

There is plenty of excellent skiing (November through April), mountain hiking, and scenic beauty to attract almost a million satisfied visitors a year—many from France. Wedged between France and Spain, it is the ward of both, but by habit and custom, Andorra is more akin to the latter. Although it functions as a republic, Andorra is, at least theoretically, responsible to the Spanish Bishop of Urgel and the President of France who are its co-princes. If this sounds like a strange situation, consider the little Spanish town of Llivia a few miles to the east. While this somewhat isolated enclave is part and parcel of Spain, it is completely surrounded by France and sole access to the motherland is via a rather poorly maintained neutral road. The cost of living is much lower than in France and, in some respects, even below that in Spain. Many articles from all parts of the world are available at practically tax-free prices which makes shopping an attractive pastime. Andorra is a two-season country as far as tourism is concerned. There is a long period of wonderful skiing and a shorter, but equally pleasant, summertime. Autumn may be clear, sunny, and crisp, but often is quite rainy.

Most overseas visitors have failed to include the tiny states and principalities of Europe in their itineraries. Some are now beginning to follow the European tourists in discovering that these often less crowded little hideaways have much of interest to offer in scenery, comfort, and costs.

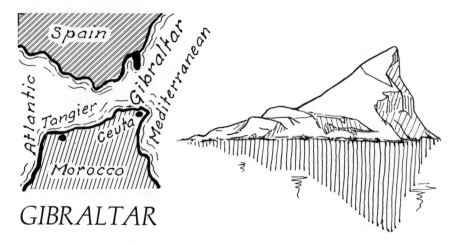

GIBRALTAR

This little 2-square-mile peninsula (not an island) jutting down toward Africa is included here because it is physically a part of Spain. The almost 1,400-foot bold drop from tip to the sea of the Rock itself is almost as familiar in profile as the Eiffel Tower. Winters are wet but mild; summertime can sometimes be uncomfortably hot, but spring and autumn are usually very pleasant. There is plenty of sunshine in summer since there is almost no rainfall during the entire season. Residents welcome the heavily moisture-laden winter clouds which replenish the 20,000,000-gallon-capacity, stone catch-basin reservoirs. The population is dependent upon rainfall as wells are inadequate and often brackish.

Water, has always been an important factor for the Rock throughout its very turbulent history. It was besieged many times and has seen nine flags flying from its peak. Forming one end of the ancient Pillars of Hercules, its strategic position at the entrance of the Mediterranean made it of great military importance.

Although total annual precipitation averages 35 inches, less than 2 inches falls from the end of April through September. Prior to the problems between Spain and Great Britain regarding Gibraltar, the military personnel favored the more comfortable nearby town of Ronda in midsummer.

Climatically, perhaps the most interesting feature is the Banner Cloud which sometimes extends from the summit of the Rock for about a mile along the ridge. This is somewhat akin to the famous Tablecloth of Capetown's Table Mountain. It is largely associated with the levanter breezes from the east which can persist for about a week, usually from August through October. When these air currents reach a high velocity, however, the Banner Cloud lifts and disappears. Few tourists will have occasion to stay more than a few hours in Gibraltar, but those who can spare the time find it an interesting experience.

Passengers heading homeward on the Italian Line or other Mediterranean ships can often pick up last-minute presents at very attractive prices. All who remain overnight will visit Caletas, the colorful little fishing village, the top of the Rock at 1,396 feet, and the miles of military tunnels cut through the mountain. It is impossible to miss the famous Gibraltar Barbary apes, which at times are overfriendly.

THE LOW COUNTRIES

THERE are both plus and minus factors relating to tourism in this little trio called the "Low Countries." Geographically, they are right in the middle of Western Europe—perhaps, too much so. Wedged between their giant neighbors, France and Germany, they are also hemmed in on the Atlantic side by Great Britain.

In spite of all handicaps, The Netherlands and Luxembourg, in particular, have inaugurated excellent tourist programs which are paying off very well. Belgium, busy with its NATO and Common Market installations and as the new European center for many large commercial enterprises, has been a bit less active with pleasure-visitor promotion.

Being generally flat country, there is little of the startling or spectacular scenery featured in many travel folders. It is a region to be enjoyed at a slower pace. There are the wide fields painted in huge patches of every shade that blossoms boast. In other directions are pleasant beaches, lovely old Dutch, Flemish, and Walloon dorps, towns, cities, and the delightful beech-covered Ardennes high country.

Who would willingly by-pass the Grand Place of Brussels, the most elegant square of its kind in Europe, or the canals of Amsterdam and Bruges? And we must remember that these were the homelands of Rembrandt, Rubens, Van Gogh, and Frans Hals. Art lovers have been known to visit Amsterdam primarily to see the famous "Night Watch" displayed at the Rijksmuseum.

But even the most enthusiastic tourist cannot live by art and scenery alone—and there is little excuse to do so in this bountiful land. The ample Dutch larder is well stocked with premium hams, dairy products —the equal of Ireland and Denmark—seafoods and fine fresh food. Be prepared for the inevitable cheese at every breakfast table which the Dutch explain with the saying, "Cheese at breakfast is gold; at lunch

100

NETHERLANDS
BELGIUM
LUXEMBOURG

Chart 19

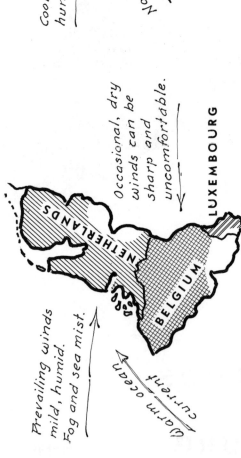

SPRING

April, May, June
Sunniest season of the year,
5½ to 7 hours of sunshine per day.
Comfortably cool, very pleasant.

Prevailing winds
mild, humid.
Fog and sea mist.

Warm ocean current

Occasional, dry
winds can be
sharp and
uncomfortable.

NETHERLANDS
BELGIUM
LUXEMBOURG

SUMMER

July – August
Light showers, many gray days,
5 to 6 hours of sunshine per day.
High temperatures very infrequent.

Cool, pleasant,
humid breezes.

North Sea

Occasional,
hot, dry wind,
generally quite
agreeable.

NETHERLANDS
BELGIUM
LUXEMBOURG

VACATION WEATHER MAPS

NETHERLANDS
BELGIUM LUXEMBOURG

Chart 20

WINTER

Quite frequent, dry, very cold winds. Very uncomfortable.

Wet, sharp sometimes severe ocean storms. Fog and mist.

NETHERLANDS

Amsterdam

North Sea

Brussels

BELGIUM

LUXEMBOURG

WINTER

November through March

Cold but seldom below freezing, heavy overcast rainy skies. 1½ – 4 hours of sunshine per day.

AUTUMN

Occasional cool, dry wind.

Cool, snappy, humid winds, fog and mist.

warm ocean current

NETHERLANDS

BELGIUM

LUXEMBOURG

AUTUMN

September – October

Cool showery but generally pleasant. 4 to 5 hours of sunshine per day. – Fog.

VACATION WEATHER MAPS
NETHERLANDS
BELGIUM LUXEMBOURG

Chart 21

silver but at dinner it is lead." The Netherlands can furnish anything in meals from the hearty Dutch breakfast to the Indonesian *rijsttafel* (rice table) which colonists brought back from the East. The friendly Hollander will interrupt almost anything to enjoy a coffee break, a cured herring, or an eel snack.

The completely underrated Brussels cuisine rivals the very best to be found in Paris, and any city that can boast epicurian spots such as the tiny L'Epaule de Mouton, the elegant Carlton, or a dozen others needs little advertising. The display of farm products in the Belgian markets is a feast to the eye: Perfect tomatoes and mammoth strawberries, each clipped from the plant with two leaves attached, are cushioned in moss or fresh, green beds. Need we mention the world-famous endive and gigantic asparagus? This is indeed a land of good living.

Topography

Lacking high mountains, plateaux, and other irregular topographical features, there are none of the widely differing subclimates which make other countries so interesting.

The land rises, but very gradually, from the sandy beaches of northern and western Netherlands and Belgium to the Ardennes highlands in the southeast corner of Belgium and the northern third of Luxembourg.

About 20 per cent of all The Netherlands is reclaimed land called *polders,* much of which, being up to 20 feet below sea level, is protected by dikes and requires constant pumping. The highest ground in The Netherlands, just over 1,000 feet, is in the southeastern province of Limburg, adjacent to the Ardennes region of Belgium.

The Netherlands has several hundred miles of coast line, much of it fine, sandy beaches, but the total length is forever changing as additional lands are diked and reclaimed. Belgium has only 40 miles of shore land facing on the North Sea, but it is an almost continuous stretch of summer resorts of every description.

Several large rivers cross or cut into these countries. There is the Rhine, which spans The Netherlands; the Meuse, or Maas, forming a portion of The Netherlands-Belgian boundary; and the Scheldt, all of which are broad slow-moving streams and important transportation arteries. The river valleys in southeastern Belgium and Luxembourg are deep and, in some cases, have very steep banks.

The 10- to 20-mile-wide coastal strip of Belgium is low and, like that of its northern neighbor, edged with sand dunes in many sections. The interior of the country is a slightly higher plateau, crisscrossed with many river valleys. The southeastern corner contains the beautiful forests and lakes which cover this hilly summer vacation land. The contours range from 1,400 to 1,800 feet elevation, the highest point being the summit of Botrange (2,283 feet) near the German border.

Luxembourg is made up of two quite distinctive geographical segments. The northern section is part of the tree-covered Ardennes plateau. The more fertile country to the south, called Bon Pays (Good Land), is at about an 800- to 900-foot elevation. This latter section forms part of the large Moselle drainage basin.

Climate

These countries are less fortunate in their weather than many others on the Continent. They are wholly contained within the large region known as the European Marine Zone. Because the land is so flat and regular, there is little to interrupt the free flow of air currents across the countries. This applies equally to the prevailing westerlies from off the North Sea and the occasional harsher winds which blow across the Continent from the east. There is a saying that when Rotterdam experiences a cold snap, every Dutchman in the country develops goose pimples.

The warm sea and air currents that maintain such an equable climate with generally mildish winters also produce many leaden overcast skies. Days are more apt to be gray and moist than clear and sunny.

The following is a 30-year composite average sunshine record of five locations throughout The Netherlands. It is reasonably typical of the western coasts of Europe. Few places in the United States register as small a total figure. The American lows average closer to 2,500 hours of sunshine per year.

AVERAGE HOURS OF SUNSHINE PER DAY AND YEAR IN FIVE LOCATIONS ACROSS THE NETHERLANDS

J	F	M	A	M	J	J	A	S	O	N	D	Year
1.6	2.4	4.0	5.5	6.7	7.3	6.5	5.2	4.6	3.2	1.3	1.2	1,570

Note that April, May, June, and July are the sunniest months. By and large, February through June are also the driest. April most often gets the lowest precipitation and is generally the most pleasant time to visit The Netherlands. The average total hours of sunshine from 1958 through 1967 in Luxembourg was only 1,430 a year.

Normally, there are few diurnal or annual temperature extremes. Occasionally, however, when the wind direction shifts and blows in from the East, after traveling across the Continent of Europe, there can be a few very hot or cold days depending upon the season. The visitor should appreciate that less frequently there can also be an almost entire subtemperature winter or a sizzling summer. Generally speaking, however, altitude and distance from the sea are the controlling weather factors.

The coast and midlands of Belgium and The Netherlands will get about 24 inches of precipitation a year; the higher Ardennes country can expect 36. Overall annual temperatures will average 50° with a January figure of 36 and July of 66°.

THE NETHERLANDS

Of the three Dutch coastal cities tabulated, Rotterdam, the world's largest shipping port, is a bit wetter and grayer than either Amsterdam or Den Helder. Note that there are 66 days with less than a ½-mile visibility, most of which occur during the winter. Rotterdam is a huge industrial and shipping complex in which we counted the ships of 64 nations in a single morning of wharf wandering. It is somewhat hemmed in and enjoys an 18-m.p.h. breeze only eight per cent of the time. The above may account for much of the heavy overcast but we suspect that, like elsewhere, man-made air pollution has contributed most to the gloom. Aside from Rotterdam being the only one of the cities where the mercury has reached 100°, there is little difference among the three City Weather Tables.

Notice the remarkably narrow range of temperatures due, of course, to the moderating effects of the sea. The spread between high and low at anytime through the year is less than 15°. By most American standards, summers here are coolish and winters rather mild. These cities rarely experience either 90° or zero weather although there will be 40 or 50 nights when the thermometer will register freezing.

We have become accustomed to Amsterdam's Christmas cards showing rosy-cheeked Hollanders skating merrily across wide expanses of ice. It is true—they do skate merrily—but not very often, except up north, toward the West Frisian Islands. Also, there is little snowfall and the ground does not stay covered for more than a few days throughout the winter.

As we have indicated, this is a damp part of the world where many days see sprinkles or some mist, but the total precipitation is not great. Twenty-six to 29 inches of rainfall a year is not particularly heavy, but along this coast, it is spread thin. It should be noted, however, that there are also heavy, but brief, showers as indicated by the 30 thunderstorms each year. These are most frequent during midsummer.

The famous Dutch flower-growing area, just behind the coastal dunes, covers a 50-mile stretch. This narrow band is parallel with, and approximately the same length as, a line between Amsterdam and Rotterdam. While there is activity throughout the year, the millions of flowers are in prime, full bloom on Easter morning. Until quite recently, almost all the bouquet and table flowers used on the Continent between January and March came from the Scilly Islands off the southwestern coast of England. Now many are being flown in from the Mediterranean area and even North Africa. April is usually the big flower month. Aalsmeer, just south of Amsterdam, is the clock auction market through which most of the cut flowers pass for shipment all over Europe.

Visitors will spend most of their Netherlands stay in these lands adjacent to the seacoast. There are many miles of fine beaches but they are generally more popular with the nationals and other Europeans than overseas tourists who find the sea quite chilly. The cold waters of the North Sea, which produce the delicious world-famous lobsters, deal less kindly with human surf dippers. Of more interest to most visitors are the fine Frans Hals Museum in Haarlem, the blue and white Delft pottery, the quiet pleasant seat of Government at The Hague, the Alkmaar Friday morning cheese market, the charming canaled Leyden, and, of course, wonderful Amsterdam.

There is plenty to see and do in this colorful Amsterdam and, starting with the excellent information booths at the airport and other points of entry, the hospitable Dutch do their best to insure that you do not miss a thing. All visitors who take advantage of the "Amsterdam Welcome Day" seem well pleased. Ask your travel agent or one of the many KLM air line offices about this very enjoyable travel bargain.

In spite of post card pictures, the famed Dutch windmills have almost been replaced by the electric motor, and there are now many more of these whirling wheels in Portugal than in The Netherlands. Only 900 of the one-time 10,000 windmills remain, with very few in active service. To forestall the complete disappearance of these distinctive landmarks, the Government recently encouraged turning over the abandoned mills for use as private residences, provided that they were reconditioned and maintained in good shape. The windmill tradition is not lost. The Dutch Windmill Society at Reguliersgracht 2 in Amsterdam supplies maps of a half-dozen areas where groups of windmills can be photographed. The one and only Windmill Museum in Koogaan de Zaan has fascinating models of the distinctive types of mills used in various parts of the country. Also those designed for pumping, sawing, drainage, or special use. It is probable, too, that you may see more wooden shoes worn in the wet farming portions of France than by the Dutch who now mostly use heavy rubber-covered shoes.

The Groningen, Twenthe (near Enschede), and Eindhoven City Weather Tables cover the eastern portion of the country from north to south.

Groningen, in the low, rich, green polder flats, is a pleasant university town and horse breeding center. There may be a subzero day from December through February but the usual low readings will be about 30° on perhaps 75 nights a year. Summers are pleasant and very hot. Rainfall, much in the form of brief showers, is lightest from February through June.

Enschede, which continues the weather pattern, is a textile industry town and a place of little interest to most tourists.

Eindhoven, in the southeastern corner of the country is noted mainly as an automobile and electrical equipment manufacturing city. Few travelers will stay here, but may pass through on their way to the more interesting panhandle section that extends down between Belgium and Germany. The highest spot in The Netherlands (1,062 feet) is at Vaals in this tongue of land which is quite a popular vacation area, and while very pleasant country, probably will not attract many foreigners. All too many tourists who fly to The Netherlands dash directly to Amsterdam and then off to the next place on the itinerary. Gouda, Delft, and The Hague are all quite close to their arrival point at the Schiphol Airport. A slight detour of even 1 day allows for a most rewarding glimpse of all three. A little preplanning often avoids missing many "almost on the way" places of interest.

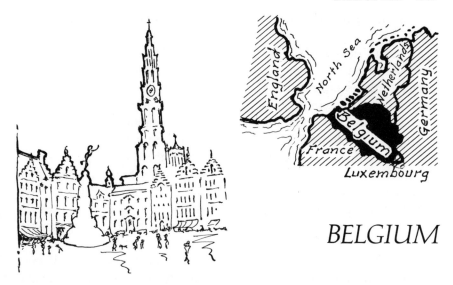

BELGIUM

Although there are three rather distinct topographical segments, the climatic variations across the country are more subtle. The differences in the five City Weather Tables of Ostend, Antwerp, Brussels, Huy, and St. Hubert are about what might be expected, based upon altitude and distance from the sea.

St. Hubert, at 1,831 feet, will average 111 nights when the thermometer drops to freezing. Ostend at 13 feet on the coast experiences only 40°. On occasion, however, there may be subzero weather during December, January, or February. February is most often the coldest month in many parts of Western Europe. Compare the high and low temperatures during those 3 months in St. Hubert, in the Ardennes highlands, and Ostend on the North Sea shore line. There is less temperature difference between the two during the summer months.

	DECEMBER		JANUARY		FEBRUARY	
	High	*Low*	*High*	*Low*	*High*	*Low*
St. Hubert	37	29	35	26	37	27
Ostend	44	35	43	34	43	33

The farther removed from the moderating effects of the ocean currents, the more frequent become the temperature extremes. The spread between the high summer figure and the low winter is 42° in St. Hubert and 35° for Ostend. Just as Great Britain has a slightly less equable climate than Ireland, inland Belgium is a bit less uniform than along the coast. At times, particularly during spring, the wind shifts to the east, bringing cold but clear, sunny days, especially in the higher country.

February through May is the driest part of the year along the coast while late autumn is the wettest. Spring is also the lowest precipitation period in the interior country, but the summer months get the heaviest rainfall. Total precipitation is generally about 30 inches in the coastal portions and 35 to 45 inches in the Ardennes highlands.

Unlike The Netherlands, where most of the tourists attractions are near the coast, there is much of interest throughout Belgium. The scant 40 miles of seashore operate on a 24-hour-a-day basis. There is golf, horse racing, and a wide variety of beach and marine activities. Ostend, Zeebrugge, and other spots combine to offer dancing, dining, a casino, and the full range of services and activities to be found in large, sophisticated resort areas.

Then there are the quaint and lovely old Flemish towns and villages. Bruges, a medieval walled city, is particularly delightful on spring and summer evenings when the old town is floodlit and reflected on the waters of the canals. Antwerp, the largest Belgian port, although 50 miles from the sea, should not be missed. People go there to see the Rubens House, the famous cathedral, float through the tree-shaded waterways, and dine well.

In common with the Dutch, French, and other marine folk, the Belgians relish the fruits of the sea. The opening of the mussel season is celebrated in Antwerp. On the first day, many restaurants will be serving large, steaming bowls of these savory bivalves to 99.9% of the patrons; that alone is almost worth the overseas trip.

Huy, the oldest commune in the Western world, was the place where Peter, the Hermit, preached the first Crusades and died in 1115. Liege is the most French of all the Walloon towns. Then there is the town of Spa, a once famous watering-place after which all others took their name.

The third big tourist magnet is the Ardennes high-country, summer vacation land. This is a fisherman's dream of crystal-clear streams, traversing a wooded hilly terrain. St. Hubert is one of the favorite spots in this region. In recent years this area has begun to attract more winter visitors, although it is not nearly as spectacular as the Alps and other mountainous regions.

LUXEMBOURG

In spite of its size, or lack of it (35 × 51 miles), this delightful little Duchy shows considerable variety in both topography and climate. There are spectacular cliff-edge views and strong contrasts, particularly in the upper segment which contains many scenic areas of savage beauty.

The northern one-third of the country (1,300 to 1,800 feet elevation) is a continuation of the Belgian Ardennes plateau. This quiet but rather rugged tree-covered, hill country is cut through with deep valleys separating high tablelands.

The larger southern portion called Bon Pays (Good Land) consists mostly of broad slightly undulating plains. Of the several rivers, the Saar which flows into the Moselle, is the most important. The fine Luxembourg white wines, both still and sparkling, are produced in the vineyard strip extending along the entire southeast border adjacent to the Saarland province of West Germany. The opposite southwestern corner of Luxembourg, a mining and industrial district, is of the least interest to most tourists.

At first glance, the City Weather Tables might suggest that the climate is quite uniform throughout the country. While there is little difference between the temperature figures, the high northern portion of Luxembourg has much harsher weather conditions during all but the three summer months. This is because that section has no protection from the winds which are more severe when there is a change in direction and they sweep in from the northeast.

Note that Clervaux, in the extreme north, experiences 18-m.p.h., or stronger, winds almost one-third of the time during the cooler months. Normally, the lowest February reading is about 26°. As we can check on the Wind Chill Factor Table, that figure, when coupled with an 18-m.p.h. wind, feels about the same as 0° on a calm day. On unusual occasions, the mercury has dropped to −10°. That is really earmuff weather but fine for outdoor activities since the higher elevations get abundant snowfall in midwinter.

Summer temperatures in this northern hill country are in the 60–70°-range with very few days over 90°. There are frequent short showers but also ample sunshine. This is a very pleasant year-round vacationland but summer is definitely the high season. It also, however, attracts a good number of fishermen and hunters in spring and fall.

The Good Land to the south can expect a 36°-temperature average in winter and 63° in summer. This lower two-thirds of the country, sloping gently toward the southeast, is shielded from much of the wintery blasts by the high mountain peaks. As a consequence, Luxembourg City and Echternach get strong winds only about one-tenth as often as Clervaux.

Total precipitation is somewhat less than in the high country, being in the 28-inch range. Much of the moisture has been dissipated from the warm sea breezes by the time they have traveled this far inland. March and April are usually the driest months. While precipitation is heaviest during late fall and winter, it is spread over the whole year. By North American weather standards, summers are cool and winters mild. Although not common, an entire season can be exceptionally hot or cold.

Luxembourg City is a charming town that was originally a medieval fortress. Perched dramatically atop a spot 1,084 feet above sea level, and surrounded by sheer cliffs, it is joined to adjacent areas by 91 bridges; several of the longer spans are 150 to 270 feet high. Another interesting feature is the 15 miles of underground passages through the rock called *casemates.* These once connected the central citadel with 50 outer forts which have long since been transformed into the delightful gardens surrounding the city proper. This spired and turreted toy city of parks and flowers is a fantastic, almost unbelievable, sight when illuminated during the summer evenings.

It is a place to visit any time of the year even though the mercury can sag to minus ten degrees on rare occasions. The modest 29 inches of precipitation is quite evenly distributed through the year. Since there are about two dozen thunderstorms a year, mostly during the summer, there will be brief but sometimes quite heavy showers. Rainfall, however, is most often not more than a light sprinkle almost every other day and is no great deterrent to outdoor activities. The mercury can climb to 100° but seldom does. While skies most often carry some clouds, the City Weather Table shows that only during December and January is visibility apt to be less than ½ mile as often as ten days a month.

The Weather Table of Echternach is almost a duplicate of the capital city. This is in the heart of a pleasant resort area and, because people love to dole out labels, it is called "Little Switzerland." Starting just to the south and extending about 20 miles, parallel with the German border, is the attractive Luxembourg wine country. The fact that the Government Calendar of Events lists activities related to wines eight times indicates the importance of the wine industry. There are many small villages in the grape-growing strip but Remich, toward the lower end, is one of the more attractive. Even though Luxembourg is not noted as a land of sunshine, viniculturists say that it takes about 100 days of sunshine to insure the proper sugar in grapes to produce good wines. And Luxembourg wines are well regarded, which is a measure of the weather.

Farther south, near the Lorraine border of France, are Luxembourg's thermal springs at Mondorf-les-Bains. This is a miniature Baden-Baden.

The following slogan is displayed in many places throughout Luxembourg, "We want to remain what we are." Each time we leave this delightful little piece of Europe in miniature, it is with the hope that neither Luxembourg nor its hospitable people will ever change.

Best Time to Visit the Low Countries

April, May, and June are generally the best months for outdoor touring. They are the periods of lowest rainfall and maximum sunshine. While March is usually the driest month of the year, temperatures are a bit low and most of the colorful spring flowers will still be in bud. The popular summer season is, of course, more crowded and peak prices prevail. Since the weather is seldom really severe, many visitors whose interests are mainly cultural find almost any time of the year satisfactory.

Tourist expenses in these three countries are not low. A recent United Nations survey indicates that costs in The Hague are 15% to 20% lower than New York City. It is said that the cost of living in The Netherlands increased 38% between 1964 and 1970. It is probable that conditions in Belgium have kept pace while perhaps only France, Germany, and Switzerland are rated higher. Overseas visitors should remember that costs in the small villages throughout the these countries are always considerably lower than in the major cities.

The KLM Air Lines, which is doing a remarkable promotion job, has been most successful with its ten-dollar-a-day program. Another pleasant feature: much like Switzerland and the Scandinavian countries, in particular, even the most modest accommodations are almost invariably extremely neat and shining. Off-season travel, where satisfactory, will usually reduce costs about 20%.

While midwinter may not be the most pleasant season for travel, many Americans and Canadians go to New York City, Chicago, Toronto, and Montreal at that time for long visits. It is often the best period for the theaters, concerts, nightclubs, restaurants, and shopping. Even with the topsy-turvy monetary situation, a jaunt to Amsterdam, Brussels, and certainly Luxembourg City can be equally enjoyable and perhaps no more costly.

Belgium maintains tourist information offices in many major cities in the United States and four in Canada. The Netherlands has offices in New York City, San Francisco, and Toronto but every KLM office is an excellent source of almost any kind of travel information. Luxembourg has a Consulate General in New York City whose office issues a wide range of literature. It also suggests writing to the local tourist office (Syndicat d'Initiative) of any particular town or city of interest.

Almost every major air line connects with some point in these three countries. KLM and Sabena (Belgian World Air Lines) have particularly attractive and, in many cases, most inexpensive package arrangements. Pan Am, TWA., Irish Air Lines and Canadian Pacific are among those that also have some very tempting combinations.

SWITZERLAND

SWITZERLAND is a tourist's dream. This 15,941-square-mile confederation is small only in physical dimensions. It boasts three distinct topographical segments, many little pieces of modified alpine climates, and four official languages. An aerial view of its surface resembles a crinkled sheet of aluminum foil.

Switzerland is all high country, made up of a wide range of elevations. The lowest point, 650 feet above sea level, is on the shore of Lake Lugano—the loftiest, more than 2 miles higher, at the summit of the 15,200-foot Monte Rosa on the Italian border.

Climate

Unlike most countries, the climatic map of Switzerland cannot be divided into large geographic segments. Latitude has much less significance than altitude, topography, or exposure to sunshine and air currents. The overall climate of Switzerland is a continental alpine type, but sharply contrasting land configurations produce many localized miniclimates.

While the annual average precipitation is on the order of 35 to 40 inches, there is a wide range as shown in the City Weather Tables. The Rhone Valley in the southwest, getting about 26 inches a year, is the driest section of the country while parts of the high mountains catch the most moisture. The lowest rainfall generally occurs during the months of November, December, January, February, March, and April. The wettest month is apt to be August.

There can, occasionally, be very great extremes—sometimes for extended periods. The 30-year record, for the whole country, shows one year that accumulated only 23.1 inches of precipitation while another was deluged with an amazing 108.4-inch total. These, of course, are exceptional and the figures shown on the City Weather Tables cover the conditions visitors can reasonably expect.

The ample precipitation at the higher elevations falls mostly as snow, of which there is plenty. Note that there may be freezing temperatures every month of the year in Santis at 8,200 feet.

In planning motor tours, prospective visitors should remember that snow drifts block the mountain passes much of the year. Of the 25 major passes, 14 are closed from October to June. A few open in May, and nine, including two tunnels, are open all year. An excellent source of this information is the little booklet *Travel Tips for Your Holidays in Switzerland* which is available through the Swiss National Tourist offices and the many very efficient Swissair desks.

There are only two general wind patterns in Switzerland. The westerlies from the Atlantic coast have lost much of their moisture by the time they have traveled over the Jura mountain ranges. While they exert some tempering influence, the sharp, dry northeasterly winds are more dominant, except in the southern Ticino country.

A very dry warm wind, peculiar to this portion of Europe, is called the foehn. It is caused when a depression to the north of the Alps pulls air from the south. These currents usually move up the south sides of the mountains until they reach a point where the temperature is low enough to condense the vapor and precipitate it as rain or snow. As the still warm but now rather dry air flows down the opposite slopes it is further dynamically heated. These winds occur mostly in springtime and can quickly raise the general temperature 20 or 30°, melting snow patches and stimulating early plant growth. The autumn foehn is usually welcomed as the elevated temperature helps ripen crops, especially the grapes. Similar weather phenomena, such as the Chinooks of western North America, are experienced in various parts of the world. A much less popular wind is the *bise* or *bize* which howls in from the north during winter. It is dry, cold and frequently accompanied by heavy overcast skies.

An element of the very greatest interest to the tourist is sunshine. The following table shows the 30-year average hours of sunshine in the five cities covered by the City Weather Tables.

CHOOSING WEATHER BY ALTITUDE IN SWITZERLAND

Much of the travel literature describing Switzerland emphasizes altitude and exposure to sun or wind as a guide to probable weather conditions. To a considerable degree, the climate in this little country

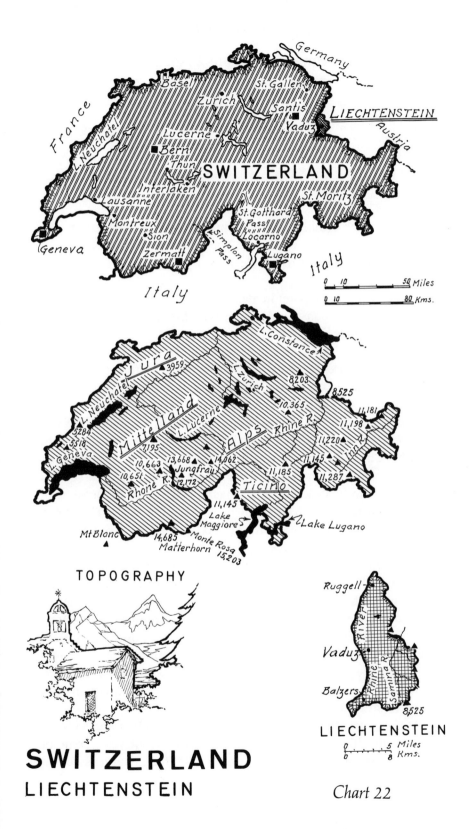

Switzerland

France
Germany

Basel
St. Gallen
Zurich
Santis
LIECHTENSTEIN
Lucerne
Vaduz
Austria
Bern
Thun
SWITZERLAND
L.Neuchatel
Interlaken
St. Moritz
Lausanne
St.Gotthard
Montreux
Pass
Sion
Locarno
Geneva
Zermatt
Simplon Pass
Lugano

Italy

Italy

0 10   50 Miles
0 10   80 Kms.

L.Constance
Jura
3959
L.Zurich
8203
L.Neuchatel
10,365
8525
5284
11,181
5518
Mittelland
L.Lucerne
Alps
Rhine R.
11,198
Geneva
7195
11,220
10,663
13,668
14,062
11,145
10,651
Jungfrau
11,185
Rhone R.
12,172
Ticino
11,287
11,145
Lake
Maggiore
Lake Lugano
Mt Blanc
14,685
Monte Rosa
Matterhorn
15,203

TOPOGRAPHY

Ruggell

Vaduz

Rhine River

Samina R.

Balzers

8,525

LIECHTENSTEIN

0   5 Miles
0   8 Kms.

SWITZERLAND
LIECHTENSTEIN

Chart 22

AVERAGE HOURS OF SUNSHINE PER DAY AND YEAR

	J	F	M	A	M	J	J	A	S	O	N	D	Year
Geneva	1.9	3.5	5.3	6.8	7.6	8.6	9.3	8.3	6.3	4.0	2.0	1.3	1,979
Bern	1.9	3.0	5.0	5.8	6.6	7.5	8.0	7.4	5.6	3.6	2.0	1.3	1,759
Zermatt	3.0	3.9	5.0	5.5	5.6	6.1	7.0	6.5	5.5	4.3	3.2	2.5	1,778
Santis	3.6	4.3	5.3	5.3	6.0	5.6	6.3	6.0	5.6	5.3	4.3	3.6	1,880
Lugano	3.7	5.0	5.6	6.1	6.2	7.5	8.6	7.6	6.2	4.6	3.6	3.3	2,101

with the much-crinkled surface can better be measured in vertical rather than horizontal distances. The following indicates some of the characteristics at the various elevations:

At 650 feet (lowest point in Switzerland) to 2,000 feet above sea level, there is generally considerable sunshine through the year. There will, however, be little, if any, on the north side of the mountains in winter, but the southern exposure can be bright and sunny. There is often high humidity but little fog, usually with somewhat more in the valleys than on the slopes. Diurnal variations are quite moderate. The rather light breezes do not vary much over the year.

Most resorts in this altitude zone operate on a year-round basis but some toward the 2,000-foot mark are more popular during the summer months. The following, generally classified as being in the alpine foothills, are south of the Alps: Brissago (800 feet), Ascona (600–1,000 feet), Locarno (600–1,300 feet), Lugano (900–1,500 feet). Some of the well-known spots north of the Alps are: Lausanne (1,200–1,900 feet), Interlaken (1,800 feet), Montreux (1,300–1,900 feet), Sonn Matt (1,800 feet), Bad Ragaz (1,700 feet).

At the 2,000- to 4,000-foot elevations, there are extended spells of fine weather with ample sunshine. The temperature extremes are slightly wider on both a daily and annual basis than at the lower elevations. Humidities are lower and there is an increased intensity of ultraviolet rays. Frequencies of fog depend upon location. Southern exposure in the valleys can be most pleasant throughout the winter, but many of the slopes facing north get little sunshine except in summer. While winds can be somewhat stronger than at lower levels, the air tends to be more still during winter than summer.

At these elevations, there are more distinct summer and winter tourist seasons although many resorts remain open all year. Some of the communities in this zone are: Aegeri (2,400 feet), Burgenstock (2,800 feet), Seewis (3,100 feet), Lenk (3,500 feet), Flims (3,700 feet), Klosters (3,900 feet).

One can label this pollution-free atmosphere alpine air or mountain air. From a health viewpoint, this could be just what the doctor ordered —and many do. But even more visitors come without prescription.

At the 4,000- to 6,000-foot elevations, changes in temperature are often sudden and extreme. The air is so dry, however, that even during the coldest weather, a bikini might not be uncomfortable in the very sunny, sheltered spots. But the almost certain mercury slump at sundown calls for woolies and a hot toddy. The sparklingly clear air allows for a strong intensity of solar radiation, especially in the ultraviolet rays.

Fog may persist for long periods but so, too, does sunshine and the sunny side of an alpine valley is usually delightfully mild even in midwinter. As indicated in the City Weather Table of Santis, sunshine is more evenly distributed throughout the year than in the lower country. Unless a place faces north or is shadowed by high mountains, it will average 3 or 4 hours of sunshine a day even in midwinter.

Normally, there is less wind in winter than other times of the year. If a valley extends parallel to the flow of strong winter winds, it will, of course, be uncomfortable during bad weather but that is not a common situation. Precipitation is very high in some of these loftier places. Santis at 8,200 feet, a more extreme case, averages a great 108 inches annually. Most of that falls as snow and explains why the winter abbreviation for the word Switzerland is "ski." The thoughtful travel agent will warn his clients from such places as New York City, Tokyo, Los Angeles, and even London about the shock to the system of becoming acclimatized. This Swiss mountain air seems to be completely lacking in the dust, allergens, and smog to which his lungs are so accustomed.

A few of the high-perched resorts are Davos (5,200 feet), Zermatt (5,300 feet), Arosa (6,000 feet), St. Moritz (5,800 feet), Silvapana (5,900 feet), Adelboden (4,452 feet), Wiesen (4,700 feet).

Some of these are open all year round but others have definite summer and winter seasons.

Western Switzerland

The Jura is a long strip of mountains, adjacent to France, separated into chains by a series of parallel valleys and lakes. The strip to the northwest of an imaginary line drawn between the west ends of Lakes Constance and Geneva is all within this zone.

Most of the country is wooded except along the valleys and a rather large flat area in the northeast which is high tableland. Mountain peaks, ranging from Mont Tendre (5,518 feet) to many others almost as high, are clustered in the lower corner between Lakes Neuchatel and Geneva. The land slopes generally to the north where the mountains are 3,500 feet or less. Although this is very pleasant countryside, especially near the south end of Lake Neuchatel, it does not attract many overseas tourists. On the whole, this northwestern portion of Switzerland gets somewhat less sunshine than down toward the Italian boundary. The Neuchatel district averages only 1,699 hours a year, ranging from a very meager 1 or 2 hours a day in December to a very respectable 8 hours in July. This is the area that produces the most widely exported Swiss wine. Some of this light, white Neuchatel is double fermented— which, as the locals say, "makes stars in the glass." This is a measure of the ample summer and autumn sunshine. Precipitation in this region averages about 32 inches distributed throughout the year, but is a bit lighter during the winter months.

Central Switzerland

The next large section, just to the east of the Jura, is the 1,500-foot elevation sometimes called the Internal Plateau but more commonly known as the *Mittelland.* This is the commercial and banking center of Switzerland and of great international importance. Who has not heard of the "Gnomes of Zurich"? The Mittelland covers only 30% of the country but contains 60% of the total population. The rich, fertile plains and rolling hills also make it the most agriculturally productive portion of Switzerland.

Zurich, Bern, and Geneva are all in this important segment. Of even greater importance to the tourists, this area also includes the famous Swiss lakes region—Interlaken, Lucerne, the fabulous views from Mt. Pilatus, and many other beauty spots. It is an easy place to find. Thirty-six air lines use the Kloten Airport at Zurich for direct flights to 106 other major cities around the world.

This central segment enjoys a very pleasant climate. The mercury seldom reaches 90° but will drop to just about freezing 100 nights a year. Total annual precipitation ranges from 39 to 46 inches and is over twice as heavy from May through October as during the other months. Although short heavy showers and thunderstorms are not uncommon, there will be at least a one-tenth of an inch sprinkle every third or fourth day. The total annual hours of sunshine are quite low—Zurich 1,693; Geneva 1,979; Lucerne 1,541; and Lausanne 1,971. The sun is mainly a summer visitor. As an example, Geneva, which is reasonably typical of the whole area, averages only 4 hours or less a day from October through February. April, May, June, July, August, and September, however, get at least 7 or 8 hours of sunshine in every 24. If that does not sound like the most ideal weather in the world, be assured that very few of the ever-increasing number of visitors are the least bit disappointed.

The Swiss Alps

The third and by far the largest zone is the magnificent Alps region. Although only 20% of the population lives in these mountains, they cover about 65% of the entire country.

The Mittelland attracts most of the "on tour" visitors, but there are probably more "long stay" accommodations scattered among these mountains and alpine valleys. A great many more Europeans than American tourists visit Switzerland—and a large percentage of them enjoy hiking and other mountain activities. Many of these high spots are equally popular in winter. It should be remembered that this region is not all mountain goat terrain. There are numerous valleys—large and small, high and low—also well-sheltered, level patches surrounding lakes of all sizes.

Perhaps, the most celebrated of all winter resorts is St. Moritz, perched 5,620 feet high in the eastern Alps toward the Italian border. It has a delightful location overlooking Lake of St. Moritz. This, too, is about as busy in summer as winter and enjoys excellent year-round weather. Of the 33-inch precipitation, almost two-thirds falls in the 5 months between June and October but mostly as rather short, heavy showers.

There are generally two off-season, or perhaps, between-season periods in this mountain country, May and November. That situation is common with most Alps summer-winter resorts. It happens that the St. Moritz weather is quite good during those months. The temperatures will range from 33 to 57° and 18 to 37° respectively, in May and November, while precipitation averages 2.8 and 2.6 inches.

St. Moritz also has a good sunshine record—1,805 hours per year. July is tops, getting almost 7 hours a day, but May, with 5½ hours, is bright and glittering. November's daily average of 3½ seems low, but absence of sunshine does not usually mean unpleasant weather. There are many clear, brisk days.

Some visitors find the occasional strong *maloja* winds a bit disturbing, but there are plenty of nearby sheltered spots where the calm air is most agreeable and never too hot. The St. Moritz weather conditions are quite typical of the many other resorts such as Silvapana and Sils, all in this upper Engadine area.

Just to the northwest of St. Moritz is the Grisons, a district where the ancient Romansh language is still spoken. But as in all other parts of this land of hospitable innkeepers, English is the second, third, or perhaps sixth tongue of these linguistic people. In this immediate section are three of Switzerland's best known mountain resorts. Arosa (5,000–6,000-foot elevation), which has one of the largest skiing schools in the country, is more popular with the younger people and family groups. Davos (5,000-foot elevation), famous as a sporting, tourist, and health center, enjoys particularly clear air with very little fog. At Klosters (3,900-foot elevation), one can ski in midsummer (at Silvietta just to the east). Like St. Moritz, many of these famous places are really complexes, each being surrounded by several little satellite communities. This allows for a wide choice since some spots are quite gay while others are quiet and peaceful. In many regions that have more regular terrain, a single weather pattern will usually extend over wide areas. That is generally not the case in mountainous country where it is more common to encounter a variety of miniclimates. Peculiarities in topography, types of exposure, and directions of air currents can cause one section to have a very distinct and different climate than other nearby districts.

As a simple example: Klosters consists of only two almost-joined communities, Klosters-Platz and Klosters-Dorf. The former, in the shadow of the mountains, gets sun only from 11 A.M. to 2:30 P.M. in winter, while the slopes facing southwesterly average almost twice as much. Klosters-Platz also has greater exposure to the valley winds, which are usually pleasant and refreshing but can be uncomfortable in bad weather. Ordinarily, the only way to get the localized conditions is by investigation on the spot. Not so in this most guest-conscious country in the world. The Swiss Government and its excellent Meteorological Institute have developed extremely complete and detailed "comfort weather" information. Perhaps one of the best examples is a most unusual booklet, *Short Climatic Guide to Switzerland* by the Swiss Meteorological Institute. Such pinpoint data can be most useful to visitors who plan to stay a month or more in Switzerland.

The Pennine Alps, to the southeast of Lake Geneva, is one of the most spectacular of the many ranges. The 14,685-foot Matterhorn, the towering Monte Rosa on the Italian border, and the even higher Mt. Blanc just across the French line as well as the popular Zermatt average a very low 26 inches of precipitation. While rainfall is heaviest from May through October, there is only about one-tenth of an inch of rainfall every fifth day. Zermatt does not have a definite rainy season. Like so many of the Swiss communities that are located in valleys surrounded by high mountains, the winter morning sun is late in arriving. On the other hand, even bad-weather winds are attenuated in the valleys unless the valley stretches out in the direction of the wind flow. There are 20 foggy days through the summer, which is about as popular as the winter season.

There is an ample 1,778 hours of sunshine a year. While summer gets 6 or 7 hours a day, midwinter will average only 3 hours. This latter figure may seem rather meager, but when the sun glistens on the surrounding snow fields and glacier-girth peaks, the whole area really sparkles and dark glasses are very much in order. The almost unfiltered rays also produce a beautiful sun tan.

Most of the Swiss people in this general area speak French. Customs and foods also reflect the close proximity to France, but whether a Swiss converses in German (70%), French (19%), Italian (10%), or Romansh (1%) every last one of these fine people is a true Swiss.

The Ticino Canton is built to order for those who prefer a milder Alpine climate. This has been optimistically called the Swiss Riviera. Its palms, architecture, and most importantly, the rather mellow climate lend a decided flavor of more southern parts.

Ticino, facing toward the Mediterranean, lies just above the Italian lake district of Maggiore and Como, while nearby Lake Lugano is almost wholly within Switzerland. Blossoms in full color from early spring to late fall, along with the ample harvest of figs, olives, and pomegranates, add to the almost subtropical atmosphere of this sun-warmed leisurely paced part of the country. Most of the Swiss in this canton speak Italian, and the foods resemble those of Italy—just to the south.

The Ticino is the warmest and sunniest part of Switzerland. Temperatures are 5 to 15° higher than other sections of the country.

Lugano's 2,100 hours of sunshine a year ranks it second after Locarno, which gets a high 2,286. Rain and sunshine are both most plentiful in summer. Since Lugano averages almost 9 hours of sunshine a day during July, the 6.9 inches of rainfall that month has to squeeze in as short, heavy showers. Note on the City Weather Tables that the very ample 68-inch total precipitation is distributed throughout the year, and while heaviest from May through October when no month gets less than 7 inches, there is plenty at all times.

This is the only portion of Switzerland where one ever thinks about heat as related to weather. Although the mercury has topped 100° on rare occasions, it is the moderately hot, but sometimes very sultry, midsummer days that cause discomfort to some visitors. This is very definitely a vacationland of all seasons and a great temptation to drop anchor and forget the remainder of the itinerary.

Tourism in Switzerland

There is no pronounced off-season in Switzerland. Some hotels and resorts either close or reduce rates slightly during the slack periods. The Swiss Government is urging these places to stay open through at least a part of the slow intervals and reduce rates 20%, but that practice is not generally followed as yet.

Costs in Switzerland are not low and rate among the top in Europe. There are few out-of-season bargains, but overseas visitors are learning that it is easier on the budget and very much more enjoyable to vacation in one area in preference to the daily pack-and-travel routine. On that basis, many of the smaller places in Switzerland are most attractive. Even with a small, rented car to explore surprisingly distant places, the overall costs can be much less than the exhausting inventory-type skirmish through vast portions of Europe at top speed.

As in most places, there are fewer visitors and more accommodations during the between-season intervals. In many parts of Switzerland, these are the months of May and November. Check the City Weather Tables for expected weather conditions; also, ask your travel agent about the high-season dates.

LIECHTENSTEIN

A natural question might be, "Why visit Liechtenstein?" To be sure, it has the same grand mountain scenery and delightfully gracious people as its larger neighbors, Switzerland and Austria. But there are two other attractions. Included in few itineraries, it is less crowded and almost completely unspoiled. Costs are considerably lower than in Switzerland but perhaps a shade above those in Austria. A triangular plot only 6 by 15 miles (62 square miles) seems a mighty small piece of real estate to hold so many enjoyable sights and interesting places. Although only about twice the area of Disneyworld in Florida, some people spend their entire vacation within its boundaries.

The southeastern edge is the loftiest part of the country and Mt. Naafkopf (8,525 feet) is the highest point. This section, and the somewhat lower area to the north, can experience quite harsh winter weather, particularly when the sharp cold *bora* winds howl in from the northeast. The bora, which is much like the famous mistral of the Rhone Valley and southern France, is often very violent. On occasion, it has reached a velocity of 100 m.p.h. but that is rare indeed. This wind usually results in cold, dry weather with clear skies. When influenced by a low pressure in the Adriatic area, however, it can sweep in over a several-day period, bringing along heavy precipitation in the form of either rain or snow, depending on the general air temperature. Overall precipitation in the highest parts of Liechtenstein is up to 80 inches a year, and since most of this falls as snow, winter resorts are being developed. Malburn, near the Austrian border, is becoming a popular skiing center. In early summer, farmers herd their cattle to the lush meadows high in the mountain valleys. Hikers, campers, and naturalists also head upward in good numbers at that time. In autumn, the sleek, fat cattle are led back to their lowland winter quarters. A bovine population explosion is celebrated by festooning the cattle with garlands and decorations, forming a triumphant procession down the mountainside. This joyful custom is also followed in Austria and Switzerland.

Liechtenstein produces very good wines and excellent beer. The vineyards extend along the south sides of the foothills, which get ample sunshine. The fertile Rhine Valley along the Swiss border is made up of fine farm lands, orchards, and rolling meadows. Total precipitation in the low country ranges from 30 to 40 inches—which is quite satisfactory for agriculture, and when coupled with lots of sunshine, is just right for vacationing. This wide variety of scenery in so small a space is one of Liechenstein's attractions.

Almost all visitors make Vaduz their headquarters, and all too few get much beyond its immediate environs. This attractive little village-capital is a very pleasant spot, but Triesenberg, a few hundred feet higher, Schaam, and the many tiny hamlets scattered among the mountains are also well worth a visit.

The official language of Liechenstein is German, the currency Swiss, and the attitude one of welcoming friendliness. This nation in miniature has retained its autonomy since 1719, although it was associated with the Rhine Confederation for a period of 50 years. Its army was disbanded about a century ago, but the two dozen genial policemen are ample protection for both the population and the 8,000 world corporations which are registered here for tax reasons.

The lucky tourist who can spare a day or two in this little principality will return home with pleasant memories—and perhaps a collection of unusually colorful stamps.

WEST GERMANY

GERMANY is a bit frustrating to travel agents who feel that it has no single focal point of tourist interest. The mention of almost any country seems to bring to mind some particularly outstanding place or feature. It is taken for granted that a visitor to Ireland will see the Lakes of Killarney and who would miss the Pyramids while in Egypt?

Although there may not be one "must" place, Germany boasts many spots that widely traveled people enjoy going to time after time: the delightful Black Forest region; Bavaria; parts of the Rhine River; Garmisch; the group of quaint little towns including Rothenburg, Bayreuth, Regensburg, Wurzburg, and a score of others. It must be admitted that, while very pleasant weather can be enjoyed at many times and places in Germany, the overall climate is far from being its chief tourist attraction.

Climate and Topography

Much of central Germany is torn between the moderating warm, humid westerlies and the harsher winds that howl in from the northeast. A shift in direction from one to the other can quickly and completely alter weather conditions. These rather abrupt changes from hot to cold or moist to dry are characteristic of Germany's weather.

The northern segment of the country, bordering the North Sea and including the low northern plains, is completely dominated by the warming ocean breezes which produce an equable marine-type climate. Summers are cool by most American standards, while winters are generally not especially harsh, except during heavy storms. The average temperatures are about 15° higher than might be expected for the latitude. As an example, Munich, at 48° 9'N, is closer to the North Pole

127

than Montreal, Canada, while Hamburg is on almost the same parallel as Goose Bay, Newfoundland. Neither of these German cities, of course, experiences nearly as cold winters as their counterparts in North America.

The south and central high plateau areas in are largely influenced by the sharp, drier northeasterly winds. The result is a typical Continental climate with cold winters and hot moist summers. Average temperatures throughout Germany, however, are more uniform than the latitudes of the various sections of the country would suggest (*see* City Weather Tables). In this case, the climatic effect of latitude is equalized by elevation. The southern portion of this climatic zone, which might normally expect to have the warmer weather, is on higher ground and temperatures, of course, decrease with altitude. This is an example of how topography can affect climate.

Even the most enthusiastic travel folder seldom emphasizes sunshine in Germany. While December, January, and February are generally overcast and gloomy, spring and autumn are quite dry and sunny. Short, heavy showers make summer the wettest period of the year, but there are also many pleasant, brilliant days during that season as shown in the following sunshine record of the southern half of the country.

AVERAGE HOURS OF SUNSHINE PER DAY AND YEAR

	J	F	M	A	M	J	J	A	S	O	N	D	Year
Freiburg	1.7	3.3	4.5	5.5	7.4	7.6	8.0	7.4	5.7	3.7	2.0	1.5	1,802
Constance	1.5	3.0	4.5	5.5	7.0	7.6	8.3	7.4	5.7	3.2	1.6	1.0	1,733
Munich	2.1	2.6	4.6	5.9	7.1	6.9	7.4	7.0	6.0	4.4	2.0	1.5	1,771
Berchtes-gayden	2.1	3.0	4.6	5.5	6.0	5.7	5.9	6.0	5.6	4.3	2.1	2.0	1,590
Regensburg	1.7	2.5	5.0	6.0	7.1	7.0	7.4	6.8	5.7	3.7	1.5	1.0	1,692
Wurzburg	1.7	2.6	5.0	6.5	7.4	7.3	7.4	6.7	5.7	3.7	1.7	1.0	1,730

It should be remembered that the above figures are averages. There are, of course, times when there will be no sunshine but other times when it will be sunny all day. November through February is the really dim period. Note that early spring gets an average of 4 or 5 hours of sunshine a day, while during May, June, July, and August there is enough to satisfy almost everyone.

Many meteorologists subdivide Germany into six major climatic zones. As is often the case, these segments coincide roughly with topographical variations. The small map in Chart 24 illustrates the

North Sea

Sylt

Baltic Sea

N. Frisian Is.

Kiel

Schleswig-Holstein

Lubeck

East Frisian Islands

Hamburg

Bremen

Luneburg

North German Plains

Berlin

Hannover

East Germany

Lower Saxony

Munster

Netherlands

Harz Mts.

Dusseldorf

Kassel

Poland

Koln

Belgium

Rhineland Palatinate

Frankfurt

Hof

Czechoslovakia

Luxembourg

Wiesbaden

Trier

Wurzburg

Bayreuth

Heidelberg

Nurnberg

Regensburg

France

Stuttgart

Bavaria

Freiburg

Black Forest

Munich

Rhine River

Alps

Garmisch

Lake Constance

9,721

Zugspitze

Switzerland

Austria

0   100 Miles

0   100 160 Kms.

WEST GERMANY

Chart 23

TOPOGRAPHY

Chart 24

Climatic zones
No.1 North Sea coast & northern plains
No.2 The Rhine Valley
No.3 Central highlands
No.4 Rhineland-Palatinate
No.5 Bavaria
No.6 Bavarian Alps

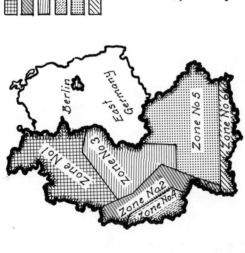

CLIMATIC ZONES

WEST GERMANY

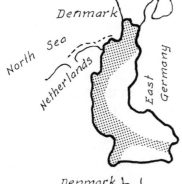

SPRING
April, May, June
Generally sunniest & driest season. North Sea & plains, cool comfortable. Rhine valley & Rhineland-Palatinate pleasant & sunny. South side of highlands & Alps valleys sunny & mild.

SUMMER
July – August
Seldom hot. North Sea & plains prime season. Rhine valley, Rhineland-Platinate, Bavaria, Alps & highlands very pleasant; heavy showers start in late summer.

AUTUMN
September – October
Rhine valley & Rhineland-Platinate, grape harvest time Many festivals. The lower Alps & central highlands very pleasant but often showery.

WINTER
November through March
Skiing in southcentral high lands. and top season in the beautiful Bavarian Alps.

VACATION WEATHER MAPS
WEST GERMANY

Chart 25

relationship between the various geographical features and the resultant climates. Air currents can also be an important factor in governing both precipitation and temperature.

Averages are usually, at best, little more than an indication but sometimes explain a general trend. By and large, the average annual temperatures in Germany are about 7° cooler in the Alps in the south (Zone 6) than at sea level in the north (Zone 1). This, of course, is due to the difference in elevations. Precipitation generally increases from north to south for the same reason. The maximum annual precipitation ever recorded in Zone 1 was 39 inches and the minimum 21, while Zone 6 has registered a high of 54 inches and a low of 37 inches in a year. In all six zones, the latter part of the summer is wettest while spring gets the lowest precipitation. This, of course, is only a general guide as there are many local miniclimates which are most often created by variations in topography.

Climatic Zone 1: North Sea and Baltic Coasts and Northern Plains

Both the seacoasts and wide plains making up this area are similar to The Netherlands in topography and climate. The windmills, but even more, the diked swamplands corresponding to the Dutch polders, further accentuate the resemblance.

All of Schleswig-Holstein, most of Lower Saxony, and the northern portion of Westphalia are included in this zone. Years before air travel, Hamburg and Bremerhaven were very important ports of entry. This was quite a popular tourist area but now most visitors who come here are concerned with commerce and industry. Perhaps the majority of present-day overseas tourists to this north country are visiting relatives in Germany.

The coasts and some interior parts of the generally pastoral Schleswig-Holstein, particularly the Frisian Islands, are still busy summer resorts but not nearly as active as they were in the past. The Isle of Sylt, with a range of attractions from nudists' colonies to prim 1880-style beach sections, is regaining popularity. While there are many pleasant sunny days, swimming is for the really rugged; the beautiful sand beaches are more popular than the surf. Average water temperatures in the North Sea at Sylt are: June, 60°; July and August, 64°; September, 65°; and October, 53°. Members of the Polar Bear Clubs should note that the December water temperature is 38°.

Many tourists (and travel agents) incorrectly assume that Hamburg is just one big, busy commercial and shipping port. Actually, the central city, including the Alster Lake District, and the many magnificent parks and fine boulevards, is one of the most attractive in Europe.

Hamburg is a place of outstanding restaurants, music, culture, and fine living.

Visit the notorious Reeperbahn, a street which offers all the after-dark attractions banned in Paris and Copenhagen. Your presence in that part of Hamburg can best be explained as having been at the Bavaria Blick Restaurant atop the St. Pauli Brewery for an excellent dinner and the panoramic view of the city and harbor.

The overall weather in this north country is neither outstandingly good nor bad. The City Weather Tables of Hamburg, Bremen, Munster, and Hannover show a medium precipitation of 24 to 31 inches. Although fairly evenly distributed among all 12 months, it is heaviest from June through October. April is most often the driest month of the year, but precipitation may start to taper off as early as January or February. There will usually be about 135 misty days a year, with gray, overcast skies. Heavy fogs roll in suddenly, especially near the water, and about 60 days a year visibility will be reduced to less than ½ mile.

Normally, the mercury will travel in a wide range from 25 to 75° during the year throughout this area. On occasion, however, it can climb into the high 90's as shown on the Munster table, or dive to − 17°. Air movement is common throughout the year in this zone. In summer, the gentle breezes are pleasant, but during the colder season when wind velocities usually increase, the weather is often sharp and biting in spite of the moderate temperatures. As indicated on the wind chill table, a thermometer reading of 35°, when coupled with a 25-m.p.h. wind, feels as cold as a 7°-temperature in a still atmosphere. It is not difficult to decide that this is not winter tourist country. Even horses and cattle wear sacking covers during the cold weather.

By and large, precipitation is of much greater concern to the tourist in this area than temperatures. Rain, mist, fog, and snow are all here in good supply. Spring and early summer are the best periods from the viewpoint of weather, even though one can expect some rain about 1 day in 4. Since there are about two dozen thunderstorms a year, most in summer, much of the precipitation at that time is in the form of short, heavy showers.

Some years, autumn is a delight with bright, clear days and rainbows of flowers. That season, however, is much less dependable than spring. More often than not, there are endless showers and overcast skies.

Hamburg is a busy commercial center which means that there are always many businessmen in town, but that should be of little concern to tourists. Their paths will not usually cross to any great extent. While the elegant Vier Jahreszeiten (Four Seasons) and Atlantic hotels must be prebooked, there are ample restaurants and accommodations in every category to take care of all.

A stay in this area should include a jaunt to the charming old Hanseatic city of Lubeck founded in 1143 and once known as the Queen of the Baltic. And for contrast, the nearby beach resort of Travemunde is worth a visit. An extra bonus is in timing your visit to be in the nearby Heide-Luneburg district just when vast areas are covered with heather in full bloom.

Although not scenically spectacular, this whole zone is an excellent place to enjoy a very pleasant, peaceful vacation. There are large wooded areas, rolling hills, and stream-traced grassy lands, interspersed with tiny ancient villages. Here one can enjoy fine food and comfortable accommodations. Prices are quite moderate at the modest little hotels and inns in the smaller villages.

Climatic Zone 2: The Rhine Valley

This has to be one of Europe's finest attractions. Tragically, like so many other rivers in Europe and America, the waters of the Rhine are fast becoming unfit for man or beast, but the lovely landscape is far from ruined. The lower reaches of this mighty 820-mile stream are bordered mostly by grain fields. The south end is one of the most famous white wine districts of the world, and sunny sections of the riverbanks are an almost continuous series of terraced vineyards. Here, too, are the legendary Lorelei Rock and many romantic old castles. No one should miss the very pleasant boat trip between Mainz and Koblenz, the most scenic section of the river. This western segment is the warmest and sunniest part of Germany. Precipitation, particularly toward the central portion of this area, is quite modest.

Wiesbaden registers a very low precipitation of 18.5 inches a year and only the month of August averages over 2 inches of rainfall. March and April receive the least precipitation when only one day in ten will get over 1/10-inch of rainfall. That is music to any tourist's ear. May, September, October, and November average not more than four or five days of rain each, but the temperatures, particularly in November, are considerably lower than late autumn.

Although there is rarely a 0- or 90°-reading, there will be about 70 nights when the mercury drops to freezing. Normally, the temperature spread is very narrow—most often less than 15° on both a monthly and annual average basis. By most American weather standards, the 28 to 40° winter temperatures are mild but the 55 to 75° summer figures are cool.

Winter days can be foggy and quite disagreeable, but there is an unusually long fine-weather season from mid-April through late fall.

There is little wind and only about 60 days a year can really be classified as rainy. Koln or Cologne, toward the north end of the valley, does a little less well but, even so, has a fairly good weather record. It gets almost 50% more precipitation than Wiesbaden—with some rain about every fifth day, although it usually comes as wet and dry spells of several days each. There is really no driest month but precipitation is often lowest in March and April. This, of course, is quite different from what most North Americans expect. They are more accustomed to April showers and glorious autumns. Most of the 18 thunderstorms occur during the summer which tend to cause short, heavy showers. Either interferes less with outdoor activities than continuous lighter rainfall. About 125 days a year will get at least a sprinkle. Each of the large German cities has its individual characteristics, but one thing they share in common is assurance of a wide and excellent list of restaurants, hotels, shops, and cultural sites.

Only 20 miles upstream from Koln, is the elegant Düsseldorf. Before World War II, it was called the Paris of the North and is still a handsome city of fine living. Just as in prosperous Turin, Italy, little effort has been made to attract tourists, which makes it that much more attractive to the lesser number that go there.

The temperatures in Düsseldorf are much the same as Wiesbaden and Koln but the 30-inch annual precipitation is higher. Here, too, March and April usually get the least rainfall, but even then, there is a little about every fourth day.

Essen is in a large complex of heavy industry and of little interest to most tourists.

Freiburg (at 938 feet), in the foothills of southwestern Germany, can be classified as either Rhine Valley or the Black Forest territory. The annual precipitation is 35 inches, being heaviest from May through August. This is not as bothersome to visitors as the figures might indicate. Much of the downfall during summer is in the form of short, heavy showers (note the 25 summer thunderstorms) which clear and freshen the atmosphere, then quickly disappear. From the viewpoint of precipitation alone, January, February, and November would seem to be the best months for tourists. There are also fewer rainy days at those times, about one day in five, but instead of short showers, those months get gray, gloomy, all-day affairs. Also note the six or seven times a month when visibility is reduced to less than ½ mile. Late spring and early summer are almost always the preferable time of year. Gamblers who pick autumn may hit a jackpot of beautiful sunny weather—or perhaps showery, rather overcast, days. The former is glorious, the latter not too bad.

Although the normal temperature spread is quite small, there are occasional extremes; over the years, the thermometer has registered from 103 to −7°. In spite of those very occasional and drastic readings, this district enjoys a mild, agreeable climate and is particularly pleasant from early spring to late fall.

Visitors, who may get bogged down a bit by rain in Koln, Düsseldorf, or Freiburg, may have better luck in nearby Wiesbaden. Freiburg gets almost twice as much precipitation as Wiesbaden—spread over about 50% more days.

Climatic Zone 3: Central Highlands

There are varying land elevations from 500 to 3,000 feet in this segment which includes parts of the Hesse and Weser highlands as well as the Harz Mountains.

The City Weather Table of Kassel (646 feet) shows what can be expected at most of the 500- to 800-foot elevations. In highly industrialized countries such as Germany, large and, sometimes, dirty industrial operations are to be found in almost every section. There may be fewer in this zone than in other parts of Germany. Kassel and Frankfurt (338-foot elevation), which share just about the same climatic characteristics, are the only really large communities and much of the unspoiled countryside is wooded with low, rolling hills. This region is favored by gliding enthusiasts and those who enjoy pleasant outdoor vacation life. Few who get to this area fail to stop at the romantic university town of Heidelberg and some even succeed in downing the oversized Red Oxen beer at that student rendezvous. This once quiet little village is now quite a busy community.

The farther removed from the coast, the less is the influence of the moderating westerly ocean breezes. Even though there is some modest tempering effect, this area is more fully controlled by the severe and erratic easterly winds that produce a somewhat modified Continental climate. Winter weather can be sharp and quite disagreeable.

Although much of Kassel's 23 inches of precipitation falls during summer, the total hours of rainy weather are limited. Most of the rainfall is in the form of short, heavy showers. Note, too, the low number of rainy days per month. There are almost no really hot spells, but an occasional 95 to 100° day is not completely unknown. The weather from April through September is generally pleasant, although April and May might be a bit cool for some.

Wasserkuppe, farther to the south at 3,035 feet, is representative of the higher portions of this Zone 3. There are two quite distinct tourist seasons for the summer and winter outdoor enthusiast. It gets a big

42-inch annual precipitation—as its name suggests. The 3 to 4 inches per month during the cold weather means 30 to 40 inches of snow. (Precipitation is always expressed as rainfall and 1 inch equals 10 inches of snow.)

The almost 5 inches of rain in July might well frighten off a potential visitor. Most of it, however, falls in nine days, six of which include thunderstorms. The hot July air can hold a lot of moisture but when released, it almost always falls as showers—often torrential but, thankfully, quite brief. Temperatures in Wasserkuppe average about 5 to 8° lower than the Kassel area at all times of the year.

Climatic Zone 4: The Rhineland-Palatinate

This is the segment of Germany lying west of the Rhine River bordering Belgium, Luxembourg, and France. It is divided into northern and southern sections by the Mosel River. The City Weather Table of Trier, at 443 feet, is typical of the lower valley portion of this area. The 28 inches of precipitation is so uniformly distributed, both by downfall and number of rainy days, throughout the whole year, that there is no distinct wet season. During October and December, every third day will experience either dense overcast or fog. The owners of vineyards measure as a bright year one getting 100 or more days of sunshine. An indication of the uncertainty of the German climate is the unique practice of picking one-third of the grapes at a time as insurance against the late summer hail or other damaging weather conditions destroying the entire crop.

Winter temperatures are intensified by brisk winds and are often quite uncomfortable even though the thermometer seldom drops much below 28°. Summer days are usually in the pleasant 70's and there is hardly ever a reading much over 90°.

The terrain on both sides of the Mosel River rises to plateaux, with heights up to about 3,000 feet. The more elevated sections have somewhat sharper winters and a fairly wide temperature spread on both a diurnal and annual basis. Hahn, at 1,649 feet, will average 5 to 8° cooler than Trier throughout the year. It will also get a little more precipitation. The valleys enjoy an overall mild, temperate climate favorable to vineyards and fruit orchards.

Climatic Zone 5: Bavaria

This single district alone has as many and varied tourist attractions as many whole countries. About the only thing lacking is a sea coast. Many of the lovely little Bavarian towns will seem very familiar to almost every visitor. Being about as photogenic as any movie star, their portraits have been as widely distributed. The climate here in Bavaria is somewhat similar to the central highland of Zone 3 just to the north. Being farther removed from the influence of the tempering warm ocean air currents, however, temperatures are a bit more extreme and winter weather can be quite harsh.

Note in the City Weather Tables of Munich (1,807 feet), Nurnberg (1,045 feet), and Stuttgart (1,300 feet) that each will average over 100 nights when the mercury will drop to below 32°. As might be expected in high country, all three will experience the very occasional subzero reading. There are also heavy foggy spells in winter but still plenty of fine, clear but not overly sunny days. In all, the season is perfect for outdoor activities since there is abundant snow through much of the winter. Munich gets about 10 snowfalls a year, totaling 44 inches. Summer weather is comfortably cool with few days over 90°, although on a very unusual occasion, there may be a 1-day 100° heat wave. A recently developing condition of grave concern to Munich is the rapid increase in air pollution. In 1971, it was described as West Germany's worst air-polluted city. Because of the topography and nature of air currents, the threat of a disastrous inversion similar to those experienced in London, Tokyo, and Los Angeles appears to be a real possibility. Fortunately, Munich is giving serious study to remedial measures.

Visitors should remember that precipitation in this region is mostly a function of elevation. Regensberg, at 1,113 feet, gets only 23 inches a year; Lechfeld, at 1,819, averages 30 inches; Landsberg (2,044 feet) gets 37 inches while Grosser Falkenstein, perched at 4,291 feet, is inundated with a massive 46 inches—mostly, of course, as snow. Munich can expect about 37 inches, being heaviest in summer when there is rain every 2 or 3 days. It gets almost 29 inches of precipitation from May through August. That may sound like water up to the kneecaps, but it is not as wet as it appears. During that same 4-month period, there are 43 short, heavy showers which means few actual hours of rainfall. Notice that the months of May, June, July, and August will average more than 7 hours of sunshine a day.

Munich has much to offer and one of the merriest events in Europe is that gigantic beer "bash," the Oktoberfest. Even the timing of this event tells something about the capriciousness of German weather. As the name suggests, it originated as an October festival, but it is not difficult to understand why the opening was advanced to the latter part of September.

One of the most delightful and satisfying vacation experiences is a few weeks' automobile jaunt through this wonderful countryside stopping off at Rothenburg, Bayreuth, Wurzburg, and a score of other quaint and some quite ancient little Bavarian towns. Each has its comfortable country inns and fine eating places. This is a most pleasant, genial country where a smile comes much easier than a frown.

Over in the southwestern corner of Germany is the charming Black Forest. After one look, many tourists forget about the remainder of their itinerary. Most would enjoy a stopover at Baden-Baden which is one big flower garden and a delightful loafing spot. This whole region is Germany at its very best.

Climatic Zone 6: The Bavarian Alps

This narrow strip across the southern portion of the country borders on Switzerland and Austria. It is a glorious land of knapsacks and *lederhosen* in summer and skis or skating in winter. The German Alpine Road from Berchtesgaden to Lake Constance passes through sections of extraordinary beauty, which makes it a grand place for auto touring. There are several resorts in these mountains, the best known, Garmisch-Partenkirchen, a year-round playland at 2,322 feet. Normally, the temperature range is not wide. Summers averaging 50 to 70° are mild to cool while winters are usually a crispy cold 20 to 40°. Records covering the past 40 years, however, show that the mercury can really travel on occasion. There have been record figures of $-17°$ up to 93°—a 110° spread, but few visitors will ever experience such extremes.

Garmisch's 51-inch total precipitation allows for plenty each month. As snow during the 5- to 8-month colder season, it is most welcome. Even the 7 inches of rain in July is not quite as bad as it sounds. As typical mountain weather, the rain most often occurs as short, somewhat violent, thunderstorms—sometimes of deluge proportions. A short distance to the south, on the Austrian border, is German's highest mountain peak, the 9,721-foot Zugspitze. Many places in these high mountains which are exposed to the sunshine and sheltered from the cold northern blasts are very pleasant and comfortable even at the lowest temperatures.

At the western end of this region, bordering Switzerland and a small section of Austria, is the beautiful and once crystal-clear Lake Constance. Like so many other world glamour spots, this is rapidly being turned into another dead marine body by man-made pollution, but fortunately, the three countries are trying to develop a joint program to restore this popular lake.

Berlin

This prewar capital of Germany is now a 342-square-mile island, completely surrounded by the German Democratic Republic (East Germany). Approximately 186 square miles of this little enclave is a political segment of the Federal Republic of Germany (West Germany).

Since most visits to Berlin are quite short, the question of weather is not very important. There are, however, good and bad seasons. The City Weather Table tells the story. Winters are rather mild, damp, and gray. Note the number of days with limited visibility. About half the time from December through February will be freezing weather but the thermometer rarely shows a 0° reading. Most visitors find this period rather disagreeable.

The weather improves in May, and from then on through September, it is generally pleasant with considerable sunshine. While there are about the same number of rainfalls every month of the year, precipitation between May and August is usually in the form of rather limited showers. Although the rainfall is slightly lower during the remainder of the year, it is more apt to be all-day drizzles. There is much more to see and do in Berlin after dark than during the daylight hours which makes weather one of the lesser considerations.

Tourist Information

Two deterring factors against tourism in Germany are high costs and climate. Large cities and beautiful luxury-class hotels can put a crimp in the heftiest purse. There are, however, plenty of small, clean hotels and inns in small towns and villages off the main thruways where excellent foods and clean comfortable accommodations can be enjoyed at a very modest figure.

In order to take full advantage of this type of fine living, it is best to use such small places as headquarters and take excursions to other nearby points of interest. Americans, in increasing numbers, are discovering that this kind of vacation can be much more satisfying than the daily pack-and-travel system.

The Lufthansa Air Lines, has developed some very attractive and inexpensive combination packages. This arrangement has done much to make European travel both pleasant and trouble-free. There are many very tempting combinations of both tours and independent travel arrangements that include a wide choice of accommodations as well as bus, railroad, auto, and air requirements. A little pretrip preparation will help insure a more satisfying visit.

The Federal Republic of Germany (West Germany) maintains very efficient and generously stocked Tourist Information Offices in Chicago, San Francisco, and New York City. We know from experience that the very helpful staff at the latter is unusually well informed.

AUSTRIA

SEVERAL of the smaller European countries seem to be special favorites of almost every world traveler. Ireland, Denmark, and Austria are three such and, while each has its individual attractions, the one common denominator must be what Austrians call "gemutlich." We have no single comparable word to describe this quality. One might say, genial, kindly, generous—but perhaps the equally undefinable Spanish "simpatico" may be the closest equivalent.

It is interesting that although sunny weather is so important a factor in attracting tourists, none of these three little nations excels in that regard although Austria is by far the most favored.

Topography

Landlocked Austria borders Czechoslovakia, Hungary, and Yugoslavia on the east and west Germany, Switzerland, Liechtenstein, and Italy on the west. It has always been a crossroads of Europe and even today, 90% of the freight on the Danube is international trade passing through Austria.

Generally high country, the average altitude of Austria is about 3,000 feet above sea level. The lowest spot in the whole country is at a 377-foot elevation on the Neusiedler See while the highest is the 12,461-foot Grossglockner.

Except for a wide section extending along the northern and eastern sides, most of the country is a combination of mountain chains and winding valleys. These pre-Alps mountains (there are over 30 distinct ranges) form a transition between the wide central European plains and the major Alps which center in Switzerland. Such topography adds up to plenty of summer hiking and winter skiing. While the Blue Danube is no longer that blue, and the once magnificent Lake Constance is fast becoming a polluted pool, there are many smaller, crystal-clear alpine

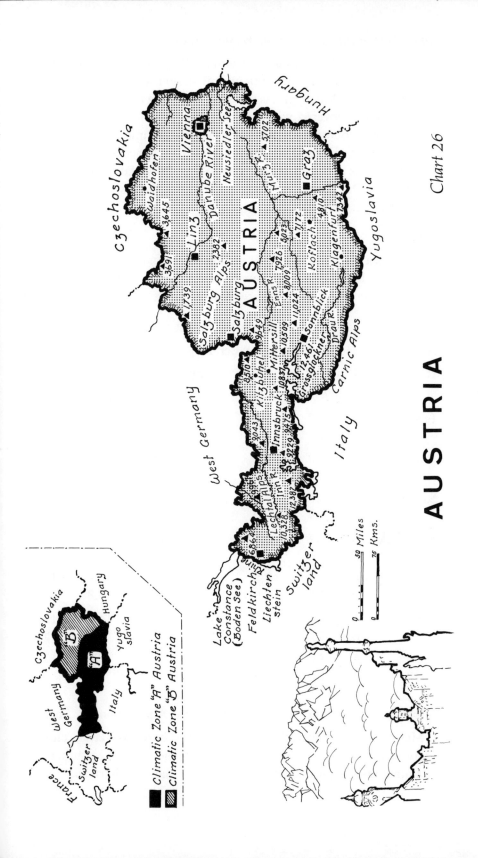

AUSTRIA

Chart 26

Czechoslovakia

Hungary

Yugoslavia

Italy

West Germany

Vienna

Linz

Graz

Salzburg

Innsbruck

Klagenfurt

Danube River

Neusiedler See

Salzburg Alps

Lechtal Alps

Carnic Alps

Mur R.

Enns R.

Inn R.

Drau R.

Waidhofen

Mittersill

Koflach

Sonnblick

Grossglockner

Kitzbühel

Lake Constance (Boden See)

Feldkirch

Liechtenstein

Switzerland

3691 1 3645

1739

2382

9549

8510

9043

9156

10,378

12,382

9229 1,752

10,837

12,461

10,509

11,024

8,009

7926

5023

7177

4810

7341

3702

50 Miles
75 Kms.

Climatic Zone "A" Austria
Climatic Zone "B" Austria

Czechoslovakia
Hungary
Yugoslavia
Italy
West Germany
Switzerland
France

"B"
"A"

lakes and fine fishing streams in almost all parts of the country.

Austria is primarily an agricultural nation although only 20% is actually farmed. Much land is devoted to grazing and about 35% of the country is still covered with forests.

The large, mountainous portion of Austria has much the same tourist appeal as Switzerland. The heaviest industrial concentration, which is rather small, is in the northeast. The rolling, hilly plateau section in the east and the Danube basin are the main source of Austria's agricultural products.

Climate

There is a good amount of summer sunshine in Austria, but since this is also very much of a year-round resort country, the winter sunshine record and hours of daylight are of great interest to potential visitors.

SUNSHINE TABLE

A—Average Hours of Daylight Per Day

B—Average Hours of Sunshine Per Day

		Nov.	*Dec.*	*Jan.*	*Feb.*	*Mar.*	*Apr.*
Feldkirch	A	6.7	5.9	6.8	7.7	9.5	10.8
	B	2.1	1.4	1.8	2.9	3.7	5.4
Innsbruck	A	7.0	6.0	7.3	8.5	10.4	11.9
	B	2.7	2.1	2.6	3.8	5.2	5.5
Salzburg	A	8.5	7.4	8.0	9.0	10.7	12.6
	B	1.9	2.0	2.1	3.3	4.3	5.0
Sonnblick	A	9.7	8.6	9.0	10.6	11.9	13.2
	B	2.6	3.4	3.6	4.6	4.9	4.4
Gras	A	8.6	7.8	8.3	9.6	10.9	12.6
	B	1.4	2.0	2.2	4.0	4.8	5.8
Linz	A	8.4	7.5	8.0	9.5	11.0	12.0
	B	1.7	1.4	1.7	3.2	5.0	5.9
Vienna	A	8.6	7.6	8.0	9.5	10.8	13.0
	B	1.8	1.5	1.8	3.2	4.6	6.1

Climatic Zone A: The Western and Central Alps Mountain Sections

Meteorologists generally divide Austria into two large climatic zones. The first of these segments includes the west and southcentral mountainous sections, shown on the small map in Chart 26.

This area can only be described as having a mountain climate, which means a great variety of local conditions depending mostly upon altitude and exposure to winds or sunshine.

There are numerous 8,000-to 10,000-foot peaks among the many mountain ranges that cover almost all of Zone A, and this territory really forms the eastern flank of the entire Alps mountain complex.

Average annual figures are very general at best, but can give an indication of climatic conditions. As an example, the average total precipitation in this zone has varied from an annual low of 27 inches to a tremendous 69. Imagine 5 feet 9 inches of water over this whole vast area! That would be quite a lake if it occurred at one time and be about ten times that depth if it fell as snow. The heaviest downfall comes during June, July, and August, but there is enough spread throughout the year to make this appear to be very damp country at all times.

The saving grace, of course, is that most of the summer rain occurs as short, heavy showers. They clear away quickly and are often immediately followed by bright sunshine. The cold-weather precipitation is mostly in the form of snow which melts throughout the summer, feeding the beautiful alpine lakes and streams.

Temperatures in this higher portion of Austria can also be very erratic, and this zone has experienced high and low annual average figures of 17 to 71°. Such extremes are, of course, most unusual.

The movement known as foehn occurs when a depression develops to the north and the air is drawn up the southern side of the mountains. As it ascends into the cooler atmosphere, the water vapor contained in the air condenses and is released as rain or snow. The almost dry but still warm air passes over the summits and absorbs more heat as it sweeps down the northern slopes. The foehn may develop at any time of the year but is most common in spring. These air currents can raise the general temperature as much as 25° in a half hour. They are usually quite welcome because the elevated temperature melts the snow very rapidly and brings on an early spring. There is, however, always the possible hazard of the sudden heat causing avalanches or floods or sometimes scorching the tender foliage.

Another characteristic of this mountain country is that cattle farmers avoid the northern mountain sides. The many vineyards are also an accurate indication of the sunny southern slopes. Winters are often long and severe in the highest mountains. While snow covers the ground 2 to 5 months in the less elevated areas, it can last 8 months and reach a depth of 30 feet in the higher portions. Is it any wonder that Austrians hold so many of the skiing records?

A quick glance at the winter sunshine table may be disappointing, but note on the City Weather Tables of this zone that 3 out of 4 days get less than one-tenth of an inch of precipitation. And even a rather small amount of sunshine can produce a bright setting when reflected on snowfields.

Feldkirch is in the small province of Vorarlberg at the extreme western tip of the country. The Rhine River forms the western boundary here between Austria and Switzerland. From this point, it flows north into Lake Constance (Bodensee). A few hours to the south are the glaciers of Sivretta. This entire section of Austria is a place of magnificent scenery, lakes, streams, and 8,000-, 10,000-, and even 12,000-foot snow-topped peaks.

Visitors to this tiny province can enjoy almost all the scenic wonders to be found in Switzerland. Prices are still very modest and Vorarlberg is less crowded than almost any other part of the country at all times of the year. Vorarlberg is also more favored by European than overseas tourists.

The City Weather Table of Feldkirch is typical of the district. The hefty 46 inches of precipitation, which is heaviest from June through August, is not of as much concern to tourists as might first appear. The total 18 inches of rain that falls during these 3 months may seem like quite a deluge and tend to frighten off potential visitors. It should also be noted, however, that while there is a substantial rainfall every third day, during that period, half occurs as brief thunderstorms of which there are about two dozen each summer. These and the other showers quickly disappear, leaving a clear, fresh, mountain atmosphere. Such showers are little interference to outdoor activities on an otherwise sunny day.

This illustrates why it is so important for readers to compare all of the columns in the City Weather Tables and why it is helpful to have as much information as possible. Since this is not a highly industrialized region, although an important textile center, the mountain air is particularly clear and free from pollution. There is almost never any disagreeably hot weather and the mercury has yet to reach the 100°-mark. The abundant snowfall makes for wonderful skiing and other winter activities. The outdoor enthusiast can normally depend upon almost perfect weather from December through March. On occasion, there may even be some snowfall in November or April.

While Feldkirch is an excellent headquarters, this little province is dotted with delightful Alpine villages which many visitors find most satisfactory. Travelers elsewhere might find it difficult to find such a spot—combining wonderful and varied mountain scenery, low costs, uncrowded conditions, fine foods, and, best of all, the gracious and kindly people.

Lovely Bregenz, on Lake Constance, is the site of one of Austria's most unusual and spectacular annual musical events. Operetta, ballet, concerts, and plays are beautifully presented on a huge stage floating in the lake. This gay festival continues from mid-July to mid-August. Midwinter and midsummer are the two very popular periods when confirmed reservations are recommended. Such attractive spots as Zurs, Bazora, and Lech are active places during the giant slalom events, the glacier race, the international ski jumping contests, and Joler Cup toboggan days. The Lake Constance area is favored in the summer.

What can be said about the Tyrol? Few places couple such sheer loveliness and spectacular beauty. The millions of travel folders, calendars, and color illustrations of Innsbruck attest to its photogenic qualities.

Perhaps the 150 cable-car and chair-lift systems tell something about both climate and topography. As indicated on the Innsbruck City Weather Table, there can be snow anytime between October and May, but heavy falls are usually limited to November through March. The comfortable 1,900-foot elevation seems to be about the ideal altitude for most people. The 1964 Winter Olympics were held here, but this whole region is equally popular in summer.

The Innsbruck Weather Table is quite similar to that of Feldkirch except that the 34-inch total precipitation is 12 inches less than that of the western city. Only 80 days a year will have more than 1/10-inch of precipitation. It occurs about 1 day in 4 during the summer and one in 5 or 6 through the other months. Most of the rain in the warm weather is in the form of short showers but, at other times, is apt to be more drawn out. Nestling down in a bend of the Inn River, there is little wind (except in March) so that even the lowest winter temperatures lack any sharp bite. Midwinter visitors will not experience temperatures much below 20°, but the records over the past 30 years show occasions when the thermometer has registered down to −22°. Readers who noted the rather low hours of sunshine should read the last column in the City Weather Table. There are only 45 days a year when visibility is less than ½ mile which spells plenty of clear, if not shining, weather. Even at the Innsbruck elevation, one can expect the normal rapid temperature drop at sundown in both winter and summer.

Tyrol attracts a greater influx of visitors than the other provinces. Such charming little towns as Lienz, Rattenberg, and Seefeld are at least part of the reason. The Austrian Government Tourist Bureau advises that there are 4,700 hotels, inns, pensions, and boardinghouses in this one mountain province. They range from the ultradeluxe to the most modest but be assured—all will be equally spotless.

A superhighway, leading directly south from Innsbruck to the Brenner Pass, is a main artery into the equally spectacular Dolomite region of northern Italy.

Salzburg, could almost substitute the Innsbruck City Weather Table for its own and pension one weatherman. The most marked difference is the high 53-inch annual precipitation which makes Salzburg a bit more showery. While it would be possible to have a slight trace of precipitation every other day, it usually occurs as a several-day, wet interval between much longer stretches of clear weather with plenty of sunshine.

Its location, deep in the Salzach Valley on the leeward side of towering mountains, provides Salzburg with protection from the northern winds. That is very important in winter while summer weather is never uncomfortably warm.

Although the Tyrol is one of the most active tourist areas, there are many small villages where both accommodations and fine foods are less expensive than in the larger cities. Nor are they as crowded as the famous name places even at top season. It is usually easier for the visitor who stays in one area to find and enjoy these attractive spots than one who makes long treks from one place to another each day.

There are still many cities in Europe where one can spend a most enjoyable unhurried vacation. Salzburg and Innsbruck are two fine examples. They are small and compact enough to cover comfortably on foot, and each is surrounded by easily accessible nearby places of interest. Dinner at either the glamorous 800-year-old Goldener Hirsch or the small, elegant Gastschloss Monchstein castle hotel in Salzburg will be an occasion. By contrast, try one of the little mountain inns the following evening for a more modest, but equally satisfying, regional meal. The periods just before or after major events are quieter and excellent times to browse at leisure.

Sonnblick, at 10,190 feet, is included in the City Weather Tables to show the conditions at high altitudes. As the name suggests, this is a grand sight with a thrilling panoramic view on a clear, sunny day. There can be snow any month of the year and there usually is. With normally about 325 nights when the mercury drops to freezing, and 40 registering 0° or lower, it is almost winter the year round.

A 60°-high reading has not been seen in recent years. The almost 60-inch precipitation, which falls mainly as snow, shows how the high peaks can drain the moist winds. The 20 or so mountain thunderstorms in July and August can be quite spectacular.

Graz, an Old World city—with a university founded in 1586—is located toward the eastern end of the mountain country in a somewhat less populated region. There is much open, green country but perhaps even more wooded land. About 37% of all Austria is forest, and it ranks as the fourth largest lumber and wood products exporter in the world. There are also grazing and farming areas and many fine orchards and vineyards.

It is a very popular holiday region with more than 200 little communities classified as vacation villages. More Europeans than overseas tourists are attracted to this quiet pleasant district. (Actually, Austria ranks close after France and Italy as a European tourist country.)

The Graz City Weather Table is quite typical of all Styria Province. Although Graz itself is only 1,100 feet above sea level, the surrounding high country is dotted with 4,000-to 7,000-foot peaks. There is plenty of snow and usually about 150 days register 32° or lower. Ten times a year the mercury will slump below 0 and, in recent years, has hit −21° on several occasions. Note, however, that the normal winter temperature range is from the low 20's to about 50°. Here also, the heavy summer rain occurs as quick-clearing showers. There are no long hot spells; in fact, even the oldest resident cannot remember a 100°-reading.

Just to the southwest is the lovely province of Carintia, the warmest and sunniest portion of the whole country. Like most of Austria, this is dotted with clear mountain lakes which are not very chilly, and many people swim here from May through September. Lake Worthersee, which may be partially fed by thermal springs, has the warmest water of any in this popular summer vacation region. The mighty Grossglockner is in the northwestern corner of the province and the 5-mile Pasterge Glacier is nearby.

Areas surrounding the large industrial centers of Europe are having much the same air-and water-pollution problems as America and Japan. Austria is generally less plagued with this grief, since there are fewer such complexes, and about 90% of the electric energy is developed from pollution-free water power.

While much of Austria's iron ore is mined in this region, almost all is shipped to the Linz and northern segments for processing. Klagenfurt, the largest community in Carintia Province, is an attractive city with many parks and flowers.

Climatic Zone B: Northeastern Austria

This segment, covering the northeastern part of the country, is made up of the two large provinces of upper and lower Austria and the much smaller eastern Province of Burgenland. It includes the high plateau area north of the Danube River called the Bohemian Massive and the wide flat valley which comprises the whole Vienna Basin.

Some parts of this region are more industrialized and less scenically attractive than Zone A. Although relatively low country (500- to 1,000-foot elevation), temperatures here are not dissimilar to those of the mountainous region to the west.

The reason for this is that, while relatively unaffected by the warm humid westerlies, it is more exposed to the sharp blasts from the north and east. These sometimes very strong winds called the bora or *tramontana* occur during the winter months. They are cold and often very dry at which times the skies are clear with some sunshine. If they have swirled over the Adriatic and approach from the southeast, there are often heavy clouds and rain or snow. The result is that this portion of Austria has a Continental climate. That means mild, dampish summers and sharp to frigid winters with sudden and, often rather violent, temperature changes.

Winds have a marked effect upon winter weather. The not uncommon 30°-days feel like 15° when coupled with an 18-m.p.h. wind. There are usually about 100 nights when water freezes and two or three that will register subzero temperatures.

Burgenland is a pleasant, green country of flatlands and rolling hills, lying between the Alps and the great plains of eastern Europe. This province borders on Hungary and the many fortresslike castles are an indication of past invasions.

It is fine agricultural and grazing country. There is an old-world flavor, accentuated by the quaint thatched houses, often enclosed in tiny white-walled compounds. The people, in national costumes and high boots, and the storks nesting on century-old chimneys add to this atmosphere. The large, rather shallow, Neusiedler See, a steppe lake in the northeastern corner is a favorite place for bird watching. The great variety of flora and fauna inhabiting the brackish waters, desolate marshes, and large beds of reeds attract naturalists from around the world.

Linz is the major city in the Upper Austria Province. The Bavarian Alpine ranges extend into the western end of this province. The remainder is rolling or plains country. Precipitation is a medium 36 inches a year, heaviest during the summer. That is also the season for thunderstorms so the 6 to 9 showery days per month do not add up to many hours of actual rainfall.

Except for the mountainous strip bordering on Bavaria, this is mostly a summer holiday area. Linz itself is not as interesting or colorful as most Austrian cities, but is a good start-off point for seeing the province. Few tourists will choose Linz as headquarters since much of Austria's heavy industry is located in this general area.

VIENNA

Too many tourists from overseas visualize Austria as a gay, glamorous Vienna surrounded by magnificent Tyrolian mountains. No experienced traveler has, of course, missed either, but neither has he failed to visit many of Austria's other beauty spots.

So much has been written about Vienna that little need be added. But one word of warning: Since this is so much a year-round tourist city, a few pre-arrival precautions are in order. In addition to booking accommodations, it is prudent to have your travel agent arrange your itinerary to include a Sunday in Vienna. Also be sure to secure tickets for that day to both the unique Spanish Riding School (established in 1580) with its magnificent, white Lippizzaner horses, and the world-famous Vienna Boys Choir in the former Imperial Chapel, just across the courtyard. Both have limited accommodations.

The weather combines with all of the other many attractions to make sparkling Vienna a place of all seasons. With only 26 inches of precipitation distributed so uniformly throughout the year, there is no really wet period but plenty of time for many pleasant, clear days. As a matter of comparison, New York City averages 43 inches of precipitation a year. Chicago gets 32 and Washington, D.C., about 41.

Winter temperatures are just brisk or snappy enough to enjoy walking and getting around to all the musical and other cultural events. The normal temperature span from November to February is about 26 to 50°, but always have in mind that a very short subzero snap is not an impossibility. To most Americans, Vienna's summer weather will seem a little cool—but very comfortable. The mercury does not often stray much beyond the 50- to 75°-marks. Normally, there will be fewer than five 90°-days all summer and, even during the hottest streak, the thermometer has never registered 100°. Vienna's fine weather is, of course, a big plus bonus, but there is so much to see and enjoy in this delightful city that a visit at any time of the year is always a great pleasure. Some physical activity is suggested to offset the irresistible Viennese *konditorei* (pastry shops). Who could leave this fascinating city without a generous sampling of Demel's goodies and, of course, some Sachertorte?

Surrounded by many diverse peoples, Austria, and Vienna in particular, offers the finest of many noted cuisines. Bavarian, Czech, northern Hungarian, and many others are all prepared with a delicate touch and flair to rival the finest in Paris.

The Vienna Woods are colorful in autumn. That is also the time of the new wines. Grinzing, Sievering, and Nussdorf, suburbs of Vienna, are gay with people stopping at each tavern and inn displaying a branch of fir or ivy, the symbol of *heurige* and the signal that the current year's young wines have arrived.

Spring in Vienna, like Paris, is a joy, with blossoms galore and an atmosphere of gaiety. No spot seems too small or insignificant to rate a box or even a tiny pot of flowers. Costs are, of course, higher than in the small villages but even so, well worth it. The top season starts from April through May when rates go up 20 or 25% to the normal summer schedules.

Tourist Information and Transportation

There are so many pleasant and surprising things not commonly known about Austria that a letter or visit to one of the Government Tourist Information Offices can be very rewarding. Austria maintains well-staffed offices in Beverly Hills, California, Chicago, New York City, and Portland, Oregon. Also two in Canada in Toronto and Montreal.

Overseas visitors should also remember the Austrian Air Line. While this excellent international carrier does not have a USA gateway, it operates in 25 countries across Europe and the Near East. Its very efficient internal service which connects Vienna, Linz, Salzburg, Graz, and Klagenfurt can often save both time and money and should not be overlooked by tourists who wish to avoid driving.

If there is any question about Austria being a particularly attractive place to visit, we can only suggest that you ask any of your friends who have been there. Almost every European country includes large industrial areas which tourists try to avoid. Except for some steel mills and other heavy industry in the Linz area, almost every square mile within the boundaries of this charming little nation can properly be classified as premium vacation country.

ITALY

ALTHOUGH Christopher Columbus arrived on our shores as a tourist almost 500 years ago, North Americans did not discover Italy in any great numbers until after World War II. Prior to that time, a grand tour might include Rome, Venice, and perhaps, Florence for the artistically inclined. Few wandered much farther afield.

But that has all changed and overseas visitors are now found in ever-increasing numbers anywhere from the secluded Cinque Terre on the Adriatic cliffs to the tiny mountain villages in sunny Sicily.

There is certainly no lack of variety in Italy, and it is also among those fortunate nations which are classified as year-round vacation lands. You need only state your preference and the Italian Meterological Bureau or "Fairweather Travel" will tell you when and where to find it.

Topography

As we know, climate is determined to a considerable extent by the physical conformation of the land. Most of Italy consists of uplands and mountainous country. The major exceptions are the great expanse of the Lombard-Venetian plains of the Po River Valley, some isolated coastal areas, and the many flat patches among the mountains. The Po Valley produces most of the nation's agricultural and industrial output. The large commercial, financial, and manufacturing complexes surrounding Milan, Turin, the large shipping port of Trieste, and the much-loved Venice are all contained within this important segment.

The massive Alpine high country, extending across the 375-mile-wide northern border boasts the famous lakes of Como, Maggiore, and

Garda which enhance so many travel folders. Here, clustered along the national boundaries, are the lofty peaks of Mt. Blanc, 15,781 feet, Mt. Rosa, 15,217 feet, and the photogenic Matterhorn, 14,780 feet.

The Apennines are an almost continuous chain of ranges extending from the Alps in the northwest down through the toe of Italy and across into Sicily. The highest points in this rugged backbone are the Gran Sasso (Corno), 9,560 feet, Amaro, 9,170 feet, about halfway down the leg, and the sometimes active Etna, 10,868 feet, in eastern Sicily.

Along much of the coast, the mountains extend almost to the shore line—sometimes in the form of sheer cliffs. This makes the fertile lowlands around Leghorn, Rome, and Naples, and the crescent-shaped coastal Foggia Plain on the Adriatic very important from an agricultural viewpoint. Most of the central portions of both Sicily and Sardinia, the two largest islands in the Mediterranean, are also mountainous.

Climate

Meteorologists generally think of a country as a patchwork map made up of segments, each having more or less distinctive climatic characteristics. The greater the variety in configuration of the terrain and degree of exposure to the sea and wind, the more weather pieces there will be. Italy experiences a wide range of conditions.

Because of its shape and location, many winds blow across Italy at different seasons. One of the most important, because of its strength, is the bora which occurs mainly during the winter months. When it blasts in from the northeast from over the land masses of central Europe, the weather will be cold and usually very dry, with clear skies. If there is a shift, however, and it blows in across the Adriatic, there may be heavy cloud formations often accompanied by either rain or snow. On occasion, the bora can tear along at a velocity approaching 100 m.p.h. Most often, however, when from this direction, it will be gusty with rather violent squalls and can persist for several days. Also known as *gregale* or tramontana, the bora is quite similar and about as unpopular as the more familiar mistral winds of southeastern France. On the western side of Italy, the *maestro* winds, blowing from the northwest, sweep around the coasts of Sardinia before reaching the mainland.

Those portions of Italy facing south are exposed to different types of air currents. The siroccos, usually dry and often dust-laden, change as they travel across the Mediterranean. Picking up moisture, they reach the Italian shores as hot, humid, and very enervating air streams. The sirocco can occur at any time of the year—it is rare in summer and most

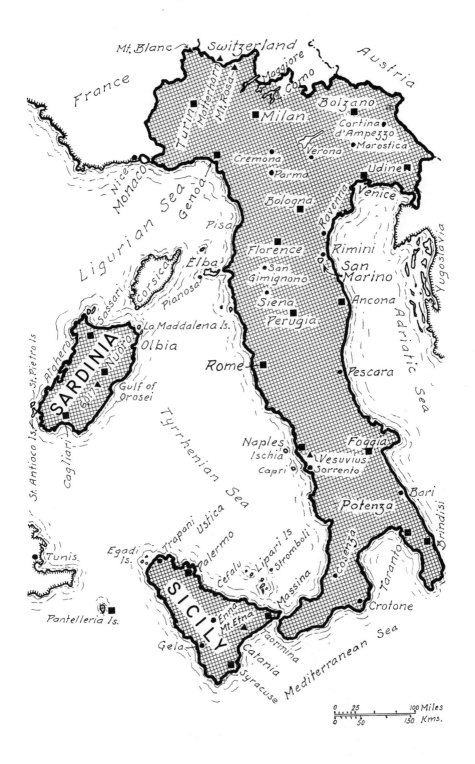

France

Mt. Blanc   Switzerland   Austria

Maggiore   Como

Mt. Rosa   Bolzano

Matterhorn

Turin   Milan   Cortina d'Ampezzo

Marostica

Cremona   Verona

Nice   Parma   Udine

Monaco   Genoa   Bologna   Ravenna   Venice

Ligurian Sea   Pisa

Corsica   Elba   Florence   Rimini

San Gimignono   San Marino

Pianosa   Siena   Ancona

Sassari   Perugia

La Maddalena Is.

Alghero   Nuoro   Olbia   Rome   Pescara

SARDINIA

St. Pietro Is   Gulf of Orosei

St. Antioco Is.   Cagliari   Tyrrhenian Sea   Naples   Foggia

Ischia   Vesuvius

Capri   Sorrento

Bari

Ustica   Potenza   Brindisi

Tunis   Egadi Is.   Trapani   Palermo   Lipari Is   Stromboli   Cosenza   Taranto

Cefalu   Messina

Pantelleria Is.   SICILY   Crotone

Enna   Mt. Etna

Gela   Taormina   Mediterranean Sea

Catania   Syracuse

Adriatic Sea

Yugoslavia

0   25   100 Miles
0   50   150 Kms.

Chart 27

ITALY

TOPOGRAPHICAL AND CLIMATIC MAPS

Alps Mts. Zone "A"

Po Valley Zone "B"

Adriatic Coast Zone "E"

Apennines Mts Zone "D"

Ligurian & Tyrrhenian Coasts Zone "C"

SARDINIA

SICILY

Lakes Area Dolomites

Po River

Chart 28

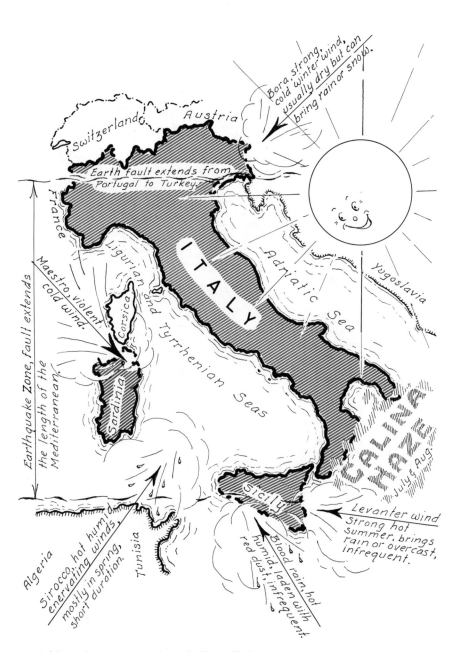

Bora strong, cold winter wind, usually dry but can bring rain or snow.

Earth fault extends from Portugal to Turkey.

Maestro violent cold wind.

Earthquake Zone, fault extends the length of the Mediterranean.

Sirocco, hot humid enervating winds, mostly in spring, short duration.

Blood rain, hot humid laden with red dust, infrequent.

Levanter wind strong hot summer, brings rain or overcast, infrequent.

July & Aug.

Switzerland Austria France Ligurian and Tyrrhenian Seas Corsica Sardinia ITALY Adriatic Sea Yugoslavia Sicily Algeria Tunisia

WINDS AND WEATHER

Italy is a most pleasant year-round vacationland with abundant sunshine.

Summer–hot, dry & sunny
Autumn–showers, mild & bright
Winter–skiing in Mts, swimming–Sicily
Spring– Eureka! almost always perfect!

ITALY

Chart 29

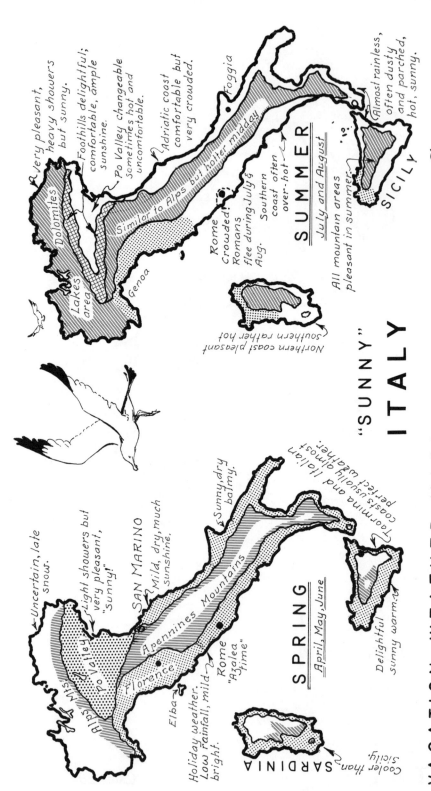

VACATION WEATHER MAPS

"SUNNY" ITALY

SUMMER
July and August

Very pleasant, heavy showers but sunny.

Foothills delightful; comfortable, ample sunshine.

Dolomites

Lakes area

Genoa

Po Valley changeable. Sometimes hot and uncomfortable.

Similar to Alps but hotter midday.

Adriatic coast comfortable but very crowded.

Foggia

Rome crowded. Romans flee during July & Aug.

Southern coast often over-hot.

All mountain areas pleasant in summer.

Almost rainless, often dusty and parched, hot, sunny.

SICILY

Northern coast pleasant southern rather hot.

SPRING
April, May, June

Uncertain, late snow.

Alps Mts.

Light showers but very pleasant, "Sunny!"

Po Valley

SAN MARINO

Mild, dry, much sunshine.

Apennines Mountains

Florence

Elba

Rome "Azalea Time"

Holiday weather. Low rainfall, mild, bright.

Sunny, dry balmy.

Taormina and Italian coasts usually almost perfect weather.

Delightful sunny warm.

Cooler than Sicily.

SARDINIA

Chart 30

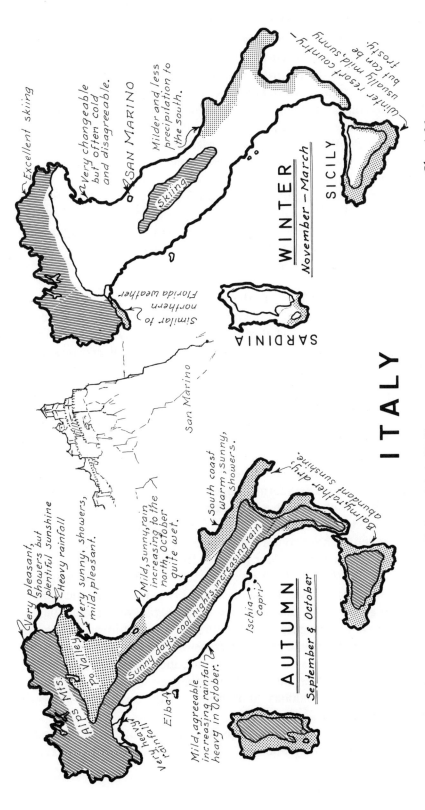

WINTER
November — March

Excellent skiing

Very changeable but often cold and disagreeable.

SAN MARINO

Milder and less precipitation to the south.

Skiing

SICILY

Winter resort country usually mild, sunny, but can be frosty.

Similar to northern Florida weather

SARDINIA

San Marino

ITALY

Very pleasant. Showers but plentiful sunshine

Heavy rainfall

Very sunny, showers, mild, pleasant.

Mild, sunny, rain increasing to the north. October quite wet.

South coast warm, sunny, showers.

Balmy, rather dry, abundant sunshine.

Alps Mts.

Po Valley

Elba

Very heavy rainfall

Sunny days cool nights, increasing rain

Mild, agreeable increasing rainfall, heavy in October.

Ischia

Capri

AUTUMN
September & October

VACATION WEATHER MAPS

Chart 31

prevelant in spring. Fortunately, it seldom lasts more than a day or two and is frequently followed by cooler air from the north. These winds often funnel into the Ligurian Sea, south of Genoa. The coastal mountain ranges forming a large arc above that seaport drain these winds of much moisture, making the lands between the mountains and the sea one of the wettest portions of Italy. Conversely, this high barrier also causes the Po Valley plains in the east to get less precipitation than would otherwise be the case. As an example, Genoa averages 52 inches a year while Turin gets a total of only 25.

Sometimes, winds from the Sahara Desert pick up fine, red dust particles which become imprisoned in the raindrops that are formed in crossing the Mediterranean. These storms, which occasionally travel as far as Great Britain are not uncommon in southern Italy. Particularly heavy showers leave a red stain on the ground and other exposed surfaces.

As indicated on Chart 28, the major climatic zones of Italy correspond pretty much to the topographical characteristics of the country. It should be remembered that while the climate within each area is generally quite uniform, there are most often small enclaves where specific conditions generate distinctive sub- or miniclimates.

Climatic Zone A: The Italian Alps

Many North Americans picture Italy as a salubrious land of sunshine, music, Chianti, and undiscovered Sophia Lorens. The wide northern segment of the country is quite a different story. The many Alps mountain ranges, shared commonly with France, Switzerland, Austria, and Yugoslavia form a huge arc across this entire region. Here it is possible to experience a 100°-summer-day but the winter weather is more spectacular. Many of the 10,000- to 15,000-foot summits are snow-capped the year round while the higher mountain passes may be closed for 4 to 6 of the colder months. In 1972, a blizzard dumped 13 feet of snow in 2 days, isolating about 100 villages and trapping 2,500 skiers for several days. That, of course, is a rare, headline weather happening. Normally, this whole region is a happy winter vacationland. Just as these towering heights discouraged military invasion from the north, so, too, do they form a barrier against much of the severe winter blasts. Actually, the southern slopes and, more particularly, the sheltered valleys are often most comfortable even in midwinter. Positive factors are the low midday humidities and the absence of smog and other man-made pollution.

But make no mistake, the Alps are prime skiing country and an overall winter vacationland. The City Weather Table of Bolzano is reasonably typical of the protected mountain valley communities. Note that, while there can be freezing temperatures and, of course, snow at any time between October and May, the winter weather is usually not at all severe. The dry air coupled with a 25 to 50°-temperature range adds up to very pleasant conditions.

This Alpine region is equally popular in summer. About a dozen days a year, the mercury may top 90°, but there is always a comfortable drop at sundown. The normal summer temperature range is about 55 to 82°. With only 26 inches of precipitation, there is no really wet period. Like most high country, this is a place of very distinct seasons with visitor emphasis on the midsummer and midwinter months.

Most overseas visitors use Cortina d'Ampezzo as headquarters because of the greater selection of accommodations, many of which were erected for the 1956 Winter Olympic Games. Much of the Dolomite mountain area was once a part of the old Austro-Hungarian Empire. This is still reflected in both the Tyrolian cuisine and wide use of the German language.

The area farther to the east, almost at the Yugoslavian border, is known as wet country. A total annual precipitation of 60 inches, mostly in the form of summer rainfall, might well frighten off many potential visitors. Actually, the heavy summer precipitation is remarkably similar to that of southern Florida, both in volume and type. Udine here in northern Italy and Miami, Florida, average up to 7 inches a month in midsummer—mostly as short, very heavy showers, often as thunderstorms. Each of these cities can expect 8 to 10 hours of sunshine a day during this warm season. Miami is a bit hotter and considerably more humid, but anyone familiar with the southern Florida weather will know what to expect during the summer in this area.

As indicated in the following table, the annual sunshine total is not particularly high. It is, however, ample throughout the year to satisfy all but the most avid sun worshippers.

HOURS OF SUNSHINE PER DAY

	J	F	M	A	M	J	J	A	S	O	N	D
Bolzano	4	5	5	6	7	7	8	7	6	5	3	3
Pian Rosa	4	5	6	6	7	7	8	7	6	6	4	4
Vetine	4	4	5	6	7	7	9	8	6	5	3	3

Climatic Zone B: The Po River Valley

Just south of the Alps lies one of the most important segments of Italy. It encompasses the wide plains of the 250-mile-long Po River Valley and includes the foothill country of the Alps to the north and the lower slopes of the Apennines on the south. This is by far the largest of Italy's many market-basket garden areas. It comprises two-thirds of all the nation's lowlands, made fertile by the rich sediment washed down from the surrounding mountainsides. This area is also responsible for most of the country's industrial output and, unfortunately, the major portion of the air and water pollution. The Milan-Turin complex seems destined to become one huge, congested megalopolis where most tourists will not spend much time. The world's most famous fresco, Leonardo da Vinci's "Last Supper," in Milan, must now be protected from the befouled atmosphere by an air-purifying system.

This zone is a region teeming with history, scenery, and architectural beauty—and some of the world's most romantic names: glamourous Venice; Bologna, which many proclaim the gastronomic center of the nation; Ferrara; and Parma, the latter too often remembered for its fine ham and Parmesan cheese rather than as the birthplace of Toscanini; the city of Guiseppe Verdi, Padua; Vicenza; Verona, home of Romeo and Juliet; and Cremona, which offers much more than memories of the world's almost perfect violins.

The delightful lake region and the lower Alps extend into this zone, and there are many little places along the lake shores of Maggiore, Garda, Lugano, Como, and the many smaller lakes. Some old-timers would not consider a "grand tour" complete without a stopover at the deluxe Villa d'Este in Cernobbio on Como.

The City Weather Tables of Milan, Venice, Turin, and Bologna cover the climatic conditions in this overall zone. The Apennines and Alps, to a great extent, prevent the mild, wet breezes off the Mediterranean from reaching this plains country. The towering Alps along the north, in turn, ward off much of the fiercest winter blasts. The quadrant facing northeasterly, however, has no such natural protection. As a result, this side is open to the cold winds from those directions, causing this whole area to have a modified Continental climate.

Temperatures are often rather uniform for extended periods in this zone with usually not much more than a 20°-spread within any one month. There are, however, almost always abrupt and rather substantial changes in weather with shifts in wind direction. The easterly summer breezes are refreshing, but the stiffer winter blasts are penetrating and most uncomfortable. The mercury can range from 0 to over 100°, but such extremes are quite unusual. Autumn and winter fogs are not uncommon, particularly toward the eastern end of the zone.

Total annual precipitation ranges from 25 inches in Turin to 39 in Milan. Although spread throughout the year, it tends to be a bit heavier during spring and autumn. As indicated in the Venice and Milan tables, there are about 40 summer thunderstorms which means that the more abundant rainfall in that season usually occurs as short, heavy downpours. The fact that there will usually be 250 to 300 days a year that get less than 1/10-inch of precipitation classifies this as a dry rather than wet tourist land.

SUNSHINE RECORD OF NORTHERN ITALY
Average Hours of Sunshine Per Day

	J	F	M	A	M	J	J	A	S	O	N	D
Milan	2	3	5	6	7	8	9	8	6	4	2	2
Venice	3	4	5	6	8	8	10	9	7	5	3	3
Bologna	3	4	5	6	8	9	10	9	7	5	2	2
Genoa	4	5	5	6	8	8	10	9	7	6	4	4
Pisa	4	5	5	7	9	9	11	9	8	6	4	3

It should be remembered that the western portion of this zone is an important industrial area. Unlike the high Alps country to the north, there are sections where one is very well aware of the heavy air pollution.

Winter is generally the least desirable time to visit this area. Although a small percentage of tourists find late July and early August a bit too warm, most suffer no great discomfort.

VENICE

Poets, painters, epicures, and historians have covered this city so completely that we need only add a word about weather. Springtime, of course, is the best season. While January, February, and sometimes December or March receive the least rainfall, the sea often compensates by flooding the Piazza San Marco ankle-deep. April is really delightful, but no month from May through August averages less than 8 hours of sunshine per day. Midsummer is the next driest period, but some visitors suggest that the rather murky canal waters would never be mistaken for Chanel No. 5 on a sizzling hot day.

Winters are most often wet, overcast, and chilly. Shoveling snow is not too unusual an occupation during that season. The weather is more uncomfortable than the normal temperature figures alone might suggest. The synergistic effect of a low temperature, plus dampness, plus a sharp wind equals one miserable tourist. On the odd occasion when the mercury drops to 10 or 12°, distress flags are unfurled. But why dwell on that small portion of the year?

While the precipitation may seem quite heavy even through the warm weather, the 40 or so thunderstorms, mostly during that period, mean that much of the rain will occur as rather heavy but brief showers. The 5 or 10°-drop in temperature that follows is a welcome relief.

Venice's 29-inch total precipitation is quite modest compared to the 52 inches that pour down on Genoa each year. There is a similarity in that each of these cities at the northern head of large water bodies averages higher rainfalls than areas to the south. Actually, New York City gets over 50% more precipitation than Venice. Although these cities enjoy about equally abundant sunshine throughout most of the year, New York gets less in summer but more in winter. There is a wide range of accommodations in Venice but whether one stays at the Danieli, the Gritti Palace, the Cipriani on nearby Torcello Island, or a modest *pensione* most of Venice's wonders can be enjoyed at small cost. If there is such a thing as as scenic *papparazzi,* this gem of the Adriatic should be a fertile field.

Climatic Zone C: Western Mediterranean Coast

This should properly be called the Tyrrhenian and Ligurian sea coasts. This segment includes many of the cities and places that are of greatest interest to visitors—Rome, Naples, Pisa, Genoa, Capri, Pompeii, and Vesuvius. The 500-mile coastal band, extending from the French border down to the country's instep below Napels, enjoys a Mediterranean or subtropical climate. Most readers could develop an accurate weather picture for themselves by observing the physical geography, winds, and sea on Chart 29.

The almost continuous chain of high mountains to the north and east ward off much of the most severe winter weather. On the other hand, these entire shore lands are exposed to the warm, humid breezes blowing in from the Mediterranean.

These air currents impinge against the mountains and ascend the western slopes the colder temperatures to condense the water vapor, causing it to precipitate as rain or snow. Meteorologists refer to this as *orographic* rainfall.

Except for the crescent extending around Genoa—which is the wettest, getting an overabundant 50 inches—most of this coast averages about 35 inches of precipitation a year. Although each month gets a fair share, September, October, and November generally experience the heaviest downpour. Summer is the driest season as indicated by the very modest July and August figures. Fortunately, from the tourist's viewpoint, there are anywhere from one to three dozen thunderstorms, mostly during the warmest weather. Those downpours will be heavy but most often very brief and are usually followed by a 5 or 10°-drop in temperature which is refreshing on a hot afternoon. Note on the City Weather Tables that there are normally about 280 days a year that get less than 1/10-inch of precipitation. This is a good yardstick to check. In the long history of statistics, this overall zone has registered up to 56 inches of precipitation in a year and as low as 24 inches. Few visitors will ever encounter such extremes over the whole area, but it is well to remember that there can always be a particularly wet or dry season.

Winters are quite mild along this Mediterranean coast, and precipitation is more apt to be in the form of rain than snow. Periodically, the whole Continent will experience an unusually severe winter—at which times there may be snowstorms clear down to the toe of Italy.

The siroccos and blood rain, both originating in North Africa, generate most of the precipitation but also account for the very uniform temperatures. While the prevailing winds are from the south and west, they sometimes shift and blow in from the east. Such a change, particularly in winter, brings on clear, dry, cold weather.

Visitors who are caught in a persistent heat wave can most often find pleasant relief by moving close to the seashore. Diurnal air movements, known as "land-sea breezes," occur in clear weather along this coast. Because land masses heat up and cool off faster than water bodies, the inland air rises at about midmorning, causing the humid but cool atmosphere to be sucked in from the sea. The process is reversed after sundown which makes for cooler daylight and warmer night temperatures near the water than in the interior. This action is more pronounced in the tropics but, fortunately, also occurs during the hottest weather in the Temperate Zone.

Zero readings are almost unheard of in this zone and even freezing weather is not too common. Overall temperatures do vary somewhat with latitude from north to south. The normal winter range is about 35 to 50° and in summer from 55 to the low 90's. There may be a few 100°-days in July, August, or September but the usual midday low humidities add to the comfort.

While most of the year, and particularly during summer, this zone can be classified as "sunny Italy," that designation is hardly descriptive from November through February. Even during winter, however, with its heavy, overcast skies, most of this zone will average 3 or 5 hours of sunshine. This, of course, means some delightfully bright, glittering days but many more gray ones.

The 7 to 10 golden hours a day throughout most of the year, however, is enough to satisfy even the most sun-starved Scandinavian. Foggy weather is rather rare, but there is occasional sea mist particularly during the cooler periods.

AVERAGE HOURS OF SUNSHINE PER DAY

	J	F	M	A	M	J	J	A	S	O	N	D
Genoa	4	5	5	6	8	8	10	9	7	6	4	4
Pisa	4	5	5	7	9	9	11	9	8	6	4	3
Rome	4	5	7	7	9	9	11	10	8.	6	4	3
Naples	4	5	5	7	8	9	10	10	8	6	4	3

This whole coast is so packed with wonders that most visitors see only a few highlights. Those who choose the overbusy months of July and August may be well advised to either have airtight reservations or perhaps seek some of the lesser-known spots. One such might be Italy's Cinque Terre (Five Lands). These are five little isolated villages, strung along the steep cliffs of the Ligurian Sea south of Genoa. Riomaggiore, in a narrow valley, and Manarola, perched on a rocky promontory, can be reached by rail or the new road from La Spezia. Corniglia in the center, Vernazza with a spectacular view from atop the sheer cliffs, and Monterosso al Mare, the largest, which boasts a fine beach and good hotels, can presently be reached only by footpath from Manarola. The La Spezia road is scheduled to be extended north to Levanto with branches into each of the five villages. That will, of course, spell the end of the quiet life for these little havens. Foreigners are still a curiosity to the residents, who will then turn from fishing to hawking souvenirs made in the Orient. So plan a visit before that happens.

Lovers of luxury will prefer the Italian Riviera just to the north— with its resorts of Rapallo, Portofino, St. Margherita, or, perhaps, San Remo, and Bordighera to the west of Genoa. The attractive little island of Ponza in the Mediterranean, midway between Rome and Naples, is a favorite of the more affluent Romans. Lucca is a comfortable and pleasant mainland town; Ostia boasts ancient ruins that rival those of Pompeii. And, of course, there is romantic Capri. Only midwinter is questionable from the viewpoint of weather. Midsummer suits most, particularly if they choose the less popular spots.

Climatic Zone D: The Apennines

Any national region that can boast so unique an art, gourmet, and scenic center as Florence is indeed fortunate. This mountain segment is not as spectacular as the Alps country nor as historically attractive as Rome, but it does have a quiet beauty and many places of interest to attract the longer-staying visitor. A month of roaming through the delightful hill towns of Italy can be a most satisfying experience. We strongly urge, however, that Florence not be used as the starting point —many of those who do never get any farther.

THE SOUTHERN APENNINES

We will begin this paper trip down in Calabria at the toe of Italy, which few tourists see except perhaps on the way to Sicily. Mother nature has not been overly kind to this scarcely populated region; perhaps its greatest current attraction is the abundant sunshine. The striking but rather forbidding terrain is agriculturally unproductive and some of the poorer people still live in hillside caves. The future, however, may hold good promise—perhaps, even a real estate boom. A hillside in the Loire River in France has recently become a spelunker's paradise; caves similar to those in Calabria are selling for $10,000, mostly for use as holiday hideaways.

The requirements of the farmer and tourist do not always coincide. Be assured that, as vacation crowds continue to increase, this presently uninviting section has the potential of becoming a very popular resort area. It combines the clear, refreshing atmosphere of the wild but majestic hill country with all the joys of marine life along the three seas that border the nearby shores.

The statistics tell the story. For instance, Potenza (2,700 feet), set in the middle of this area, averages only about 10 inches of precipitation a year. The four or five showers a month during winter seldom add up to more than 1½ inches. While summer is not exactly bone dry, the light sprinkle about once every fortnight is no deterrent to outdoor life. This is indeed a land of sunshine and thankfully, no man-made air pollution. At this 2,700-foot elevation, there can be a 100°-day, but the normal summer temperature range is 60 to 80° with a slight drop at sundown. The low midday humidities almost preclude any heat discomfort. The wintertime mercury usually roams between 34 and 50° with few freezing nights and almost never a 0 reading. The best accommodations in this district are of the Jolly Hotel classification, clean and pleasant but not deluxe. Temperatures tend to be progressively lower while precipitation increases as one travels northward up through the center of Italy.

THE CENTRAL APENNINES

The high country between Potenza and central Italy is pleasant but scenically spectacular in only some sections.

L'Aquila (the Eagle) enjoys a dramatic setting in the midst of snow-capped mountains. The rugged high country down to Sulmona is also strikingly beautiful. Viterbo, the ancient walled city, is an inviting place to wander with a camera or sketch pad.

Umbria, just to the north and on the route between Rome and Venice, is one of the most blessed of all Italy's 17 regions. The hillside town of Assisi is associated with the much-beloved St. Francis. Perugia (1,677 feet), known in America for delicious chocolates, is highly regarded by Italian gourmets for its savory ragouts, suckling pigs, roasted pigeons, and other fine dishes. Orvieto, on a high plateau surrounded by steep cliffs, has lent its name to a grand line of white wines ranging from flinty dry to sweet dessert varieties. Spoleto's recent fame rests on the famous art and musical festival but has long been noted for its architectural treasures. Every epicure is well aware of Scheggino, in the center of the black truffle country. The more fragrant white variety is found in Piedmont to the northwest. Trained dogs have largely replaced the hog in searching among the oak forests in autumn for these black nuggets (which are a tuber, not a fungus).

The City Table of Perugia is reasonably typical of this area. Note that the rather high 36 inches of precipitation is distributed throughout the year but is heaviest in October and November. The next highest is during the April-May-June period while the summer, particularly July and August, each get only 4 or 5 showery days. This leaves plenty of time for sunshine which is quite abundant in this whole area. While gray skies may limit December to about 3½ hours of sunshine a day, that figure increases rapidly to 11 hours in July. No month from April through September averages less than 7 of those golden hours a day which should satisfy almost everyone. There are about 280 days that get less than 1/10 inch of precipitation.

Temperatures are quite equable. The normal 36 to 45°-spread in January and the 64 to 83°-average during July and August suggest pretty good conditions. Like almost all higher country, there can be quite wide extremes and readings from 17 to over 100° have been registered. Such hot and cold spells are almost always of very short duration.

FLORENCE

Very deservedly, there have been volumes lauding the wonders of Florence, the jewel of all Tuscany. Of the millions of words, few have been devoted to describing the weather. The City-Weather Table statistics are very inviting. The ample 33-inch precipitation is spread over the year and, except in winter, often falls as rather brief showers. July and August are particularly dry and sunny with hardly one sprinkle a week and 9 to 11 hours of sunshine a day. If one adds to this the normal 65 to 90°-summer temperatures and generally low humidities, Florence is very appealing. During the occasional 100°-spell, the experienced traveler moves up into the hills. Even nearby Fiesole is usually quite comfortable. Heat-sensitive people should be prepared for 90°-midday-temperatures about half of the time during July and August. Although most visitors find the weather conditions from early spring through autumn very pleasant, springtime is almost always the prime season.

Between Florence and Siena is the famous Chianti region, which is very attractive countryside. Many overseas visitors recognize the little black rooster against the golden background on the label, which proudly certifies a bottle's contents as the élite Chianti *Classico*. Vineyards supplying the choice grapes for this premium wine are limited to a very small acreage. A step down in prestige is the bottle seal showing the chubby, white *putto* or Della Robbia angel. While the familiar wicker flasks are very attractive, the finest of the Chianti wines are bottled in the conventional Bordeaux-type bottle rather than the straw-covered *fiaschi*.

For those who crave excitement, few places draw more wildly enthusiastic crowds than the July and August horse races in the huge central piazza of Siena. On these 2 days, this normally quiet place becomes a medieval town, hung with a fantastic array of crimson and gold banners and with a display of colorful pageantry that defies belief.

While in the Apennines country, no one should fail to visit tiny San Gimignano, midway between Florence and Siena. This ancient, walled town has 15 high towers, all that remain standing of the original 72. The purpose of these monumental structures was twofold, prestige and safety. They were family fortresses and are quite common in Italy. Like the abandoned windmills in The Netherlands, many of the towers along the coast have been converted into summer homes.

A more complete tour of Italy's hill towns would include some of the most attractive in the northern part of the country. Because of the great revival of chess, the town of Marostica will probably be on many itineraries. Every second year, during September, a colorful pageant is staged. A gigantic chessboard is laid out in the main square of the town and beautifully costumed players, each representing a chess piece, go through the moves of a complete game. The knights in shining armor are, of course, mounted on magnificent horses. Huge grandstands are erected along the full length of the square, and reservations are required.

Climatic Zone E: The Adriatic Coast

This side of the peninsula is somewhat dryer and cooler than the shore lands along the Tyrrhenian and Ligurian Seas. Again, we can examine the topography, winds, and seas for the answers. Since the prevailing warm, moist air currents are from the south and southwest, the Apennines form an effective barrier and prevent their reaching this zone. Conversely, this coastal strip is exposed to the drier winds which have traveled over the large land masses to the north and northeast. During the summer, this air is warm but can be more than chilly in winter.

THE LOWER ADRIATIC COAST

From the Gargano promontory, or spur of land opposite Foggia, down through the heel is noticeably hotter and drier than lands to the north. Note on the City Weather Tables that Taranto, Brindisi, and Foggia can expect temperature extremes of over 100° anytime from June through September, while Ancona rarely experiences such readings. Also compare the columns showing the number of 90°-days each month. The summer temperatures usually range from about 65° to the upper 80's and in winter between 40 and 55°.

The total annual precipitation is a modest 18 to 23 inches which does not interfere with the sun. This is particularly true from April through September which average 7 to 11 hours of sunshine per day. During the short winter season, skies are often overcast, with the sun showing perhaps 3 to 5 hours daily. Like most of the less elevated parts of the Italian peninsula, this zone enjoys a Mediterranean climate with very dry, bright summers and the heaviest percipitation during autumn and into winter. The fact that at least 300 days a year get less than 1/10-inch of precipitation is perhaps the most reliable gauge.

This segment of the country, along with Sicily and Sardinia, is, in a large measure, responsible for the designation "sunny Italy." April through September, in particular, is bright and sparkling. The following table will allow sun lovers to pick their spots.

AVERAGE HOURS OF SUNSHINE PER DAY

	J	F	M	A	M	J	J	A	S	O	N	D
Terminillo	4	4	4	5	7	7	9	8	7	5	3	3
Pescara	3	4	5	7	8	9	10	10	7	6	4	3
Brindisi	4	5	6	7	9	10	11	10	8	7	4	4

The lower portion of the Adriatic coast is made up of both attractive and rather barren, uninviting sections. This is a place for people looking for a quiet, inexpensive vacation spot unspoiled by tourism. Most will probably find the small fishing villages of greater interest than the larger communities. Perhaps Lecce, with its rich heritage of baroque architecture and pleasant atmosphere could be an exception. It also boasts a clean, attractive Jolly Hotel. The regional dishes are simple but tasty and satisfying. Bari, Brindisi, and Otranto are busy ports not only for freight but also for passenger service to many ports in Yugoslavia, Greece, Turkey, and the Near East.

THE MIDDLE ADRIATIC COAST

The Abruzzi region, just to the north, is far more scenically attractive. The highest ranges of the Apennines form a dramatic backdrop, topped by the snow-capped 9,500-foot summit of Corno Grande. There are always many more Italian than foreign tourists both along the shores in summer and on the ski lifts in winter. Aside from being a bit less hot in midsummer, there is little difference between the climate here and to the south.

The long stretch of coast from Abruzzi almost to the Gulf of Venice includes those active beaches and waterfront areas sometimes referred to as "the poor man's Riviera." During July and August, it is "standing room only" but just off-season can be an excellent spot for an inexpensive stay. The City Weather Table tells what to expect in this whole section. It is seldom either too hot or cold and the low 27-inch precipitation is spread over every month with July and August usually the driest. Summer humidities are rather low which adds to the comfort.

SICILY

After a week in Sicily, many tourists almost forget that they are still in Italy. This feeling is echoed by the islanders who commonly use the expression "over there in Italy."

Sicily is unique in several respects. Note its location. Little wonder that it has long been a major crossroads of invasion and travel. Sitting almost in the middle of the Mediterranean, it forms a division between the eastern and western basins of that sea. It is also a stepping stone between North Africa and Continental Europe. Some of the uninvited early guests were the Greeks, Carthaginians, Phoenicians, Romans, Goths, Vandals, Byzantines, Saracens, Arabs, Normans, French, Spaniards, and Italians. Many have left evidence of their presence. Taormina, Syracuse, Catania, and many other communities were settled before 500 B.C.

Topography

This island, the largest in the Mediterranean (9,925 square miles), is a continuation of the Apennines complex. The highest portions are in the northeastern corner and include Mt. Grande, 4,508 feet; Mt. Soro, 6,060 feet; Conserva Peak, 6,480 feet; and, of course, the famous Mt. Etna that towers 10,705 feet. Its summit is snow-capped about 6 months of the year.

The entire center of the island is an elevated plateau crisscrossed with valleys and flat patches. Toward the eastern end of Sicily is a massif dotted with several 5,000-foot peaks. The short, steep rivers flowing to the north flood during the wet winter season and dry up almost completely in summer. The two longest rivers (about 70 miles) in the southwestern segment, the Platani and Salso, have similar characteristics.

The 646-mile-long coastal strip encircling the island accounts for most of the agricultural output. The important vineyards and many orchards, however, are located on the lower mountain slopes. Along the northern coast, the steep mountains crowd the shore line but the coastal road is a continuous succession of beautiful seascapes and striking mountain views. The west and southeastern coastal lands are wider with extensive plains in the Palermo and Catania areas.

As previously mentioned, Sicily is on the series of earth faults extending east and west the full length of the Mediterranean. The northern portion of Italy is also situated over a long chain of earth fractures which extends from Turkey to Portugal. There is much evidence of the many communities that have been damaged over the ages in both of these areas. Tremors are not uncommon but perhaps less frequent than in such places as California and Japan. Almost 90% of all earth tremors and actual quakes occur around the perimeter of the Pacific Ocean.

Climate of Sicily

The lower shore lands and the foothills country enjoy a pleasant Mediterranean or subtropical climate. Rain is heaviest in winter, while summers are almost bone dry and sunny. The climate on the plateaux and mountains, like most elevated terrains, varies with altitude and exposure to sea and air currents.

AVERAGE HOURS OF SUNSHINE PER DAY

	J	F	M	A	M	J	J	A	S	O	N	D
Messina	4	5	6	7	8	10	11	10	8	6	4	3
Gela	5	6	7	8	9	10	11	11	9	7	6	5
Trapani	5	5	7	8	9	10	12	11	9	7	5	4
Ustica	5	5	6	8	9	10	11	10	8	7	5	4
Pantelleria	4	5	6	7	9	9	11	10	8	7	5	4

The City Weather Tables of the coastal cities, of Palermo, Syracuse, and Massina, at the tricorner points of the island, show very similar weather conditions. The 30- to 38-inch annual precipitation is heaviest in winter. It is a season of a dozen or so thunderstorms with short, heavy showers, occurring about once every 3 or 4 days. There is no 0°-weather and even freezing temperatures are not common in the lowlands. The normal winter thermometer readings are from 40 to 65°.

Most of the winds of concern to Italy were described earlier in this chapter. Sicily is, of course, more affected by the hot blasts from the Sahara than the frosty ones from the north. There are, however, two others which are much less often encountered in other parts of the country.

The eastern coast of Sicily is exposed to the levanter winds, named for the direction of origin. These hot and sometimes very strong blasts, occurring mostly during the summer, often bring rain or overcast skies. They are generally known to the Italians as *solano*.

Even less frequently, the almost constant midsummer sunshine of southeastern Sicily is blotted out for short periods by clouds of finely powdered dust known as calina haze. This seldom occurs in other than July or August.

Palermo

Only distance and location has delayed Palermo from becoming one of the important tourist spots of Europe. It can offer just about everything that a visitor to the southern climes looks for. The elegant old Villa Igea in a magnificent, sea-front garden setting is the prize, as it would be in many larger cities. Palermo has excellent eating places, and its architectural and scenic wonders will exhaust the most ample film supply. In addition to Palermo itself, no one should neglect to visit nearby eye-filling Monreale on a hill overlooking Conca d'Oro Bay.

April, May, and early June is the choice season or a bit earlier if you want to enjoy the almond blossoms. At that time, there is plenty of sunshine, very little rain, and usually 60 to 80°-temperatures. Midsummer is even brighter with 10 or 11 hours of sunshine a day. While that season is not quite bone dry, precipitation might be labeled meager. It is not unusual, however, to have a 100°-hot-spell at any time from late June through September. The rather low humidity and usual sea breezes make outdoor life comfortable for most although heat-sensitive people may find spring or fall more to their liking.

Circular Tour of Sicily

The most interesting parts of Sicily at present are around the perimeter. Your travel agent will probably suggest the scenically exciting Nastro d'Oro road tour which encircles the island.

Traveling counterclockwise, many windmills soon show up on the Trapani skyline. Their almost constant churning is an indication of conditions at this westernmost point on the island. This ultra sunny spot, which one day may be a major resort area, gets only 18 inches of rainfall a year, with almost none from May through August. The weather conditions along the long southern coast clear around to Syracuse are very similar to those in Trapani. There are many interesting sights scattered across this long stretch, and everyone will recognize the town of Marsala as the home of that familiar cooking wine. There is much evidence that the Greek architects and builders were here, but the real glory of this segment is the marvelous sunny weather and the sparkling waters of the Mediterranean. Just how long it will take for this to be developed into a busy vacation land is anyone's guess but make no mistake—it is sure to happen.

Taormina is perched 670 feet above the Ionic Sea, with Mount Etna, the largest active volcano in Europe, as a backdrop and the pleasant Mazzaro beach at its feet. This delightful city boasts a hotel classified as luxurious by most travel guides. The view alone, however, from the San Domenico Palace Hotel, a landmark converted from an old monastery, is worth the price. Although sometimes nippy in midwinter with chilly swimming, Taormina is popular all year round and, compared with many places farther north, is a weather paradise.

Interior of Sicily

Enna, at 3,163 feet, is fairly typical of the inland mountain country. There may be snow from November through April, which is also the period of highest precipitation. Normal winter temperatures are not particularly low, 30 to 45° with the mercury dropping down to about 15° on occasion. Summers are dry and sunny with an average temperature range of 65 to 82°. The thermometer has never registered 100°, but a few days each year, it comes very close to that figure.

Sicily is another of those places favored by nature that is certain to become a very popular vacation and resort area. The people, the scenery, and perhaps most important, the wonderful climate are all in its favor. If the proposed bridge (or less likely tunnel) to the mainland near Messina is ever built, the floodgates of tourism will be opened wide and at least some parts of this happy peaceful island will quickly resemble other busy vacation spots.

Sicily's Satellite Isles

There are always those intrepid Americans in search of rare undiscovered hideaways. No matter how remote or uncharted, very few places really fall within that category. The English and German tourists, in particular, have a nose for digging out such Shangri-las and invariably make their discoveries *sans* timetable, guidebook, or compass. They generally fit in with customs of an area and leave conditions undisturbed. There is a general complaint and with some little justification, that all too often prices boom coincidentally with the influx of American visitors, some of whom insist that everything from the orange juice, ham and egg breakfast, to the twin-bedded double with bath be so standardized that if it were not for the language, everything would be exactly as at home. An overexaggeration to be sure, but with at least a modicum of truth. But there are still many relatively unknown spots, some of which are mentioned throughout this volume. One might be the Lipari Islands, just to the northwest of Messina. They can be reached by hydrofoil from several ports in Sicily or by boat from Naples. This little archipelago includes the often-active, volcanic isle of Stromboli which ship's passengers watch for at night when passing between Sicily and the toe of Italy. There are a half dozen more islands but visitors are most apt to stay on Lipari, which boasts a very attractive hotel, rugged scenery, and enough real estate to allow one to enjoy a happy holiday.

Pollution has become so widespread that the pure, clear water that surrounds these islands is irresistably inviting to underwater swimmers.

The sunshine history of tiny Ustica, which is just to the west of the Lipari group, is on the sunshine table. The annual 2,694 hours is not a particularly high total, but the average 7.5 hours per day in April to the 8.4 in September, with the tops of over 11 hours a day in July, adds up to more sunshine than most doctors recommend being exposed to. The 4 to 5 hours per day in midwinter allows for many gray, overcast skies. There can also be gusty winds and violent sea storms during that season. Costs are still very low, but Romans and a few other Europeans are beginning to find their way to the Lipari Islands, particularly during July and August. We have included the City Weather Table and sunshine record of Pantelleria, the tiny island which is a bit closer to the coast of Tunisia than Sicily. The scant 11 inches of rainfall and the 335 days a year that get less than 1/10-inch fall could be a great temptation to folks from the north. Then there is the ample sunshine and still clear, beautiful water, although the Gulf of Tunis area is one of the relatively few heavily polluted water bodies along the North African coast and the current flows from that direction toward Pantelleria.

There is also the tiny Egadi Island group just off Trapani at the western end of Sicily and a few other isolated isles in the general area, but they do not, as yet, have much to offer in the way of accommodations.

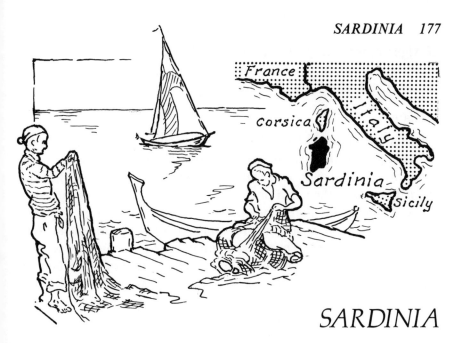

SARDINIA

This 9,300-square-mile island may be second in area to Sicily, but it includes more real estate than the state of New Jersey or the nation of Israel. Sitting out in the middle of the Mediterranean, it also was an ancient crossroad of travel. While relatively unknown to North Americans, it has long been a favorite of some Europeans. Sardinia is primarily agricultural land but not particularly bountiful. In general appearance, it is less green and lush than Corsica, its French cousin, just to the north.

Topography

Almost all of Sardinia is classified as high country. Mountain peaks dot the island but few are over 3,000 feet. The large southeastern segment is mountainous with high plateaux. The central and western portions are a complex of massifs, broken by very wide valley plains which form a large "V." One arm extends from Oristano to Cagliari and the other from Oristano to the northeast, short of Nuoro. The large, low area just to the northwest of Sassari is mostly agricultural and grazing lands. There are also many small, flat areas among the mountains and along the coasts. In many sections, however, the mountains crowd down almost to the water's edge.

The entire eastern coast of Sardinia is quite steep and straight with few indentations. On the opposite side, which is generally more popular with visitors, the mountain slopes are rather gentle, and the irregular shore line is a continuous string of beaches, coves, bays, and, just south of Alghero, there are high, rugged cliffs. The Tirso, the longest river in Sardinia (93 miles), has been dammed to form Lake Omodeo, in the center of the island.

Climate

Sardinia experiences a modified temperate climate. As usual, weather in the higher sections is generally governed by altitude and exposure. Winters are usually not severe but can be very uncomfortable when the winds blow from the north or there is a sea storm. Summers are very dry and sunny; there are times when the midseason can be very hot.

Most of the notes concerning the winds of Sicily and mainland Italy also apply to Sardinia. There is one situation of particular interest for sea travelers heading toward the island from the west. The famous mistrals which sweep down the Rhone Valley in France during the cooler weather end violently in the Gulf of Lions. Ship passengers traveling to Sardinia during that season, arrive rather groggy after being tossed around in those turbulent waters. Italians know these windy blasts as the maestro. We mention these less than pleasant conditions because they can occur, but few visitors, will encounter them, since, normally, there are only the gentle sea-land breezes.

The sunshine record of Cagliari is typical of the southern end of the island, while the figures for Alghero apply generally to the west coast. The record of Olbia shows what visitors to the famous Emerald Coast resorts can expect.

The tiny isle of Pianosa is about 10 miles south of Elba, the 94-square-mile island to which Napoleon was exiled in 1814.

AVERAGE HOURS OF SUNSHINE PER DAY

	J	F	M	A	M	J	J	A	S	O	N	D
Cagliari	5	5	6	7	9	10	11	10	8	6	4	4
Alghero	5	5	6	7	9	10	12	11	9	7	4	3
Olbia	4	5	6	8	10	9	12	11	9	6	4	4
Pianosa	5	5	6	8	9	10	11	10	8	7	4	4

General Information

Sardinia owes its increased popularity to several factors. First, the malaria which had plagued the lowlands was eradicated—mainly due to the efforts of the Rockefeller Foundation. Then more recently, the Aga Khan (Prince Karin) and the Guinness family of Ireland gave the island widespread publicity by establishing a fabulous luxury resort complex along the northeastern coast.

This glamourous segment, the 35-mile-long Costa Smeralda (Emerald Coast), extends northward from Olbia. There are actually a half dozen individual resorts each sitting in solitary splendor, isolated each from the other—and from the general public. Of the three Aga Khan resorts, only Cervo, the largest, stays open all year—the others from April through October. The jewel, Pitrizza, consists of only six villas accommodating 25 guests and serviced by a marvelous staff of 125. The rates paid by Princess Margaret and lesser guests may not cover the operating costs but that is probably of little import. The overall enterprise is a real estate development project. It would take some careful looking to find a more relaxed and elegant spot in which to enjoy a quiet vacation.

The City Weather Table of Gagliari is typical for this coastal strip. Combining the modest 21 inches of precipitation with an average of 8 to 12 hours of sunshine a day from April through September just about eliminates the need for additional information. Note that the mercury does not often climb above the comfortable 80's but the visitor must be prepared for the possible few 100°-days in midsummer.

Winter is usually quite mild with considerable sunshine. When the strong, cold wind howls in from the north or a sea storm kicks up, however, the prudent refuge is indoors. Yes, winter is somewhat of a gamble with the odds a bit adverse.

Some underwater fishing is done along the eastern coast, but apart from the beaches and a few resorts around the Gulf of Orosei, there is not much tourist activity until you reach the Cagliari area. The shore line here is a series of sharp cliffs, steep mountain slopes almost to the water's edge, open beaches, and small coves. Cagliari, which predates Rome, is a busy place—the terminus for steamship, railroad, and air lines. The immense salt works are also interesting. The City Weather Table is almost a duplicate of Olbia. There are refreshing sea breezes throughout most of the warm weather, and this area also qualifies as a segment of sunny Italy.

The same general weather patterns continue up the west coast past Alghero to the northernmost point opposite Corsica. This west side is much more of a vacation playland and the area where most Italians and other Europeans stay. It will be no great surprise to the experienced traveler a few years from now when the front covers of travel folders start to feature such charming little places as Stintino. Note that there is practically no rainfall along either coast during the summer. The heaviest of a rather modest precipitation occurs through the autumn and well into the winter months. There are usually over 300 days that will get less than 1/10-inch of precipitation around the entire perimeter of the island. Visitors will quickly notice that compared to Sicily, which is a very densely populated place, many parts of Sardinia appear almost deserted.

Nuoro, at just under 2,000 feet, is reasonably typical of the elevated inland country. Even here temperature extremes are seldom encountered. Normally, the thermomenter will not register much over 90° or drop below 30° throughout the whole year. Lengthy statistics, however, show an occasional 100° and 18°-figure. There can be some snow anytime from December through March, but this is not skiing country.

As time goes on, many more overseas visitors will discover Sardinia. We have already mentioned the Emerald Coast glamour spots but there are fine accommodations that are clean and comfortable, although not deluxe. There are, of course, the pleasant modestly priced Jolly Hotels. The French Club Méditerranée has a vacation village, and the English, who come in the greatest numbers, are represented by a fine 1,500-bed holiday village. A recent survey of several resorts showed that only 1% of the guests were from the United States.

Tourist Information for Italy, Sicily and Sardinia

There are so many conditions and climates throughout Italy that it is obvious there is no period when the visitor cannot find wonderful weather, whether his aim be skiing or scuba diving. The vacation maps show at a glance the most pleasant areas during each season of the year.

An institution not familiar to most overseas visitors is the *Diurno* which can be found in the larger Italian cities. As the name implies, this daytime comfort lifesaver is a quick stopover station offering many useful services. The leaden-footed tourist, far from his hotel, can duck into the nearest Diurno for a siesta followed by a shower, Turkish bath, hairdo, pedicure, or to have his clothing pressed.

There are discounts up to 30 per cent on some forms of transportation, accomodations, and services from January to May in parts of Italy and Sicily. In addition, there are many package tours run by airlines. Ask your travel agent to check.

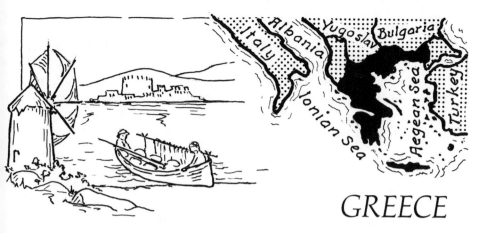

GREECE

The Greeks were among the earliest of the Cook's tourists. They sometimes appeared in the role of teachers and sightseers but more often as warriors who coupled civilization with conquest. Today, all the world seems intent on returning the visit.

In spite of its generally delightful climate, Greece is not really that far south. Athens, at almost the extreme lower tip of the mainland, is about as far away from the equator as San Francisco and Richmond, Virginia. The southern portions of the country in particular have a relatively dry climate and most often average about 300 sunny days a year.

Greece is slightly smaller in area than either England or the state of Florida. The many islands comprise about 20% of the 50,550 square miles. Little wonder that Greece is known as a nation of sailors and fishermen. It boasts 9,335 miles of seacoast whereas its total frontiers bordering Albania, Yugoslavia, Bulgaria, and Turkey measure a bare 621 miles. The generally equable climate is due in no small measure to the mellowing effects of the Aegean Sea along the east, the Ionian on the west, and the Mediterranean to the south.

There are many attractions for the fortunate tourist whose ticket includes Greece. There is tremendous scenic, cultural, and climatic charm and its strategic location makes it a convenient spot for radiating in all directions to the many nearby places of interest.

These exciting little side trips are usually made by air. If time permits, the visitor can enjoy an added pleasant and comfortable minivacation by traveling on the numerous ferries and steamers to a dozen countries that border the Mediterranean and Black seas. In addition, there are cruises of various lengths that make stops at some of the most exciting ports in the world. A word of caution, however, is in order. Do not sign up for these cruises from newspaper advertisements. Most of the ships are smart and sparkling, but a few would require generous applications of DDT to become almost habitable.

A glance at a single travel poster is probably as convincing as reams of text in triggering a visit to Greece. All that is necessary after that is the time, money, and information telling when and in which parts to find the best weather.

Topography

In order to get a more complete preview of what the weather is most apt to be, it is prudent to take a look at the topography of the country. Greece, including most of the islands, is predominantly mountainous. More than 40% of its area is higher than 2000 feet while only about 30% is less than 700 feet above sea level. The remainder includes the flat coastal plains, some river valleys, and the low, flat areas extending inland from the numerous coves and inlets.

In spite of its size, there are four recognizable geographic segments:

1) The damp, mountainous, somewhat isolated region in the northwest;
2) the dry sunny plains and lesser mountain ranges in the northeast;
3) the central or lower mainland portion of the country with Athens at its southernmost tip.
4) the mountainous, four-fingered Peloponnese, separated from the mainland by the Gulf of Corinth and the Canal. The numerous islands including Crete and Rhodes also fall within this zone.

Mt. Olympus, at 9,550 feet, is the highest mountain in Greece. This lofty home of the ancient gods is in the northeast almost overlooking the Gulf of Salonika. The Pindus chain, extending north and south, is the most important mountain group and contains sharp peaks over 8,700 feet high.

The 6,000- to 7,000-foot ranges along the Bulgarian and Yugoslavian borders are generally softer, with rounded or flat tops. They also include many fertile alluvial plains and agriculturaly productive valleys. The Grammos Mountains, between Greece and Albania, are more rugged and range up to 9,000 feet high.

The rivers are not navigable and, although most dry up by late summer, they are rushing torrents during the spring floods. Some have cut deep gorges through the mountains. About 30% of Greece is arable and, perhaps, 45% suitable as rough pasture for sheep and goats. Most of the rich productive sections are in valleys such as the flatlands bordering the Peneus River.

There are both coastal and inland lakes. Kastoría and Ioánnina are

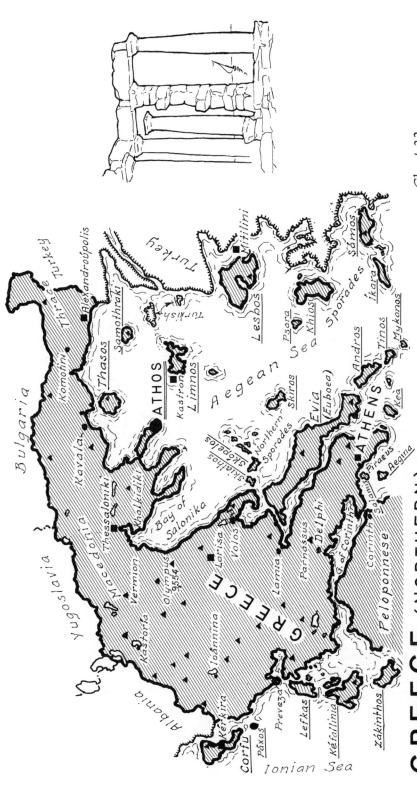

GREECE (NORTHERN)

Chart 32

Turkey

Rhodes
Castellorizo
Symi
Rhodes

Karpathos
Kasos

Tilos
Kos
Kalymnos
Leros
Patmos
Sámos
Ikaria

Dodecanese

Astipalaia
Amorgos

Aegean

Andros
Tinos
Mykonos
Delos
Náxos
Syros
Paros
Ios

Iraklion
Mirabello

Evia

ATHENS

Kea
Kythnos
Serifos
Sifnos
Milos

Santorini
Thira

Cyclades

Sea of Crete

Chánia

Crete

Kithira

Corinth
Argos

Sparta

Peloponnese

Olympia
Andravida

Patrai

Kefallinía

Zákynthos

Ionian Sea

GREECE (SOUTHERN)

Chart 33

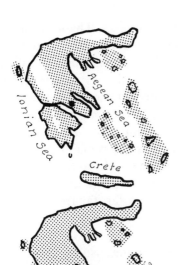

SPRING
April, May, June
9–12 hours of sunshine per day everywhere. Light showers. Cool in north & high country. Particularly delightful in southern portions.

SUMMER
July & August
Most favored vacation season popular areas quite busy. Very dry, sunny, & clear Almost no rain.

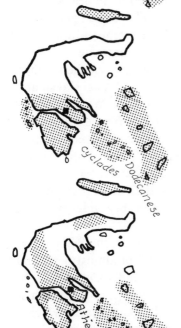

AUTUMN
September-October
A bit less sunshine & heavier showers than spring. Quite wet in northwest. Southern areas & islands very pleasant.

WINTER
November-March
Precipitation generally heaviest in winter, particularly in the north. Skiing in Mt. Vermion & high areas. Sunny with lighter showers in south.

VACATION WEATHER MAPS
GREECE

Chart 34

two of the most beautiful of the mountain lakes. The western coasts of both mainland Greece and the Peloponnese are rather regular but the Aegean shores are indented with deep bays, inlets, and natural harbors.

Just for the record, we should mention that Greece is on the long series of earth faults that extend continuously across southern Europe from Turkey to Portugal. Earthquakes and tremors appear to have been more frequent in ancient than more recent times. At least, the many references throughout history, as well as the remaining physical evidence of towns that were destroyed, seem to substantiate this belief. As mentioned previously, about 90% of all earth disturbances occur around the perimeter of the Pacific Ocean far, far from Greece.

Climate

Although there are many local miniclimates, usually caused by topographical irregularities, Greece is made up of four main zones having distinctive weather characteristics. To a great extent, they conform to the changes in land configuration.

1) The western slopes stretch from the Ionian Sea across and up into the foothills of the Pindus Mountains. This region gets more abundant rainfall than any other part of Greece. The temperatures are rather mild and the seasonal variations not generally extreme.

2) Thrace and Macedonia, stretching between the northern coast of the Aegean and the borders of Bulgaria and Yugoslavia, are subjected to a more changeable Continental climate. Winters are often quite severe as compared to the south. This area frequently comes under the influence of the cold north winter winds. Conversely, summers can experience hot spells when the mercury occasionally reaches 100°. At such times, the refreshing north breezes are most welcome.

3) In eastern and southern mainland Greece, precipitation is lower, but the range of temperatures is considerably wider, with higher summer and lower winter readings.

4) The lowest portion of mainland Greece, the Peloponnese, and the many islands south of Limnos is an area of almost perpetual sunshine and delightful weather. This is a year-round vacationland. Most of the modest rainfall occurs during the 3 winter months, when some areas are rather chilly and uncomfortable. In many places which experience an unpleasant winter, the hotels close for several months.

The most important element of weather from the viewpoint of the tourist is usually sunshine. Greece fares very well in that respect. Individual segments of some other countries experience more hours of sunshine per year, but the following table shows how high the total figures are throughout all of Greece.

AVERAGE HOURS OF SUNSHINE PER DAY AND YEAR

	J	F	M	A	M	J	J	A	S	O	N	D	Year
Límnos	3	4	5	8	10	11	12	11	9	7	4	3	2,666
Lesbos	3	4	6	8	10	11	12	12	10	7	5	3	2,744
Thessa-loniki	3	4	5	7	9	9	11	10	8	6	4	3	2,377
Corfu	4	4	6	8	10	10	13	11	9	7	5	3	2,698
Athens	4	5	6	8	10	11	12	11	9	7	5	4	2,788
Náxos	3	4	6	7	9	10	11	10	9	7	6	4	2,611
Khania	4	4	6	8	10	12	13	12	10	7	6	4	2,874
Iráklion	3	5	6	8	11	12	13	12	10	7	6	4	2,929
Rhodes	5	5	7	8	10	12	13	12	11	7	6	4	3,076

While Greece is not free of pollution, conditions are very much better than throughout North America, Japan, and other parts of Europe. The three locations where sea water contamination is most serious are the Salonika, Vólos and Athens areas. These are all on the eastern shore, which, to some extent, is also affected by the outfalls from the Istanbul industrial zone.

Some scientists predict that unless effective remedial steps are taken promptly, the huge Mediterranean will become a dead sea in a few generations. The most severe pollution occurs along the north shore, west of Greece. The only substantial supply of clean, oxygen-bearing water is through the Straits of Gibraltar.

Fortunately for Greece, this incoming flow follows the north coast of Africa. It then forms a large loop, flowing counterclockwise as it passes the Near Eastern countries, Turkey, and Greece before it reaches the highly polluted west coast of Italy and the French Riviera. If the flows were reversed, Greece would be in trouble.

Because Greece is not a highly industrialized country, air contamination is not of too great concern—certainly not for the visitor. There are a few spots that are less than pleasant, but they can be easily avoided.

The Northwestern Segment of Greece

This includes the mainland area from the Albanian border southward and between the Ionian Sea and the high sections of the Pindus Mountains. It is rugged, but generally quite spectacular, country and is a region of oaks and pines rather than olive groves, grapevines, and citrus trees. Winters are very wet but not usually severe. It is generally a delightful summer vacationland although spring and autumn can also be very pleasant.

The two most important communities in this wet zone are Ioánnina and Préveza. The former is built on a 1000-foot-high rise overlooking the beautiful mountain lake of the same name. Although this district gets an ample 48 inches of precipitation, most of it falls during winter. July, August, and September are dry and sunny, but no other month averages less than 3 inches of precipitation. About half the summer days get temperatures in the low 90's with the occasional 100°-hot-spell. At a 1,600-foot elevation, Ioánnina is fortunate in having refreshing breezes during the day and a comfortable drop in temperature after sundown.

Préveza, at sea level, is at the mouth of the large lakelike Gulf of Amurakía. Much of the land in the immediate area is marshy plain. It gets fewer hot days than Ioánnina, but the precipitation pattern and other weather conditions are quite similar.

The coastal strip and Ionian Sea islands enjoy a much softer climate with fewer of the temperature extremes of the mountainous interior country.

Off the northwestern coast is Corfu, one of the most attractive and better known of all the Greek islands. Unlike many of the Aegean islands, it is green with wooded mountains, olive groves, and fertile lands. It is not on the usual crossroads of tourist travel; most visitors come here either to spend their full vacation or as passengers on cruise ships making a short stop. This has long been a favorite of European royalty and the more affluent. It has retained many cultural and sympathetic ties with places to the west. There are numerous architectural reminders of its years as an outpost of Venice's ancient prowess. The Corfiote is quite different from other Greeks. In attitude he somewhat resembles the Sicilian who is apt to remark "over there in Italy . . ."

Although only 3 to 15 miles wide and 36 miles long, Corfu has plenty to keep one occupied for a long comfortable stay. The choice weather periods are spring and autumn. As indicated on the City Weather Table of Kérkira, summer may be a bit hot for some although it is the most popular season. Winters are rather mild but quite showery. Most of the 46-inch annual precipitation occurs from October through February. Summers are almost bone dry with blue skies and 12 or 13 hours of sunshine each day. A measure of the weather is the 25-year record which shows that Corfu averages only 55 days a year which are classified as cloudy. Three times that number are clear; the remainder experience partial sky cover.

Here, too, will be found figs, grapes, lemons, and perhaps kumquats. Also Glyfada and other fine sandy beaches, a lively casino, a 1-to-4-o'clock siesta period, and even a Sunday afternoon game of cricket. This is a place to loaf and play—not to catch on the fly. If yachting along this coast, there may be the temptation to cast anchor off the tiny isle of Skorpios owned by Aristotle Onassis.

Northeastern Segment of Greece

This, of course, is a big, important area and includes most of the major cities.

Thrace, a somewhat wild and, certainly, imposing, area has retained a flavor of the East, as well it might. All of European Turkey and this province of Greece combine to form the overall region known as Thrace. Komotiní is typical of many communities near the Turkish border which are most reminiscent of the Orient. In these colorful places, one can enjoy the many Turkish delicacies served by natives in baggy trousers and wide, bright sashes. Roaming the narrow winding streets lined with mosques, minarets, and ancient buildings is a very pleasant pastime.

In some places, such as Langadhás and Aghia Eleni, not far from Thessaloniki, an extraordinary religious rite is performed. After sacrificing a lamb, the devout dance barefoot on a blazing bed of charcoals.

Thrace is mountainous from the Bulgarian frontier on the north to the shore of the Aegean Sea. The good bags of wild boar, hunted in this country, are an indication of its rugged character.

The climatic conditions of the whole region are quite similar to those of the meteorological station at Alexandroúpolis. The annual precipitation is a low 23 inches with only about 60 days a year getting more than 1/10-inch of rainfall. November and December average 4 or 5 inches each while summers are bright and almost bone dry.

Winter can be shivery with cold blasts from the north, although temperatures are usually in the 35- to 55°-range. Lest we have made all Greece sound like an eternal land of milk and honey, be warned that your warmest woolens are fashionable winter wear in some of these northern areas and the high mountain country. Alexandroúpolis may have a dozen snowfalls a year while Thessaloniki and Límnos can expect 5 or 6. Even Rhodes and Crete occasionally wake up to a light frost in January or February, but the sun soon wipes it away. The high mountain areas, of course, welcome that beautiful white fall and the Greek government is encouraging the development of skiing centers and winter resorts throughout the country. While the mercury does not reach 90° more than two dozen days between June and September, there can be the occasional, short, midsummer heat wave of 100°. Nights are most often very comfortable and even on the hottest days, the sea and mountain breezes offer pleasant relief. This is prime hideaway country for those allergic to the Athens summer crowds.

This beautiful country is generally mountainous but in Macedonia many of the ranges are separated by wide, fertile valleys. Two peaks of particular interest in this region are the fabled Olympus, home of the 12 Gods in Greek mythology and the lesser 6,200-foot Mount Vermion. The latter is fast becoming a very popular winter resort and skiing center. Another attraction is Kastoría, one of the loveliest of the mountain lakes near the Albanian border.

This whole northern section, now the least visited by tourists, cannot fail to, some day, become an important resort area—a summer, winter, spring, and autumn vacationland.

As indicated on the Thessaloniki (Salonika) City Weather Table, the climate here is a bit milder than that of Thrace, but the low precipitation pattern is much the same. Only the period from November through February averages less than 5 hours of sunshine per day. The other 8 months enjoy from 6 to 11 golden hours in every 24. The clear, pollution-free mountain air on a sunny day coupled with delightful people and wonderful scenery is not easily topped.

MOUNT ATHOS—THE THEOCRATIC REPUBLIC

Not many have ever heard of this tiny independent state, wholly within Greece. Fewer still, and males only, have visited this holy place. Measuring a bare 6 by 20 miles, it could be lost in a corner of Texas' King Ranch. It has functioned as an autonomous entity for 1,000 years but that status was formally recognized by the Greek Constitution only in 1927. It now contains some 20 monasteries, all Greek except three which are Russian, Serbian, and Bulgarian.

At one period, the hermit population rose to 40,000, and the region became known as the Holy Mountain. Today, there are about 3,000 monks who live by their own code of laws and forms of authority. The peace and serenity is due to prolonged prayer and meditation. Only males with passports and proper credentials are admitted. They are welcomed on arrival with a glass of *ouzo*, a spoon of jam, and a glass of water. Those who choose to remain overnight are furnished with simple food and shelter in the form of a bare monk's cell.

This little mountaintop treasury of icons, religious relics, and beautiful works of art is perched at the extreme southern tip of the Khalkidkiki Peninsula overlooking the Aegean Sea, about 90 miles from Thessaloniki.

Southern Mainland Greece

Farther to the south on mainland Greece are some of the more industrialized areas. Vólos, a city of 50,000, is a busy seaport with many factories. Lárisa, about the same size and a bit inland, is also a busy commercial and industrial town. But that does not produce the heavy banks of smog and catastrophic water pollution all too familiar at home. Even this more populated area is made up mostly of quaint or attractive little mountain villages and fishing ports. They dot the miles of seacoast and the many pleasant valleys and high areas devoted to agriculture and grazing. As mentioned previously, only the immediate environs of Thessaloniki, Vólos, and Athens exhibit flagrant evidence of man-made contamination. The lucky visitor who arrives in April or May finds the highest mountaintops still glistening with snow. The peach, almond, and other orchards will be in blossom and the fields covered with bright, red poppies.

The effects of latitude are evident from north to south of the mainland. Lárisa is considerably warmer than Alexandroúplis with more 90°-days and fewer 32°-nights. Being somewhat less subject to the cold northern winds, the winters are milder than the rather slight temperature difference might suggest. Many think of Greece as being semi-arid. That is the case, particularly toward the south and east, including some of the islands. Much of the northwestern mainland and numerous valleys throughout the country, however, are lush and green.

The southern tip of the mainland peninsula holds much of interest. Who would fail to pause beneath the summit of Mount Parnassus? The many lesser gods and nine muses have long departed, but the grand panoramic view is reward enough.

Then turn toward Delphi with problems of the mind or body. We associate the oft-quoted maxims, "Know thyself" and "Moderation in all things," with the Oracle of Delphi. This region is also the haven for those who have overindulged in food, drink, or endless sightseeing. The mineral, thermal, and radioactive springs, spas, and sanitoria are here in sufficient number and variety to cope with any medical problem. Even if completely mentally and physically sound, a jaunt to these parts to enjoy the abundant sunshine and spectacular scenery is a pleasant rejuvenator.

MAINLAND EAST COASTAL ISLANDS

There are a number of islands along the eastern and northern coasts which are more akin to the mainland than the various island groups.

The attractive island of Samothráki, a place of peace and solitude, is to the south of Thrace. There are a number of small hotels, including one of the excellent Xenia group (hotels operated by the government).

The 150-square-mile Thásos is only minutes away from Kaválla by ferry or a pleasant sail on a *caique*. Unlike most other Greek islands, its 25-inch annual precipitation and controlled goat-grazing makes Thásos a lush, green spot covered with pine forests and aromatic plants. There is almost no rain during summer and with less than 50 days a year getting even 1/10-inch of rainfall, there is plenty of sunshine. It is also called the "honey island" because of the thousands of apiaries sustained by the great abundance of flowers and blossoms. Linena, the largest of the 11 villages, boasts a comfortable "B" category Xenia Hotel. For the person seeking a quiet, pleasant, inexpensive hideaway to write that book, this could be the ideal place.

ATHENS

Some enthusiastic travel writers declare that the long journey from America just to view the magnificent Parthenon atop the Acropolis would be ample reward for the time and money. Perhaps few tourists would agree, but all too many make the trip and never get beyond the environs of Athens.

It is necessary to have quite complete information in order to get a reasonably accurate picture of the climate in a strange place. For instance, the annual 2,788 hours of sunshine which Athens normally gets rates high but is topped by Miami's 2,900-hour figure. That alone, however, is far from the whole story.

By comparing the City Weather Table of these two glittering sun spots, we find that the Greek capital enjoys many more clear days. Miami averages a 60 inches of rainfall a year, mostly during the summer as against Athen's meager 16 inches, which is heaviest in winter. There will be just over 100 days a year that Miami will have 1/10 of an inch or more of rainfall while Athens will usually manage with a bit under 50 such days.

Athens' 45- to 60°-average winter temperatures are some 10 to 15° lower than Miami usually experiences. Both cities can get a bit chilly but an earmuff salesman would starve in either. Summers are generally just a trifle warmer in the Greek city, but short, 100°-spells are a possibility in both places. Athens is agreeable during all 12 months. Almost without exception, however, it is perfect in springtime and only a mite less than that in autumn.

There is little need nor sufficient space to list even half the wonders of Athens. Our favorite procedure is to mix part-days roaming the fascinating remains of ancient Greece and museums with short tours both within and outside of the city. Rambling through the flower market, the older streets of the city, or along the Piraeus wharfs and waterfront can occupy many delightful hours. All of this puts one in the proper frame of mind to sit in the sunshine at an outdoor table in any of the numerous inviting *tavernas*. Alternating from feet to seat is a most satisfying routine in sunny Athens. Although there is always the temptation to feast at the fine establishments in the main city, the lure of all these seafood restaurants strung along the sweeping Tourkolimano crescent is mighty strong. They are equally inviting during the day or at night. Time enough for the traditional lamb and eggplant dishes when in the north. The menu of these waterfront places comes directly from the many colorful fishing boats in the Old Harbor.

ISLANDS CLOSE TO ATHENS

Those on limited time schedules need not return home disappointed at not having stepped onto even one of the 1,426 Greek isles. A few were conveniently dropped into nearby Saronic Gulf by the benevolent Greek gods and they are only a ferry stop away from Athens.

There is Salamis, venerated as the place where, in 480 B.C., the Athenians were victorious against the Persians; Aegina, the site of the first Greek government; dramatically stark Hydra, which attracts so many artists and is the only major Greek island which does not allow automobiles. Poros and Spetsai boast pine-shaded beaches where eyes, weary from museums and ancient works of art, can be soothed by equally wonderful but far more glamourous modern works of art—in economy-size bikinis.

During off-season, these islands are often quiet and peaceful. In July and August, it is surprising they do not sink under the load of visitors.

The Peloponnese

Before the Corinth Canal was cut, this large region was a continuation of the mainland. It is an ancient country, and Árgos is said to be the oldest city in Greece. In the 8th century B.C., long before Athens and Sparta gained prominence, this was a center of civilization and government.

It is generally rugged and the scenery, while often spectacular, does not boast quite the breathtaking grandeur of the Swiss Alps or Norwegian *fjords*. In some of the more remote mountain villages, the inhabitants still speak the ancient Doric dialect which is incomprehensible to people from other parts of the country. There are two chains of mountains, extending from the northwest to southeast, with peaks over

7,000 feet only in the upper portion. There are also wide alluvial deposits, particularly along the Evrótas River in the South. Tobacco, vegetables, and grain are abundant in the north, while the south produces figs, olives, cotton, citrus fruits, and grapes.

Precipitation is highest along the western side, especially toward the north. The islands of Zákinthos and Kefallinía get an ample 44-inch annual rainfall. Little occurs from April through September, while June, July, and August are dry and sunny. Although there are seldom any 32°-nights, winters are wet and chilly, with many overcast skies.

Precipitation diminishes toward the south and is even lighter along the east coast. Andravidha gets about 32 inches while the island of Kíthira in the extreme south accumulates only 25. As we have noted earlier, the Athens area rarely averages much over 16 inches a year. In all cases, winters tend to be wet, but the remainder of the year is sunny, clear and dry. Although the mercury may very occasionally mount to 100°, the normal midsummer temperature range is 70 to 85°.

The Aegean Islands

There are almost 1,500 of those glamorous sun spots sprinkled around the edges of the Greek peninsula. They range from the 3,207-square-mile Crete (equal to the combined areas of Delaware and Rhode Island) down to tiny specks inhabited only by the birds and bees. We have already mentioned the Ionian islands strung along the western side of Greece. Aside from Corfu, the most popular groups are in the blue Aegean. Each of the Greek islands has something to offer, but since it would require 4 years to spend one night on each, most visitors must choose a few and be happy.

THE NORTHERN SPORADES

The largest in this archipelago is the elongated Euboea, Evia, or Evvoia, depending upon which map you use. The overseas visitor need not be disturbed by the overgenerous use of names in Greece. The worst that can happen is that the visitor will end up in a wrong but equally attractive place.

Although Evia is only 2 hours from Athens by car and ferry, this pleasant island is relatively unknown to overseas tourists. Stretching along the northeastern flank of the Attica Peninsula, it forms almost a part of the mainland and contains about the same kinds of places, sights, and attractions. Chalkis is the port of entry and the largest town from which all roads radiate.

One of Evia's Northern Sporades neighbors is the fashionable Skiathos which offers both luxury and quite modest accommodations. This is a rather green island with miles of sand beaches and large areas of pine-scented woodlands.

Nearby is the plum island of Skópelos. Apart from its 400 churches, this is one big fruit and vegetable garden. Alonissos or Khilcodramia is the silent one—as delightful as the others but even more peaceful with many fruit orchards and pleasant, uncrowded sandy beaches.

By good fortune, this little group of islands is seldom harassed by the often strong *meltemi* or *etesian* winds, which sometimes carry a heavy dust burden. When clean, however, and of light breeze velocity, these air movements, although hot, can be very refreshing in summer. Except for midwinter, which is often quite showery, most of the year is dry, sunny, and pleasant. The mercury has registered up to 108° on these islands, but the sea breezes are refreshing during these short, hot snaps.

THE EASTERN ISLANDS

This is a group of pleasant, green, fertile land patches that are more charming than scenically spectacular. It includes mountainous Lesbos which ancient scholars revered as the birthplace of Sappho and Aesop. Current visitors appreciate it more for its fine beaches, two beautiful bays, olive trees, petrified forest, and unusual architecture. The younger ones gravitate towards the town of Molybdos which threatens to become a little St. Tropez. Prudent older tourists arrive just before or after the slightly hectic midsummer rush.

There is also the wooded Sámos, famed for excellent wines. It is mountainous but includes much fertile land which produces plentiful crops. Límnos and Ikaría are perhaps just a bit less interesting.

As indicated on the City Weather Tables of Kástron (Límnos Island) and Mitilíni (Lesbos Island), the moderate 24 to 28 inches of rainfall occurs mostly in November, December, January, and February. The 6 months from April through September are very dry, sunny, and generally gorgeous. There is an average of 8 to 12 hours of sunshine per day throughout this most agreeable season. If we add to that plenty of excellent beaches, hills and mountains, bountiful fruit orchards and vineyards, the total must add up to the makings of an almost irresistible travel folder. Little reason to ask why all of the Greek islands are so popular.

THE KIKLADHES (CYCLADES)

The word means circular, and legend claimed that the 24 islands making up this archipelago are satellites surrounding the sacred isle of Delos. This tiniest of the group, and uninhabited, was the birthplace of Apollo and his twin sister, Artemis—but that is going back a bit—about

425 B.C. Because it is almost completely covered with a fantastic display of ancient ruins, Delos is often called the Pompeii of the Aegean.

The largest of the islands is Náxos, which, like most others, has mountains and elevated areas down through the center. Apart from the marble quarries, most of the population is engaged in farming and fishing.

The reader can either read 1,000 pages describing the climate of the Greek islands or take a quick look at the City Weather Table and sunshine statistics of Náxos. Either will quickly convince the most skeptical that this is a summer vacationland not easy to match.

The days are dry, sunny, and comfortable, with gentle sea breezes and plenty of inviting beaches and mountains to suit both the gregarious and the seeker of solitude.

One of the most intriguing sights in the Aegean is the Island of Thíra or Santorini. A volcanic eruption in 1500 B.C. changed it from a round to a crescent-shaped island. The crater now rises in the large bay caused by the upheaval. The multicolored cliffs which edge the sea are topped with little white and blue chapels and homes. By contrast, the other side of the island is a series of gentle green slopes, patched with orchards, vineyards, and small farms. All of that plus a sampling of the island's heady wine and some beach loafing make this a worthwhile jaunt.

Other islands of the Cyclades are Paros, with its pine forests and attractive little villages, and Mílos, made famous by the unearthing of many beautiful works of art, the most renowned being the statue of Aphrodite, the Venus de Milo now in the Louvre.

The best known and perhaps most visited of the Cyclades is Mykonos. This draws both the so-called high society and the beatnik. One attraction is watching the weaving of handsomely designed fabrics on old-fashioned hand looms. There is quite a sizable colony of foreigners, many of whom live there the year round. The active, churning windmills are another feature of Mykonos and an indication of air conditions. Characteristic of Tínos are the thousands of towerlike dovecotes, while Sífnos boasts the largest of the many beaches in the Cyclades.

THE DODECANESE (SPORADES)

Including Kastellórizo, off by itself a few miles south of Turkey, there are 13 islands in this group. The most important, by far, is delightful Rhodes whose climate is so friendly and inviting that the hotels stay open the year round. There are dozens of reasons for a visit, but the happy statistics shown on the Sunshine and City Weather Tables are more than enough.

If the island seems to sink an inch or 2 during July, August, and

September, blame it on the visitors. One might suspect that every Swede not in the Canary Islands is here and wonder who is minding the shop at home. Note that May, June, and October are also almost perfect from the viewpoint of beautiful blue skies and sunshine. April, too, is about as good but, by November, there are usually quite brisk showers that add up to over 5 inches of rainfall.

Do not overlook this spot that the Knights of Rhodes found so pleasant and rewarded with such beautiful and imposing evidence of their stay. This is a place one should absorb slowly—not catch on the wing. Visitors are too late for a view of the 107-foot-high Colossus of Rhodes which was toppled by an earthquake in 224 B.C. They can, however, roam the fascinating old city—the world's largest medieval restoration. Also visit the unique Valley of Butterflies or take a 1-day jaunt to Turkey, only a dozen miles away.

Kalymnos, the blue island, rates next after Rhodes in tourist appeal. Its town of the same name, with a setting of little blue-trimmed white cottages surrounded by gardens and orchards, is the home of the sponge divers. At the springtime fiesta, the parade of decorated caiques is blessed, marking their departure for a 6-month's stay along the coasts of Libya and Tunisia.

Pátmos is probably the next best known of the Dodecanese Islands as the place where St. John wrote the Book of Revelations. Women's Libbers will be pleased to hear that on the little isle of Léros nearby, the ancient matriarchal custom of having the eldest daughter rather than son inherit the family wealth is still observed. All associated with the medical profession should stop at Kos and pay their respects at the home of Hippocrates; perhaps visit the Asclepeion, an ancient sanitarium, and gaze at the statue of the old pillmaker himself, dating from the 4th century B.C.

CRETE

This 160-mile-long island is the place where Zeus, the father of ancient Greek gods, was born. One of its present-day claims to fame is a most delightful climate.

The forests of palm trees, particularly in the south, and the 8 to 13 hours of sunshine per day during more than half the year are fair indications of what to expect. As indicated on the City Weather Table of Iráklion, there are about 28 inches of precipitation a year but two-thirds of that total falls between November and February. More than 300 of the 365 days get less than 1/10-inch of rainfall. The Greek farmer is sad but the tourist joyful that most precipitation is in the form of short, heavy showers.

With only a dozen days a year when the mercury reaches 90° and almost no nights experiencing frost, there is little need to discuss weather further. The hard to please, however, will point out that while the normal midsummer temperature averages about 85° there may be the occasional hot spell with readings over 100°. The answer to that is —sit in the comfortable shade or in areas exposed to the refreshing sea breezes. After all, there are 20,000 windmills on Crete, several times the number remaining in The Netherlands.

There are four mountain groups extending the length of the island. The 8,058-foot Mt. Ida, almost in the center, is the highest, but there are several others in the rugged hogback almost as lofty. A curious feature is the unexpected fertile basins encountered almost at the summit of some of the high country.

Khaniá, the commercial center, and Iráklion, the major seaport, are on the northern coast which is indented with bays and natural harbors. The Gulf of Merabéllo is striking. Agriculture, grazing, and fishing are the chief occupations on Crete which has very little industry or man-made pollution. There is an atmosphere of well-being and markets are noticeably better provisioned than on the mainland. This abundance is reflected in even the most modest taverna.

Adding to the eye and ear appeal, this fortunate island lies directly on one of the great bird-migration routes. There are huge splashes of floral coloring everywhere, and it has been said that in order to build even an outhouse, the first step is to clear away the flowers.

El Greco was only one of the many famous people this proud island has produced and, indeed, Cretans take for granted that they are a bit different from their mainland cousins. Some say they tend to accentuate traits of the uninvited visitors of the past—the ruggedness and vigor of the Frank and the Venetian's appreciation of the finer things. Others attribute these characteristics to their artist forefathers and their days of fighting the Turks. Whatever the reasons, you can tag them as either super-Greeks or just wonderfully hospitable and self-confident people.

DENMARK

MOST foreigners picture Denmark as a little, toy-soldier country. One could, of course, point out that this mighty kingdom of the north, comprising some 850,000 square miles, is twice the combined area of Great Britain, Italy, and France. That measurement, however, would have to include all the acres of ice cubes that make up Greenland as well as the 19 fog-bound Faeroe Islands. Actually, Denmark proper, about the size of Switzerland, is the smallest of the five Scandinavian countries.

There are so many interesting sides to this little country that only a fraction could be covered in any single travel guide. One area of interest to almost all tourists, and certainly the Danes themselves, is food. These people really eat well and meals here assume an importance matched in few other parts of the world.

All Scandinavians seem to agree that Aalborg's aquavit is the world's best. The vote is also unanimous that the tiny Danish shrimp, which number 40 to a little, mounded, open sandwich, are also unbeatable. The same is true of the two world-famous beers. And this generous little market-basket nation does not hoard its gastronomic delights; it serves as gourmet delicatessen for Britain and nearby neighbors. There are cheeses of every color, type, and size; premium hams and bacon; eggs and dairy products; sausages to fit any occasion; and fish in a dizzy array. This latter is, of course, not unexpected, for where in all Scandinavia can one travel very far and not come face to face with the sea?

But in order to enjoy all of these places and things at the most favorable times and with the greatest comfort, it is well to take a look at Denmark's weather story.

Topography

As usual, we start with the shape of the land—that so often is an important key to the climate. To describe a country as having been formed by sedimentary deposits with an average elevation of 100 feet and the highest spot only 566 feet above sea level might suggest The Netherlands or perhaps the flatlands of Florida.

Such is not the case. The terrain of Denmark is quite irregular, with only a small portion of the country being a low, level plain. Most of the land is undulating and broken with ridges, small steep hills, moors, bogs, downs, and marshes. This is ideal for bicycle touring. There are also many short rivers and inland seas. Denmark is made up of the large mainland, Jutland Peninsula, and almost 500 islands of various sizes.

The entire 4,600-mile boundary, except the 42 miles bordering on West Germany, faces the sea. Aside from the port of Esbjerg, the western side of Jutland is made up mainly of a chain of sandy beaches and dunes. The eastern shores are very irregular and indented with fjords, bays, and inlets.

Climate

Although meteorologists are inclined to divide the country into three or four climatic areas, there actually is not enough difference to be of much concern to the average tourist. A combination of the prevailing westerly winds plus the fact that no part of Denmark is more than a few miles from the tempering effects of the sea tends to produce reasonably equable temperatures.

There is generally not a great variation within any one day, but when the wind shifts and blows from the northeast, there can be an abrupt and often sharp change in the temperature. That is not of great importance in summer, but during the winter season, it can drive people indoors very quickly. As we know from the Wind Chill Factor Table, a 35°-temperature will feel like 12° when coupled with a 25-m.p.h. wind.

Normally, the day-to-night summer temperatures range from 70 to 55°, while the winter figures hover between 40 and 25°. Even during the hottest days, the mercury seldom tops 90°. In the winter, however, there have been occasional − 12 or 14°-readings in almost every section of the country. There is considerable overall uniformity and the average temperature in southern Jutland is seldom more than 4 or 5° higher than Skagen at the northern tip.

In general, precipitation is rather low throughout Denmark. The annual total in the north and eastern segments averages a modest 24 inches. This figure increases to about 31 inches in the southwestern

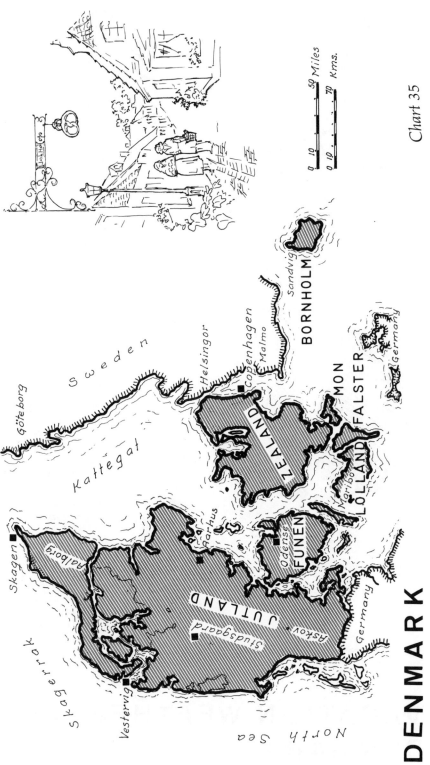

DENMARK

Göteborg

Sweden

Skagen

Skagerrak

Kattegat

Ålborg

Helsingor

Copenhagen
Malmö

Sandvig

BORNHOLM

Vesterdig

JUTLAND

Studsgaard

Århus

Askov

ZEALAND

FUNEN
Odense

MON

LOLLAND

FALSTER

Maribo

Germany

Germany

North Sea

50 Miles

70 Kms.

0 10

0 10

Chart 35

SUMMER

July and August
Showery, $2\frac{1}{2}$–$3\frac{1}{2}$ inches of
  rain per month
Temp.: Days 65-75°. Nights 55-65°.
Sunshine 7-8 hours per day
  This is vacation time in
Denmark. A good time to visit
but reservations necessary.

Copenhagen

AUTUMN

September & October
August wettest month. Sept.
  next with $2\frac{1}{2}$–3 inches
Days 50-65°. Nights 40-50°
Sunshine {Sept. 5-7 hrs per day
         {Oct. 3-6 " " "
Copenhagen very pleasant
and not too crowded.

SPRING

May and June
Delightful weather
Low rainfall $1\frac{1}{2}$-2 inches per month
Temp. Days 65-70°. Nights 40-50°.
Sunshine 8-9 hours per day
Much less crowded than summer.

WINTER

Nov. thru April
Low precipitation
  Days 35-45°
  Nights 25-35°
Sunshine :
  March 4 hrs per day
  April 6 "   "   "
  Remainder 1-2 hours
    per day

Shaded maps show
most desirable vacation
periods.
There are many fine
beaches for sunbathing.
The sea water does not
reach the 64° maximum
until August.

VACATION WEATHER MAP
DENMARK
Chart 36

portion of Jutland. Rainfall is heaviest during late summer and fall, and almost without exception, August gets the highest total precipitation of any month in the year. Although the 5 to 15 thunderstorms, occurring mostly in summer, would usually indicate short, heavy downpours, much of Denmark's precipitation is in the form of extended rainy days. As a rough gauge for comparison, the total annual precipitation is about two-thirds as much as the normal fall along the New England seacoast.

As a result of that 23°-tilt in the world's axis, the hours of daylight vary greatly by season in northern Europe. There is a difference of 10 hours between the longest and shortest days of the year. Along the 55°-parallel of latitude, which cuts across the center of Denmark, June 21 has more than 17 hours of daylight while December 21 is limited to about 7. During winter, the Dane drives without headlights only from 9 a.m. to 4 p.m.

Visitors are generally pleased with the average 7 to 9 hours of sunshine per day during the 4 months of May through August. April and September sually get about 6 hours per day but midwinter is a bit gloomy as indicated in the following table.

AVERAGE HOURS OF SUNSHINE PER DAY AND YEAR

	J	*F*	*M*	*A*	*M*	*J*	*J*	*A*	*S*	*O*	*N*	*D*	*Year*
Skagen	1	2	4	6	9	9	9	7	6	5	2	1	1,799
Aarhus	1	2	4	6	8	8	7	7	5	3	1	1	1,693
Askov	1	2	4	6	8	8	8	7	6	3	2	1	1,719
Odense	1	2	4	6	8	9	8	7	6	5	2	1	1,756
Skaelskor	2	2	4	6	8	9	8	7	6	6	2	1	1,825
Copen-hagen	1	2	4	6	9	9	8	7	6	3	1	1	1,794
Maribo	1	2	4	6	8	9	8	7	7	5	1	1	1,793
Bornholm	1	2	4	6	9	9	9	7	8	3	1	1	1,856

The following swimming information is included mostly for the benefit of native Vikings and very healthy polar bears, since the sea water does not reach its maximum temperature of 64° until August. Meteorologists, but few others, are interested in knowing that the February reading is about 34°. However, in spite of all the statistics, the sand beaches and dunes are a delightful place to loll away a sunny afternoon.

Zealand, Falster, Mon, and Lolland Islands

Zealand is, by far, the largest of all the islands and the site of the delightful Danish capital. The Copenhagen City Weather Table is quite representative of this island group. Spring is a gay, bright time of year here, which averages 8 to 9 hours of sunshine per day. There are 15 to 17 hours of daylight in each 24 throughout this whole period. Temperatures are rather consistently in the 48 to 68°-range. There may be the occasional 90°-day or the just-below-freezing night, but neither is too common.

Rainfall is hardly ever a problem and seldom exceeds a low 1.5 to 2 inches per month. Only 10 days in this whole 60-day-period will get more than 1/10 of an inch of rainfall. These are almost ideal conditions for reasonably active sightseeing, while the comfortably lazy fellow can sit in the warm spring sunshine enjoying the pleasant surroundings.

One can see quite a good cross-section of Denmark without ever leaving the Island of Zealand.

COPENHAGEN

If ever a single city was the mainspring and hub of a nation, Copenhagen is that place. Although it sits on the northeastern edge of Zealand, the largest of the Danish islands, it makes a wonderful headquarters for visiting many nearby villages and places of interest.

Tivoli Gardens, the fabulous Danish Disneyland smack in the center of the city, offers an amazingly wide range of pleasures and amusements. Carousels and games to be sure, but also fine foods, music, dance halls, concerts, cabarets, a pantomine theater and entertainment—and much of it at no cost. This is one of the biggest tourist bargains in all Europe. No matter how tight the budget, every visitor should splurge just a bit on an enjoyable dinner at the pleasant Belle Terrasse, the Divan II, or one of the 20 other fine restaurants in Tivoli. As everywhere in Denmark, food is abundant and good, whether in a deluxe restaurant or at one of the many little open stands, tidbit vendors, coffee and pastry shops, bars, or cafes.

Copenhagen is one of those perfect "walking cities"—a place where it is easy to spend an entire day just strolling with no awareness of time or distance. Few fail to travel the mile-or-so length of the famous Stroget, an area from which autos are banned. It is actually made up of three interesting old squares and five streets, none of which are named Stroget.

Bing and Grondahls or the Royal Copenhagen Porcelain showrooms resemble fine art galleries more than commercial establishments, and these will be only one highlight of the inevitable shopping spree. The well-informed tourist, however, will first spend a few spellbound hours at Den Permanente (Permanent Exhibition) where generous samples of all Denmark's remarkable arts and crafts are on display. When the tourist tires of shopping, a little detour to either the Carlsberg or Tuborg breweries will show what a Danish welcome really means. Most already know why these two fine beers are rated among the best in the world.

This largest of all Scandinavian municipalities also has many art galleries, museums, and a university founded a dozen years before Columbus arrived in America. Any camera fiend who discovers the colorful old "New Harbor" will not leave until the last roll of film has been shot. Its main attraction is Hans Christian Andersen's "Little Maiden," who looks just as young and graceful as when placed on her rock in the harbor in August 1913. This little bronze ranks as one of the best-known and most beloved statues in the world.

Visitors will also want to see Copenhagen's world-famous zoo, the nearby 3,500-acre Deer Park, and dozens of city and area tours.

Let us not forget that unlike its Scandinavian sister cities, Copenhagen enjoys a rollicking nightlife second to none in Europe. On one occasion years ago, a cab driver sadly related the story of the newly elected "blue-nose" mayor. He explained that since he took office, there was no drinking spot or place of amusement open between 5 A.M. and 6 A.M. In those days, there were three groups of establishments that operated about 8 hours a day each. More recent visits confirm that the 5-to-6-A.M. problem has been cured; the droopy-eyed tourist wonders if the native population ever sleeps.

North of Copenhagen is a land of gently rolling countryside. There are tiny coastal fishing villages and inland hamlets. All is peaceful and a delight to the eye, particularly when the countryside is alive with spring flowers and flocks of migrating birds. At this time of year, which is a bit off-season for most vacationists, the little country inns, known as *Kros* are not too busy. The fact that most of these buildings do not have private baths is much less a problem than in the popular July-August period.The almost national monomania concerning cleanliness means that even the most modest accommodations will be bright and spotless.

Not to be missed is Helsingor, a few hours' drive north of Copenhagen. William Shakespeare's moody Hamlet no longer prowls the ramparts of Kronborg Castle, but the castle, with its commanding view of the sea, is well worth a special stop.

Roskilde, the medieval capital of Denmark just to the west of Copenhagen, should not be overlooked when touring Zealand. Ringsted, Soro, and Slagelse are attractive towns that can be seen on the route westward to Funen Island. For those interested in crafts, the Kahler pottery works, the glass operations, and several others in the lower part of Zealand may be worth a detour.

The islands of Lolland, Mon, and Falster, all south of Zealand, are still mostly pleasant pastoral countryside. While very attractive, they are of less interest and most overseas visitors are most apt to see them only on the way to the boat ferries at Gedser, Rodby, and Nakskov.

Funen Island

The usual route from Copenhagen to the mainland is across Funen, which is referred to as the island of chateaux, the Hans Christian Andersen island, or one of the popular vacation islands of the Danes.

The City Weather Table of Odense, the most important community on Funen, shows how pleasant the weather is apt to be in May and June. The sunshine table, usually the most reliable gauge, indicates 8 or 9 hours per day. The reader should also note that there are almost constant light breezes across most of the country at this season. That suggests a sweater or light topcoat, even though the air will be clear and the sun shining most of the time. Americans will find the coastal segments of northern Europe somewhat cool compared to home.

The Jutland Mainland

This great peninsula comprises almost two-thirds of Denmark and includes the largest fjords and bays. It is an interesting land of rolling countryside, lakes, busy market towns, and ancient villages. Except for the short section adjoining West Germany, the peninsula is edged with sandy beaches, dunes, some sharp cliffs, and many bays and coastal indentations. Scattered across Jutland are many hamlets resembling that of Hansel and Gretel.

The City Weather Tables of Skagen, Vestervig, Studsgaard, and Aarhus all confirm how wonderful the weather is during May and June.

Bornholm

This island, about 100 miles east of Copenhagen in the Baltic Sea, is a bit off the crossroads of travel for the overseas visitor. That is unfortunate as it is one of the most attractive parts of Denmark and a good vacation area. It is about the only place in the country where rock formations project above the surface of the ground. It is fringed with fine coastal scenery. Bornholm is also one of the driest areas with only about 20 inches of precipitation a year, which means that there is no really wet season. Every hotel and inn on Bornholm is filled to capacity in July—the favorite vacation time throughout Denmark.

The Best Seasons to Visit Denmark

This nation does not enjoy the wide range of climates that many of the more weather-fortunate countries in Europe do. Although visitors arrive every one of the 12 months, the choice is not always the most satisfactory. As ever, there is a best season which varies somewhat throughout the country.

The vacation weather maps of many countries resemble a patchwork quilt. That is particularly the case where there are wide variations in the terrain. Altitude often determines temperatures while a mountain range may act as a climatic barrier, effectively separating two areas which experience very different weather conditions. Marked irregularities in the topography also very often produce miniclimates in local areas which may differ quite markedly from the surrounding countryside. These conditions prevail to a much lesser extent in Denmark than is the case among many of its neighbors.

It takes very little research to decide that May and June are tops for good weather throughout just about all of Denmark. That is generally the time of lowest rainfall, maximum sunshine, and cool, but not uncomfortable, temperatures.

Although far more Danes and Europeans generally vacation during July and August, the weather is rarely quite up to the excellence of May and June. While the summertime sunshine is quite abundant, averaging 8 hours per day, note on the various City Weather Tables that precipitation increases substantially over the 2 previous months. August, which almost always gets more rainfall than any other month, usually accumulates almost twice as much as the May-June average. That, of course, does not label summer as a poor time to visit; it only means that the spring season is usually even better. Notice in the "extreme temperature" column that there need be little concern about heat waves. Throughout the spring, summer, and autumn seasons, temperatures are just right for walking and sightseeing in all parts of the country.

The Scandinavian countries attract more tourists every year and in spite of massive building programs, hotel operators are hard put to keep pace with the demand, particularly in July and August. Prebooking of accommodations is strongly advised. So the conclusion is—come in summer if that is the most convenient time, but be prepared for the combined native and foreign tourist peak load. Also remember that August, in particular, will almost always experience more rain than May or June. All parts of the country will be busy, but Copenhagen gets the greatest influx of visitors. Jutland's many clean west coast beaches are also one of Denmark's busiest vacation areas during July and August. There are, however, many tiny hamlets, fishing villages, beaches, coast and peaceful countryside areas which are not crowded even during those two popular vacation months. The Danish Tourist Bureau can help you locate them on off-shore islands and away from the places featured in tourist travel guides and folders.

The Danish autumn is unpredictable, and the odds are that spring will be more pleasant. In the northern latitudes, the days become noticeably shorter toward the end of summer. Note also in the table that the hours of sunshine taper off in September, with a very sharp drop by the first of October.

Next after August, September and October are almost always the wettest segment of the year. Even factual descriptions sometimes tend to be quite misleading. For instance, New York City averages a healthy 4 inches of precipitation in September and 3 inches in October as against Copenhagen's 2 and 2.2 inches for those 2 months. But it is necessary to delve further into the statistics to get a true picture. While New York averages 271 and 235 hours of sunshine during those two months, Copenhagen gets about 230 and 182. This means that there is not a wide difference, although the American city is a bit sunnier but wetter. Its autumn rains, however, are most often in the form of rather short showers, while Copenhagen's very much lighter rainfall tends to be somewhat more persistent.

In 1972, most of Western Europe experienced one of those unusual years with a gray, cold, wet spring and a generally dry, sunny, late summer and autumn. Meteorologists say that particular situation was caused by an unusually heavy crop of icebergs which did not melt until early summer.

Copenhagen's autumn is very pleasant. There will be more than 1/10 inch of rainfall only about 1 day in 6 and it is too early to worry about snow. Temperatures are in the 50 to 60° range and, although October will average only 3 hours of sunshine a day, September gets a glittering 6. Another plus factor is that as the crowds diminish, one again sees more Danes than tourists.

Yes, autumn is a good time to visit Copenhagen. That is the time when cultural activities approaching a peak, Tivoli is still in full swing, and restaurants are festive with service back to normal. And a nip of good *schnapps* will thwart even the imaginary cold.

Copenhagen is a glowing exception during the drab winter months. Shops are less rushed, restaurants, symphony orchestras, opera, the famous Royal Dutch Ballet and, of course, the "lids off" nightspots operate at top pitch.

As an indication of what to expect with regard to weather—there will be about 50 foggy days, mostly from December through February. The first snow of the year arrives about the middle of November and the last fall may be as late as April 15. There are usually about 40 snowfalls each winter. In spite of the considerable sunshine in spring, summer, and fall, Denmark is not a bright country. Copenhagen is lucky to average 1 hour of sunshine a day in November, December, and January. The country as a whole can expect only 35 completely clear days a year. There will be about 140 cloudy ones and the other 190 may be mostly one or the other. So come to wintertime Copenhagen for merriment, not suntanning—but very few winter visits are a disappointment.

The Faeroe Islands

Even experienced navigators deny that there is such a place, but a careful search in the fog banks due north of Scotland will prove them wrong.

There are several excuses for the jaunt to these 19 earth specks which were kicked up from the ocean bottom and scattered over a 50-by 70-mile area by volcanic upheavals in the distant past. The pride of one-upmanship—and to have your puzzled neighbors ask "Faeroe who?" A better reason might be to see the countless flocks of marine birdlife equaled in few parts of the world. Perhaps some come in the interests of science—to note the reaction while inhaling completely pure nature-made atmosphere. People have traveled farther with less justification.

TOPOGRAPHY

There is a velvety greenness but there are almost no trees on the islands. The irregular stone outcroppings, peat bogs, lakes, and moorlands give the landscape a peculiar beauty. The coasts are indented with small fjords, sometimes walled with 1,200- to 2,000-foot sheer cliffs. Many of these serve as "high-rise condominiums" for the limitless number of terns, gulls, puffins, guillents, and other beautiful, Arctic water birds. The highest point in the archipelago is less than 3,000 feet above sea level. Although few visitors will get to more than two or three of the islands, which total 540 square miles in area, most will want to see several of the many tiny fishing villages that continue a by-gone way of life that has almost completely disappeared throughout Scandinavia.

CLIMATE

It does not always rain in the Faeroes—just most of the time. There is some precipitation more than 200 days a year, which adds up to an ample 56 inches. Sunshine is less than plentiful, but there are some beautiful days when the skies are not completely overcast or lost in fog and mist.

Temperatures are surprisingly moderate and, as shown on the City Weather Table of Thorshavn, the average normal low even in midwinter is just above freezing. The minimum recorded in 50 years was only down to 8°. The relatively mild climate is due, of course, to the North Atlantic Drift.

Nor is the weather in the Faeroes ever very hot unless the August maximum 55°-high is bothersome. Perhaps, it should be noted that the mercury zoomed up to 70° one July afternoon several years ago. The combination of the ocean streams and the strong tidal currents surging between the islands tend to keep the harbors ice-free throughout the winter.

LANGUAGE AND GOVERNMENT

Although Danish is the official language, many people commonly speak an old Norse dialect which is somewhat similar to Icelandic. Many learn English in school. While there is an undeniable attachment to the homeland, there is also an atmosphere of rugged independence, rather akin to that of the Sicilians, Basques, and Bretons who strive to retain their identities. The Faroese have their own flag and enjoy almost complete home rule. They are, nevertheless, an integral part of the Kingdom of Denmark and two island representatives sit in the Folketing in Copenhagen.

Ninety-eight per cent of the male population are engaged in fishing and the remainder raise sheep—the name Faeroe comes from the old Norse word meaning sheep. Since no spot on any island is more than 3 miles from the ocean, it is easy to understand why so many of the rather sleek, high-bow, high, stern Viking-type of fishing boats put out to sea each dawn and return laden to the gunwales with cod, herring, or haddock.

TOURISM

Although the Faroese are the most gracious hosts, little is done to promote tourism. The answer is that, while not classed as an affluent region, incomes are relatively high and the standard of living is adequate to take care of all requirements. This is a somewhat remote, but most certainly not primitive, piece of the world. By measurement, it is 400 miles to Norway, 500 to Iceland, and 850 to its national capital, Copenhagen.

Air and ship service are bringing in a few Europeans each year. No doubt the numbers will increase and they will, as usual, be followed by the American who has been everywhere. There are about ten modest hotels but private accommodations can also be arranged. Most overseas visitors are apt to stay at the Hotel Hafnia in Torshavn, the capital, on the island of Stremoy. Those arriving by air will land on Vagar, the most westerly large island and the only one with suitable flat land. It takes about 3 hours by car and boat to get from Vagar to Stremoy, where about a quarter of the 40,000 total island population live. The one drawback to a Faeroes Islands' visit is that the air back home may not seem as fresh as before the trip.

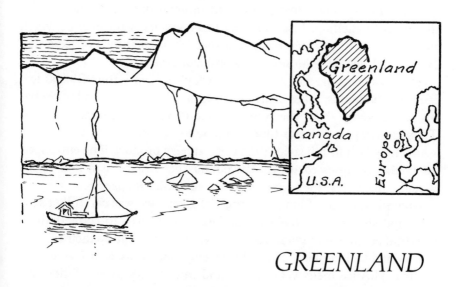

GREENLAND

Some may be surprised that this remote and largest of all islands is included in a volume on Europe. Many Americans first learned the name in a hymn book which gave some indication of its climate. They may still know it only as a huge ice slab somewhere in the Artic north of Labrador.

The Danes, however, think of it as being just across the Denmark Strait from Iceland and used by their Viking ancestors as a convenient filling station on jaunts to the New World. While substantially autonomous, Greenland has actually been an integral part of the Kingdom of Denmark since 1953, when two representatives were chosen to sit in the

Folketing at Copenhagen. The culture is more akin to Scandinavia than America and transportation (except from the United States base at Thule) is almost entirely from Denmark and Iceland.

Topography

This vast island is a spectacular region. Great Britain, Italy, France, and a few more could be slid under the 2-mile-thick icecap and not stick out along the edges. The entire 840,000 square miles, except a 50-mile wide strip bordering some of the shore lines, is frozen solid permanently. The 10,000- to 12,000-foot high, black granite mountain ranges near the coast are one-sided. That is, only the ocean side is exposed as the whole center of the island is mounded solid with a permanent icecap. The glaciers creep slowly down the fjords, and on reaching the ocean, calve or break off with a deafening roar and often create giant tidal waves. These mammoth chunks, frequently a half mile long and 300 feet above water, float southward in great floes until they finally disappear by melting. Just watching this awesome sight may be worth the long trip. It has been estimated that 20,000,000 tons of ice a day are spewed out into the sea at Jacobshavn alone.

These icebergs are not simply yesterday's frozen water. Glacial ice, like gem diamonds, takes centuries to form. What we see today probably fell as snow about the time that Nero was taking music lessons. Tremendous pressure and time combine to convert it into a dense ice which may be a fifth heavier than the ordinary product. The entrapped air is also compressed up to 250 pounds per square inch and forms wirelike markings, the shape of which identifies its age.

Recently a cargo of lumps or chips of glacial ice was carefully packaged and shipped to Copenhagen. The thought was to convert these beautiful crystals into gold, not directly, of course, but by their sale for use as highball coolers! The scheme may not be as weird as it first sounds. This ice melts very slowly and causes very little dilution. As the compressed air is released, it pinks in the glass as does sparkling water. It is also said to improve the taste of any drink—iced tea, coffee, or highball. In addition, there is an ecology angle. Smog and other such man-made pollutions had not been invented a couple of thousand years ago when this ice started forming. At any rate, the Danes are regarded as very astute businessmen and trial shipments have also been made to a dozen other countries. Glacial ice from Alaska, used in highballs at an Explorers Club dinner in New York City some years ago, vied with many strange and exotic foods as a conversation piece. Who knows? At least, they are not apt to run out of raw material.

Climate

Before plunking down hard cash on the barrelhead for tickets to Greenland, the prudent traveler asks, "What's the weather like?" Well —in summer, it is cold, and in winter, even polar bears wear fur coats. Precipitation ranges from a meager 2-inch total for the entire year at Perry Land in the extreme north to a 76-inch deluge at Torgilsbu some 1,200 miles south, at the lower tip. The average tourist is, of course, interested only in the midsummer weather conditions.

The City Weather Table of Godthaab is reasonably typical of the whole island but Ivigtut in the south averages about 5° warmer and Thule, in the north, about that much colder. Since almost all of Greenland is north of the Arctic Circle, the sun is above the horizon—not necessarily visible—at all times during most of June, July, and August.

Tourism

Travel from one village to another is usually most conveniently done by boat. There are no deluxe foods or accommodations but the Grouland Hotel at Godthaab and the White Falcoln at Jacobshavn are rated first-class. Greenland's population is made up of about 7,000 Danes and 40,000 Eskimos. There has been some admixture of white and Indian blood in the latter group. Civilization conquered almost all tuberculosis but, in exchange, introduced drunkenness and venereal disease. Most Greenlanders depend upon fishing, but in addition to generous budgetary assistance, the Danish Government has sponsored the development of fish processing plants and other suitable industry. There is also a little mining activity, but the great hope is that petroleum and gas may be found as was the good fortune in Alaska.

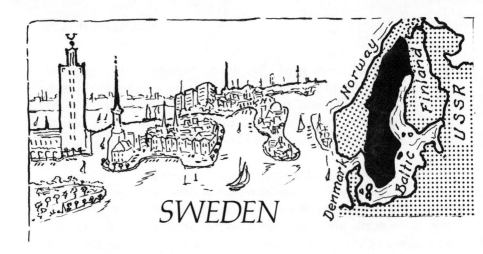

SWEDEN

ONE does not usually think of Sweden in terms of a single spectacular tourist magnet such as the fjords of Norway, the Swiss Alps, or the Lakes of Killarney. Rather, it is mainly a pleasant land of quiet, peaceful countryside. The largest of the Scandinavian countries, and indeed ranking fourth in size among the nations of Europe, there is room for a great variety of both scenery and climate. Its midnight sun is as bright as other Arctic regions while the lovely lake district, the long coast line of beaches, coves and rocky shores, the mountains, the canals, the seascapes, and Baltic Sea islands make a prize vacation package.

Topography

The Kjolen Mountains extend down along the Norwegian border from the northwestern tip of Sweden southward to just below Särna. This 60-mile-wide band of high country averages 3,000 to 4,000 feet above sea level. The loftiest peaks are in the northern segment where the 6,965-foot Kebnekaise tops all others. The midsection drops somewhat, but the elevation again increases at the southern end where the Helagsfjaelet (5,892 feet) looks down on its shoulder-high neighbors.

A huge plateau, which slopes from the mountains down to the Gulf of Bothnia on the east, occupies the major portion of the country. There are a great number of short, swift-moving rivers which cut their paths southeastward to the coast. Almost all of these streams flow from the thousands of finger-shaped lakes dotting the higher western side of this entire plateau region.

The segment immediately south of the mountain and plateau areas, is fertile lowlands containing a number of very large lakes. It is said that 9% of Sweden is covered by lakes, and Vänern, the second largest in Europe, is one of the estimated 96,000 in the country. This 2,140-

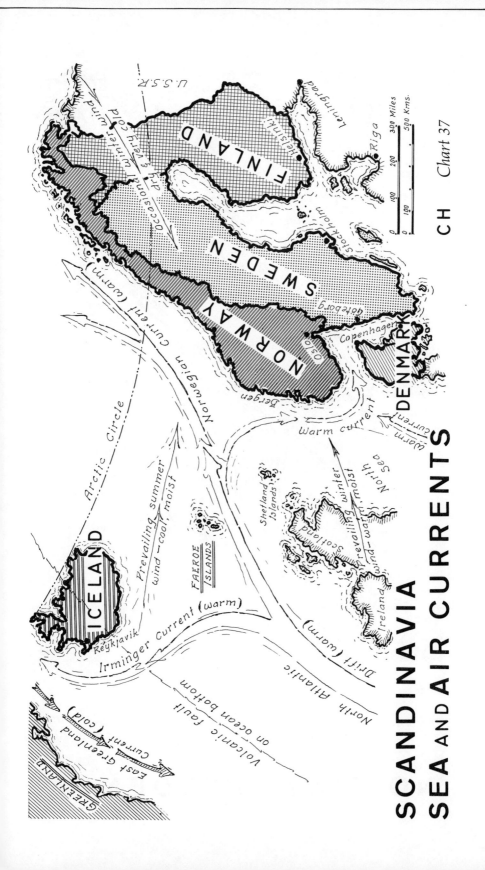

SCANDINAVIA
SEA AND AIR CURRENTS

CH   Chart 37

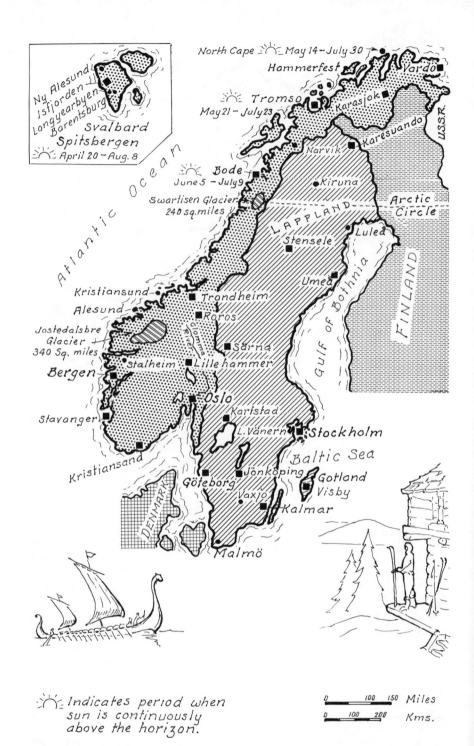

Ny Alesund
Isfjorden
Longyearbyen
Barentsburg
Svalbard
Spitsbergen
April 20 - Aug. 8

North Cape ☼ May 14 - July 30
Hammerfest
Vardo
☼ Tromso
May 21 - July 23
Karasjok
Narvik
Karesuando
U.S.S.R.

☼ Bode
June 5 - July 9
Kiruna
Arctic
Circle

Swartisen Glacier
240 sq. miles

LAPPLAND

Stensele
Lulea

Atlantic Ocean

Umea

FINLAND

Kristiansund
Trondheim
Alesund
Roros
Jostedalsbre
Glacier
340 Sq. miles
Bergen
Stalheim
Lillehammer
Särna
Glomma River

Gulf of Bothnia

Oslo
Karlstad
Stavanger
L. Vänern
Stockholm

Baltic Sea

Kristiansand
Jönköping
Gotland
Visby
Göteborg
Vaxjo
DENMARK
Kalmar

Malmö

☼ Indicates period when
sun is continuously
above the horizon.

0   100   150 Miles
0   100   200 Kms.

NORWAY SWEDEN

Chart 38

square-mile body of water is the source of the Gota River which empties into the Kattegat at Göteborg.

Just south of this low lake region is a large patch of 1,000- to 1,500-foot elevation uplands known as Smaland. The nub of land at the southernmost tip of Sweden, called Skane, is the finest agricultural section in the country. It is also the warmest. While the bikini-clad Swedish sirens may regard this sandy coast as a Riviera, visitors hardly classify it as subtropical.

Climate

There is a drastically sharper temperature gradient in Sweden than in Norway. Norrland, the province in the upper portion of the country, will average at least 10° lower readings than the south. The northern area can also expect more than 7 months of winter and less than 3 of summer. Skane, way down in the south, enjoys over 4 months of summer and a short 2 months of cold weather.

The prevailing westerly air streams, which have traveled over the warm ocean currents, temper the climate of Sweden to some extent. Winters, particularly in the southern half, are much less severe than the latitude might suggest. When the winter wind shifts and blasts in from the northeast, however, it is high time for bundling. These uncomfortable conditions are much less pronounced in Norway than Sweden.

Precipitation is generally quite modest throughout th country. Few areas average much more than 20 inches annually and most do not experience over 1/10 of an inch precipitation more than 60 days a year. None equal Bergen's 79-inch deluge. The wettest segment of Sweden is along the southwestern coast facing on the Kattegat. The total here will range from 30 to 35 inches with 75 to 85 days getting over a 1/10-inch of precipitation.

As we know, the farther north, the longer the summer day. Sunshine is surprisingly plentiful during April, May, June, and July in these polar regions. Actually, the sun never drops below the horizon during the whole month of June above the Arctic Circle, although it may be hidden behind clouds or fog. In the southern part of the country, there is usually less sunshine in April than August.

There has been a great flood of newspaper and magazine articles recently concerning many serious air- and water-pollution problems throughout Europe, comparing conditions with ours. The Scandinavian countries are seldom mentioned. There are current disturbing indications that this situation may change. The southern portions of Sweden and Norway have been experiencing increasingly heavy banks of polluted air. Swedish scientists have traced the origin to the Birmingham area of England and the belching smokestacks of the Rhur Valley, 1,000 miles away. If that sounds a bit far-fetched—the United States National Weather Service claimed that the east coast of Florida was engulfed in a heavy dust haze which had traveled 5,000 miles from the far-off Sahara.

But there are dangers closer to home. The rapid industrialization of Russia, Poland, and the Scandinavian countries themselves threaten to produce major pollution problems. The huge petroleum development in the North Sea, if not carefully controlled, could seriously contaminate not only the long, exposed coast of Norway but also seep into the Baltic on the small branch of warm water from the Norwegian Current. Both Norway and Sweden enjoy one tremendous natural advantage. Most of the energy required by the railroads and industry is supplied by clean water power generating stations.

The Mountain Climatic Zone

The City Weather Table of Karesuando, in the extreme north, and that of Särna, at the lower end, indicate the varying conditions that may be encountered in this 600-mile-long stretch of high country. Note that only about 3 months of the year do not get at least a taste of freezing weather in Karesuando. The mercury will drop below 0° about 65 nights during the long, dark winter. In this region of rather changeable weather, it is important to observe not only the normal average figures but also the radically different extreme recordings. The −42°-temperature might drive even the most rugged Laplander into hibernation. On the basis of latitude alone, this should be a mighty chilly place. Karesuando is actually a hair closer to the North Pole than Verkhoyansk in frigid Siberia. On at least one occasion, this Russian city registered a −90°-reading. The relatively warm air currents from the Atlantic help save the day for this part of Sweden.

The total 13 inches of precipitation looks really meager but that translates into about 11 feet of snow.

The Swedes are almost as avid skiers as the neighboring Norwegians. Both slalom and cross-country are popular. The midwinter tan that some of these handsome people sport was probably acquired in the Swedish mountains and high country rather than on the French Riviera.

On whiff of the pure, pine-scented Lapland air should restore life to even the most wilted tourist suffering in a hot, overcrowded midsummer European resort. How to get to this almost unknown northland inhabited by 10,000 Lapps and 200,000 reindeer? There is a wide choice, but the well-planned SAS Midnight Sun Flight out of Stockholm is the quickest and most popular with Europeans. This very efficient air-line has developed several package jaunts which are quite inexpensive. Driving is possible but a bit tiring for some unless there is plenty of time. For economy and adventure, the young at heart may be tempted to try the unique mail-bus route. This network of clean, comfortable, yellow vehicles covering 5,000 miles of roads carried some 5 million (mostly Swedish) passengers last year. Accommodations are not deluxe but every last boardinghouse, cottage, and hotel is immaculate. That, of course, is no surprise since even the most modest housing in every corner of all the Scandinavian countries is invariably spotless and shining. Complete information in this service is available from the Postal Administration, Mail-Bus Section, P. O. Box Stockholm 1, Stockholm, Sweden.

For the even more vigorous, there is a camp near Kiruna where Lapp huts can be rented very inexpensively by the week. There are no dining rooms but food is available and your dinner of cold-water char and trout is waiting at the nearest stream or lake.

The immediate Kiruna environs, where the richest iron mines in the world are located, are of lesser interest to most tourists.

The method best suited to many who want to see one of the largest remaining untouched wildernesses in Europe are the several Swedish Railway tours. These are beautifully arranged and the service, accommodations, and food will come as a pleasant surprise to overseas visitors familiar with passenger railroads in the United States. One of the most comfortable and interesting itineraries rambles across the Arctic Circle to Riksgransen near the Norwegian border. Those who wish can make the hairpin-turned trip from there down to Narvik on the Atlantic Coast and back. The roundabout trip includes most of the finest scenery and interesting towns and can be arranged with stopovers for sightseeing. It is an unusual experience that will be long remembered.

While even summer nights can be nippy, the daytime temperatures are just right for this type of travel. The 7 to 9 hours of sunshine per day is enough for the most enthusiastic camera bug while the 18 to 24 hours of daylight call for drawn curtains at bedtime.

Särna (1,500-feet elevation) at the southern end of the mountain chain follows much the same pattern. On the average, temperatures are normally about 5° higher but precipitation drops to a modest 22 inches a year. May, early June, and September offer the best combination of temperature and precipitation; however, there is somewhat less sunshine during the latter month. In all of Lapland and the mountain country, thermal underwear is *haute couture*. Although the night temperatures do not usually drop below the 40's during the summer tourist season, a touch of freezing weather is not unknown.

The Northern Plateau Climatic Zone

This is a vast region of countless crystal-clear lakes and dashing streams, some of which have cut deep chasms in their rush to the sea. It is mostly a forest wilderness of pines and other scented evergreens. There are also some bogs and fertile lands as well as stark, rocky patches. It could be the answer for outdoor people and all who enjoy an undisturbed natural environment. This entire region slopes from the mountains in the west down to the sea which is fringed with sections of very rich agricultural lands. Luleå, at the head of the Gulf of Bothnia, is an important export center but is often icebound in winter. At that time, commodities are shipped by rail to Narvik, which, although farther north, is a year-round open port thanks to the benevolent Gulf Stream. One of the interesting features in Luleå are the several hundred tiny huts or cabins which rural folk once used when they journeyed into town to attend church.

The City Weather Tables of Stensele and Umeå indicate that there is not a great difference between the coastal and inland weather conditions in this vast region. Unlike the warm Atlantic streams on the coastal areas of Western Europe, the wide Gulf of Bothnia does not materially affect the climate of the adjacent shore lands. Winters are dark and cold, sometimes quite severe, with rather low precipitation. Summers are sunny and warm during the day, with many hours of daylight and long evenings.

Stockholm, The Lakes, and Southern Climatic Zone

From the plane window, it would seem that one has been skyjacked to Amsterdam or Venice. Stockholm has miles of intriguing waterfront and many canals and waterways separating the numerous islands. The quaysides seem endless and the city fathers claim that the Stockholm archipelago comprises almost 24,000 islands. It is said that although every resident may not own a car, he most certainly will have a boat, no matter how small.

In addition to all of that, there is the Gota Canal winding southward, while to the west spreads a whole region of glittering lakes. The varied segments of Stockholm make one appreciate why so many visitors never get much beyond the city limits. There is the grand central section of town with many colorful parks and handsome wide boulevards lined with swanky shops. The "old city," which seems to have remained almost untouched for centuries, is a joy to stroll on a quiet Sunday afternoon. No one would fail to visit the magnificent city hall which is like none in America. Any pleasant, sunny morning or afternoon invites a stroll in the spacious Skansen, a most unique park on an island. There is everything there from a glass-blowing factory to the things found in the world's largest parks.

Although the climate of this general area is the most pleasant in all Sweden, weather is certainly not the principal attraction. Average temperatures throughout the year are cool but moderate. Midsummer days normally register 65 to 70°, with nights in the middle 50's. In winter, the mercury ranges from about 25 to 35°. On occasion, however, there can be wide extremes. Sometime during the past 30 years, there was a recording of −26 degrees, but no Stockholm hot spell has ever reached 100°. There are almost sure to be about 130 nights a year with readings of 32° or lower, but the chances are only one in 100 that the mercury will zoom above 90°.

The average precipitation of 23 inches is quite modest; New York City gets more than twice that figure and is not considered an overly wet place. Note on the Stockholm City Weather Table that almost 300 days a year get less than 1/10 of an inch of precipitation. Rainfall is usually highest in July, August, and October and distributed fairly uniformly during the other months. Much of the summer rain is in the form of showers or heavy thunderstorms which are soon finished.

Although records show that Stockholm usually gets only about 40 totally clear days as against 180 cloudy ones, no month from April through July averages less than 7 hours of sunshine per day. There will usually be about 45 foggy days mostly spread from October through February. Prevailing breezes are from the south or southwest but only in winter and very early spring are they forceful. Some Americans associate snow with all of the Scandinavian countries—we should note that there will be at least some ground cover 90 to 100 days a year in Stockholm. Snow does not usually appear until October but can continue through early April.

There are countless lakes scattered all over Sweden but the largest are to the southwest of Stockholm. Vänern and Vättern, the biggest, are very popular summer vacation areas but there are many others equally attractive. This is a region of pleasant, rolling countryside and pine-covered hills. One might look far and long to discover a more inviting pastoral setting for a long, lazy vacation.

The City Weather Tables of ancient Karlstad and Jönköping, at the north and south interior sections of this district are quite typical of the entire area. There are never any really hot spells but nighttime readings reach the freezing point about 6 months of the year. The winter temperatures generally range between 25 and 35°, but on rare occasions, the mercury slumps into subzero figures. Total precipitation is a low 20 to 25 inches and, while distributed throughout the year, is heaviest in late summer and autumn. The average 9 + hours of sunshine per day during May, June, July, and August may have given rise to the Swedish expression, "More solar than polar."

The town of Växjö is a place of particular interest to visitors of Swedish extraction. It is estimated that more than 1 million Swedes emigrated to North American from this district between 1850 and 1930. That period is sometimes referred to as "the years of American Fever." There is a unique building called "The House of Emigrants" devoted to the history of that period. Växjö is also in the center of the famous glass industry with about 30 well-known glass works within an hour's drive. The Glass Research Institute and glass museum, with one of Europe's most extensive collections of old and modern pieces, are ample reward for coming to this district.

There are two very interesting tours, both originating in Stockholm which can be made by auto or bus. The one to Oslo, Norway, goes first through Uppsala. This town of great historic interest, with its 500-year-old university, has long been a cultural center of Sweden. After skirting the lakes of Malaren and Hjälmaren, the route heads through pleasant, rolling country to the charming old town of Karlstad on Lake Vänern. The run from Fryken Lake and over the Norwegian frontier to Kongsvinger is across a ridge of pine-covered hills. It follows the foaming Glomma River and on in to Oslo.

The other equally enjoyable jaunt is a 2-day trip from Stockholm to Malmö in the extreme south, just across from Denmark. This drive starts westward through prosperous farmlands past ancient manor houses to medieval Nyköping. The road winds along the Gota Canal and across the lovely, undulating Gotaland plains. This trip follows the shores of Lake Vättern to Jönköping, the "match city," then southward with perhaps a stop to watch glassblowing at Växjö. The route to Malmö is through the chateau country of Skane and some of the most productive farmlands of Sweden. Malmö is only 18 miles across the narrows from Copenhagen.

We should also mention the trips from Stockholm to Göteborg. There is the choice of car, bus, train, or the more leisurely canal boats. Two-thirds of the 350-mile boat trip is on the canal, with its 65 little locks, the remainder on the lakes and rivers. Many restless Americans find this 3-day water trip a bit monotonous but enjoy the combination motor coach and steamship tour.

Göteborg and the Sunny Coast

Göteberg, the second city of Sweden, not only dominates the country's summer playground but is also the center of its major manufacturing activities. It is the home of the giant SKF ball-bearing complex and Volvo, Sweden's largest automobile company, which is discarding the Henry Ford production line theory for group assembly of cars.

As shown on the City Weather Table of Göteborg, this area enjoys the mildest and most pleasant weather in Sweden. Temperatures are surprisingly uniform throughout the year with never a scorching spell and seldom a 0 reading. The moderately ample 31 inches of precipitation is spread over the 12 months but, in common with most of Western Europe, is highest in late summer and fall. There is a snow ground cover about 50 days a year and 100 or more nights when the mercury will drop below 32°.

Although there are only 75 fully clear days and twice that number of cloudy ones during the year, this part of the country is bright and sparkling much of the time, and gets 7, 8, 9, and even 10 hours of sunshine per day, every month from April through August.

AVERAGE HOURS OF SUNSHINE PER DAY AND YEAR IN SEVERAL SWEDISH CITIES

	J	F	M	A	M	J	J	A	S	O	N	D	Year
Karesuando	—	1	4	7	7	9	9	5	4	3	1	—	1,568
Luleå	1	2	5	7	9	10	10	7	5	3	1	—	1,818
Storlien	1	2	3	4	6	6	7	6	4	2	1	1	1,322
Sundsvall	2	2	5	7	9	9	9	8	5	3	2	1	1,854
Karlstad	2	2	5	6	9	10	9	9	5	3	1	1	1,862
Stockholm	1	3	5	7	9	11	10	8	6	3	1	1	1,973
Göteborg	2	3	5	7	9	10	9	8	6	4	2	1	1,988
Visby	1	2	4	7	9	11	10	8	6	4	1	1	1,932
Olands S	1	2	4	6	8	10	9	8	6	3	1	1	1,833
Malmö	1	2	4	6	9	9	9	7	6	3	1	1	1,738

There is much of interest in Göteborg, a city of gardens and canals. Although Stockholm's elegant Restaurant Riche and deluxe hotels may well dispute the claim, Göteborgers boast that their city is the center of fine cuisine, particularly seafood. The wonderful Park Avenue Hotel of Göteborg lives up to the challenge. Do not be dismayed to discover that New York City may now be the smorgasbord capital of the world

—it certainly is not to be found in Sweden. Few restaurants there serve this once national "board of plenty" except on special occasions. In its place has appeared a miniature individual tray called *assietter.* Tasty? Of course, but not the famous Swedish smorgasbord.

The tourist with time on his hands and pleasure in his heart may break away from Göteborg's attractions to enjoy a jaunt along Sweden's "Sun Coast" just to the north. He may also have a scientific curiosity in learning if the famous Swedish beauties really wear the world's skimpiest bikinis.

The bright and lively coast line bordered with 3,000 islands and skerries is packed with Swedish and Norwegian vacationists during the month of July. The waters of the Kattegat are warmed by a tiny branch of the Norwegian Current but official records show a typical year's temperature readings as: June 65°, July and August, 67°, and September 65°. The Swedes point out that the English Channel averages only 65° during that period. To a Floridian accustomed to almost 80° December-swimming, this would feel more like a polar bear sauna.

There is quite a wide choice of both scenery and accommodations along this popular summertime coast. Brastad and Stromstad are rather busy popular resorts while Marstrand, the only town in Sweden with no motor traffic, has a bit more of the Old World staidness. Bovallstrand and Grebbestad are typical of the many tiny fishing villages. There are also industrial centers. Here also is Gullmarsfjorden, Sweden's one and only genuine fjord and Kinnekulle, a unique plateau with strange flora and many medieval churches and manors. The south is primarily an area to come to for a pleasant vacation with reasonable assurance of fine spring and summer weather.

Gotland

Visby, the charming little capital of the Isle of Gotland, is included in all too few itineraries. It is less than 1 air hour from Stockholm and a pleasant 7 hours by water. Gotland is popular with Swedes in July; other Europeans are more apt to arrive in August. Few overseas tourists discover this 80 × 30-mile Baltic island which means another choice find for the alert traveler. Visitors spend most of their stay in Visby, the only completely walled medieval town in northern Europe. Like all other such strategically located islands, Gotland received many invited and more unwanted guests over the ages. There is evidence that both the Greeks and Romans were among the early "Cook's tourists." As the name might suggest, this was a one-time stronghold of the early Goths. The well-preserved ancient ramparts are studded with many fortress towers. At least for the present time, this is still a delightful place for a comfortable and pleasant vacation. When picture folders appear in the travel agents' racks, a modern invasion will be on.

The Visby City Weather Table shows how inviting the weather is. With a very low 20 inches (less than half that of New York City) of precipitation, there are 300 days a year that do not get more than 1/10-inch of rainfall. May averages only about 1 inch and even the wet month of August gets less than 3 inches. In addition, much of the summer precipitation occurs as showers, sometimes thunderstorms which can be heavy but very short.

Normal vacationtime temperatures are just cool enough for walking while heat waves are almost unknown. The crowning glory of Gotland's weather are the 7, 9, 11, 10, and 8 hours of sunshine per day, respectively, for the months of April through August.

The Best Seasons to Visit Sweden

Many people limit the scope of their travel enjoyment by associating a country with a single, supposed characteristic. The French Riviera stands out in many minds as a winter resort when the prime weather period is spring and early summer. Who would dream of spending July in Turkey although such spots as Foca are much more comfortable than most parts of Europe at that time? And so it goes—there are similar examples in almost every country.

If one picture is worth 10,000 "I told you so's," the vacation weather maps of Sweden should tell a complete weather story of that country at a glance.

Except for the skiing areas, April, May, and June are the most desirable months to visit much of Sweden. The hours of sunshine table shows a sharp rise between March and April and an equally steep slump at the end of August.

Even the foothill portions of the mountain country are at least as pleasant in spring as any other time of the year. The air is clear, cool, and quite dry. Sunshine ranges from 7 to 9 hours per day. Precipitation increases from 2 or 3 showery days during the month of April to about 1 day a week in June. The following months get more. Those who prefer rather warm days may find the spring weather in the foothills a bit chilly but it seems to suit most active sightseers. There is, of course, always the possibility that it may be a bit muddy underfoot but a pair of rubbers takes care of that.

Springtime would not be the first choice for the huge northern plateau region. Although sunshine is plentiful and precipitation low, the days can be nippy and the nights a lot colder. There are usually large patches covered with snow, but the countryside is generally green with flowers everywhere. We suggest that prospective spring visitors first study the weather statistics carefully.

Without question, springtime is the top season in the important southlands. Precipitation ranges from 1.5 to 2 inches per month, being higher in June than April. There is usually light rain about once a week. The 50 to 65°-daytime-temperatures refer, of course, to readings taken in the shade. Sunshine increases from about 7 hours per day in April to 9 or 10 by mid-June. June 21, with about 18 hours of daylight, is the longest day of the year. These conditions also apply to Visby and the Isle of Gotland. As a word of caution, North Americans should remember that the general weather patterns of Europe are not as uniform as at home. More often than here, there will be exceptionally hot, cold, wet, or dry seasons. The odds, however, are that the visitor will experience the conditions which we describe.

The mountains are not very high—3,000 top 4,000 feet—and are very pleasant in summer. Precipitation increases over the springtime figures, and August is most often the wettest month of the year. Even then, it rarely exceeds 3.5 inches of rainfall which means a shower every 3 or 4 days. This is the best season in the high northern plateau, although there, too, August gets the most rainfall. Sunshine is a plentiful 8 or 9 hours a day and the crystal-clear lakes and streams are teeming with hungry fish. It is the perfect time to see Lapland and come home with tales of moose and game, footprints of bear, wolverine, and other "wild beasties" but also 200,000 reindeer.

July and August rank next after the spring months in the south and lake region. Sunshine is almost as plentiful but rainfall is somewhat heavier than the previous months. Temperatures are as high as they ever get in Sweden but no need to fear heat stroke. Swedish thermometers would be just about as useful if the portion above the 90°-mark were sawed off. Overseas tourists would do well to remember that most Swedish workers get a month's vacation and just about every last one of them decides that July is the ideal time. It is necessary to book accommodations well in advance in order to have any choice. More of the English vacation in August while the French divide their time between those 2 months.

Few Americans realize that autumn ranks well below springtime as a vacation choice in most of Western Europe. There is much less late-season sunshine but also usually lighter rainfall than in August. Overcast skies are the rule with, sometimes, a tinge of winter cold filtering in. Occasionally, Sweden will enjoy one of those glorious dry, sunny, New England autumns with a lavish display of foliage colors, but pleasant springtime weather is a much safer bet in this part of the world. There is an outstanding redeeming feature when excessive August rainfall keeps one indoors. Every person who has been in Stockholm on the eighth of that month will remember the reason. That is the first day that *kraftors* may be sold or served in restuarants. This delicacy is a succulent, little, fresh-water lobster (not a shrimp, crevette, or crayfish) and is ample justification for a most enthusiastic annual celebration. The time-honored ritual of comsumption is one kraftor followed by a single gulp of aquavit and a chaser of excellent Swedish beer. That procedure is repeated many times, often by going from one party to the next. Anyone who thinks of a Swede as "old sobersides" should witness a kraftor happening. (Incidentally, August 8 is also the beginning of the kraftor season in Norway, but for some reason, the date in Finland is July 25.)

Skiing in the mountains or a visit to enjoy Stockholm after the tourist bustle has quieted, are the choices for winter in Sweden. The mountains are cold, clear, and dry although dim. Every Swede finds at least a little time for skiing and this upland country is full of skiiers all winter long.

Visitors enjoy Stockholm at all seasons and winter is no exception. It averages about halfway between New York City and Montreal in winter temperatures. Precipitation is not heavy and will occur mostly as snow. November, December, and January will not average more than 1 hour of sunshine per day, and daylight will extend only from about 10 A.M. to 4 P.M. That, however, offers little problem as much of the time will normally be spent indoors. The shops ranging from the most elegant to very simple are very tempting. Restaurant service is at its peak and musical activities are in full swing.

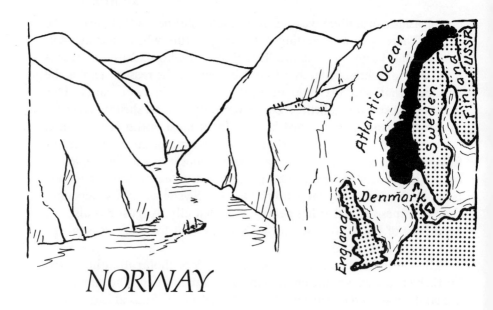

NORWAY

NORWAY is rather easy to describe—3.3% of its area is neatly cultivated farmlands and the other 96.7% is a tourist wonderland. It is a fantastic travelogue in living color—majestic mountain scenery, spectacular fjords, and startling seascapes.

And there are a few additional bonuses that help make it one of the most popular tourist spots in Europe. First must be the bright, cheerful people who are, perhaps, a bit more akin to the fun-loving Danes than their more serious Swedish cousins.

A pleasant surprise awaits the visitor who assumes that cuisine of excellence starts and ends with the French, Chinese, Italians, and Hungarians. Ptarmigan in cream, *multer* or cloudberries, delicately smoked salmon, and the famous northern trout from the icy streams and fjords, that require 3 months of home processing to become prime *rakorret*. Those are but a sampling of the unusual palate pleasures.

Topography

Because the sculpturing of the terrain has so important an influence on climate, it is well to first have the geography of a country in mind.

Norway's 1,100-mile north to south dimension is greater than that of any other nation in Europe except Russia. About one-third of the country lies north of the Arctic Circle. And just to orient its position on the globe—Vardö, the northernmost town in Continental Norway, is just a bit closer to the North Pole than Point Barrow, at the top of Alaska. Kristiansand on the Skagerrak, way down at the southern limit,

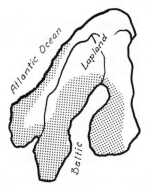

SPRING
May June
Glorious weather almost
everywhere. Driest & sunniest
season. Cool but comfortable.
Still snow in north & high
country but no skiing. All
other areas delightful. Wild
flowers cover countryside
Spitsbergen season.

AUTUMN
September October
Usually cool with overcast
skies & often quite wet.
Some years (1972 was one of
them) are sunny, dry &
crisp with magnificent
fall foliage.

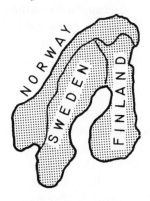

SUMMER
July August
More showers than spring
but sunshine plentiful. Rarely
hot. Days still very long. Lap
land & north very pleasant
also fjord boat trips. This is
favorite vacation time for both
Norwegians & tourists. Some
places quite busy.

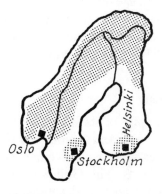

WINTER
November March
The winter choice is skiing
in every form or three
wonderful capital cities—
many enjoy both. Oslo is
milder, gets a bit less snow,
but no more daylight
sunshine.

VACATION WEATHER MAPS
NORWAY SWEDEN
FINLAND

Chart 39

is on almost the same parallel as Juneau, Alaska. That may explain the absence of any reference to a "Norwegian Riviera" in travel folders. The sea water at Kristiansand does not reach a shivery high of 66° until August. Bikinis are for show and suntan except for a few of the most hardy.

The surface of Norway is dominated by mountain chains extending the full length of the country. Only 20% of the land is less than 500 feet above sea level and the average overall elevation is about 1700 feet. Most of the country is about 60 miles wide and only down in the southern bulge does it exceed 250 miles across. Rivers are short and generally swift-moving. Glåma, the longest (380 miles), flows from north to south parallel to the Swedish border.

It is claimed that if the coast lines, including fjords, inlets, and islands, were stretched out, they would span 17,000 miles. Is it any wonder that the Norse have always been a seafaring people? A common expression is, "We have been sailing the ocean since the world was flat." While the mountains extend the full length of Norway, the loftiest masses are in the northern and southern portions of the country where the average altitude is about 3000 feet. In the central Trondheim area, the land drops to 1,500 feet. Although the sharp, snow-capped peaks are scenically spectacular, they are not very high. Few of the summits top 5,000 feet and only Galdhöpiggen, north of Bergen, reaches 8,400 feet.

Climate

In spite of its northerly location, the climate of Norway is fairly temperate. As is commonly the case, the almost continuous mountain barrier and the ocean currents play a major role in controlling weather conditions. The Gulf Stream, which in turn, has become the Great Atlantic Drift and subsequently the still warm Norwegian Current, greatly tempers the climate of the low coastal areas. (*See* Chart 37.)

On the basis of latitude, there is a considerable difference in temperature readings between the northern and southern segments of Norway. The coastal lands generally experience temperate winters and cool summers, with the heaviest precipitation in autumn and winter. The interior portions normally have warm summers, cold winters, and moderate to low rainfall.

The prevailing, humid air currents blow in from the Atlantic. As might be expected, the highest precipitation occurs along the coastal lowlands and the western slopes facing the ocean. These rather constant winds can be very strong, especially during winter when they may reach gale force. Prudence suggests some indoor activity during heavy sea storms. Since large water bodies heat up and cool off more slowly than land masses, proximity to the sea usually spells equable temperatures. To the good fortune of the Norwegians, the path of the westerlies across the Atlantic brings these winds in from a somewhat southern direction in January. They gradually shift and blow from a more northerly quadrant by midsummer.

The higher inland districts experience lighter breezes and accumulate a lower total precipitation than along the shore. The temperature range in the interior is far wider between the extreme high and low recordings. This is generally true on both a diurnal and annual basis. As indicated in the following table, Norway is not a bright, glittering region.

HOURS OF SUNSHINE PER DAY AND YEAR

	J	F	M	A	M	J	J	A	S	O	N	D	Year
Oslo	1	3	5	6	7	9	8	7	5	3	1	1	1,649
Stavanger	1	3	4	6	6	8	6	6	4	2	1	1	1,483
Kristian-sand	2	3	4	6	7	9	8	7	5	3	2	1	1,749
Trondheim	1	2	4	6	7	7	5	6	4	2	1	—	1,359
Bodo	—	1	3	5	7	7	5	6	3	2	1	—	1,238
Tromsö	—	1	4	6	7	7	6	6	3	1	—	—	1,273
Bergen	1	2	3	5	6	6	5	5	3	2	1	—	1,214

As is the case in most of Western Europe, April, May, June, and July get the most sunshine. August, particularly the first half, is also usually quite bright and clear. The annual totals are, of course, very low since the sun hibernates during most of the winter in these northern parts. The average maximum possible sunshine (ratio of hours of sunshine to hours of daylight) ranges from about 26% to 40%. In spite of that, the southern portions of the country do quite well.

Norway has three major climatic zones: the northern and southern halves of the low country adjacent to the coast lines, and the much larger high and mountainous interior. Within these, there are two small districts which experience rather low precipitation. One is the inland area in the extreme north, from Kirkenes to Karasjok, which averages a scant 13- to 15-inch total annual precipitation. The other dry spot is the section around Röros and Tynset, inland southeast of Trondheim, which normally accumulates 16 to 18 inches a year.

The Northern Coastal Section

This narrow strip, north of the Arctic Circle, contains such well-known places as North Cape, Vardö, Tromsö, historic Narvik, and Bodo. All are either on the ocean or a fjord. This coast line is very irregular and dotted with islands; the largest fjords are in the southern segment between Trondheim and Stravanger.

There are almost continuous mountain chains quite close to the shore line which include many 4,000- to 6,000-foot peaks. This barrier causes the humid ocean breezes to drop most of their moisture on the western slopes. As a result, the larger glaciers also from close to the coast.

North Cape is a favorite stopping place for cruise ships where passengers get a kick out of reading the newspaper and taking snapshots by the midnight sun. Just because the sun is above the horizon, bright, glittering weather is not guaranteed. Sometimes, fog makes it as scenic as a trip through the New York subway. On a clear day, however, North Cape rises above the Arctic Ocean, its sheer cliffs 1,000 feet high. That you will remember. The sun is above the horizon continuously from the middle of May until the end of July. Conversely, it is one long night for over 2 months in winter. In between, there are varying degrees of brightness to twilight. The City Weather Table of Vardö, just along the coast to the east, gives a fair idea of what to expect. While the temperature is below freezing two-thirds of the year, there is seldom much subzero weather. Fortunately for summer visitors, there is rain (or sometimes snow) less than 1 day a week. Temperatures normally range from about 35 to 55°, but there have been occasional 80°-days and 14°-nights during that period. What is the excuse for traveling so far afield? It could be to say that you have been to the top of the world or even to get away from smog and pollution. The best reason is the magnificent coastal scenery.

It is not necessary to journey all the way to North Cape to witness 24 hours of continuous daylight. The time period that the sun remains above the horizon decreases gradually as you travel southward. The whole disc is continuously visible above the horizon for certain periods at several locations as this table shows.

THE MIDNIGHT SUN

North Cape	May 14 to July 30
Hammerfest	May 17 to July 28
Tromsö	May 21 to July 23
Narvik	May 26 to July 19
Bodo	June 5 to July 9

All of this is the result of the 23.5°-tilt in the world's axis. If it were vertical, there would be neither midnight sun nor change of season. There are overnight SAS air flights from both Oslo and Stockholm to Bodo (Latitude 67° 16′ N) for those on a tight schedule. Droopy-eyed passengers squint at the midnight sun, eat a meal of reindeer meat, and receive a certificate proving travel above the Arctic Circle. When time permits, most visitors go to see nearby Svartisen, the second largest glacier in Norway. Covering an area of 240 square miles, this gigantic ice cube could chill considerable aquavit.

An extra sweater is all most tourists wear. The summer temperature is most apt to be about 45° at night and perhaps 60° during the day, but because of the almost constant light wind, many find a windbreaker or raincoat over a sweater more comfortable. Lacking that coverage, use newspaper under the jacket. Odds are about one in five that there will be a little precipitation.

Temperatures along the coast are mostly a function of latitude. The 40 to 50° summer readings at Vardö and North Cape are about 10° lower than those at Bodo. Precipitation increases from an annual 24 inch total at Vardö to 42 inches at Bodo.

Narvik, remembered from World War II, is an important seaport and shipping center with perhaps less interest for tourists. Hammerfest, not far west of North Cape, is a pleasant, modern fishing town and an excellent harbor which, thanks to the benevolent Gulf Stream, is ice-free all year.

Tromsö, the capital of this northern section, was the starting point of many polar expeditions. It is an attractive place which, like many coastal towns, has a dramatic backdrop of magnificent mountain scenery.

This section of Norwegian coast is, of course, primarily summer vacation country. It is interesting, however, to note that the December, January, and February temperatures in Bodo and Tromsö are almost identical with those of Chicago, which is about 1,500 miles farther away from the North Pole. Such is the power of the Gulf Stream.

Finnmark, the northermost province is a unique land. This vast area is one of Europe's last few "wild tracts." A short stretch of its boundary at the north is common with Russia and another at the south with Sweden, but the major portion borders on beautiful Finland. This trackless *tundra,* an untouched wilderness, is interspersed with heavily stocked streams and lakes. Some portions are bare, stark, and rugged; others are dense forest affording sanctuary to the lynx and wolverine. The Laplander moves across the land, following the annual movement of his reindeer herds up the mountains in spring and summer and back to their grazing grounds in the forest and closer to the coast in late autumn and winter.

But such freedom is not for the average tourist. Facilities are not generally available and the only community of consequence in the inland portion is Karasjok, close to the Finnish border. This interesting little place, at only a 430-foot elevation in the valley of the Karasjok, is in a most attractive setting of 2,000- to 4,000-foot mountains.

There is little doubt that this currently unknown Finnmark region will one day become a major summer and winter vacationland. Its pure, invigorating mountain air alone is almost sufficient magnet.

The modest height of the Norwegian mountains comes as a surprise to many who equate this area with snow scenes and skiing. The snow line (boundary of perpetual snow) is determined mostly, but not entirely, by latitude. Near the Equator it varies from 16,000 to 20,000 feet, in the Alps and Pyrenees about 8,000 to 10,000 feet, while in parts of Scandinavia, it is down below the 4,000-foot mark. Noting the skimpy 13-inch total precipitation in the Karasjok-Kirkenes area would cause some to question the quantity of snow. Since 1 inch of rainfall equals approximately 10 inches of snow, it would be possible for this region to accumulate about 11 feet. This, of course, would apply only in those places where the mercury remains below the freezing point—such as high in the mountains.

While there are roads and a railroad extending all of the way up to North Cape, most tourists will prefer the marvelous water route. In addition to the cruise ships that make several stops along the coast, perhaps a more exciting adventure is aboard one of the many spotless and pleasant passenger-mail-cargo boats which are operated by five Norwegian lines. They sail year round and may make as many as 30 stops in the 2,500-mile trip, lasting 11 days. The Bergen S.S. Co. has complete information on all these delightful, low-cost trips. Anyone who has enjoyed the Alaskan Inside Passage trip will have some idea of what to expect.

The Southern Coastal Section

This is the segment of Norway which overseas visitors know best—and not without reason. Fantastic chasmed fjords, charming spots such as Oslo and Bergen, and a breathtaking 3-day mountain trip between these two cities is a scenic experience which, even the morning after, is hard to believe. Little wonder that colorful posters of Norway occupy such prominence on travel agency walls around the world.

These southern coastal lands continue the marine climate with wet, not very low temperature winters, but almost constant moderate winds (which raise the chill factor). There is also the not infrequent violent sea storm. Summers are cool and, while the ample precipitation is distributed through the year, it is generally lightest during spring and early summer. That also is the period of greatest sunshine which may be anything from 5 to 7 hours per day.

There is a continuous maze of spectacular fjords from the seaport of Trondheim to Stavanger. Trondheim, Norway's third largest city, is on the Nid River—where it empties into the extensive Trondheimsfjord. Because it is back some miles from the ocean and in the lee of several small mountain chains, it gets somewhat less precipitation (34 inches) than most of the coastal towns in the southwest. Note on the City Weather Tables that temperatures continue to increase from north to south. The overall climate, particularly the precipitation pattern, is quite similar to the mid-New England coastal cities such as Boston. The combination of proximity to the sea and protection from the harsh northeastern winds by the high mountains to the east produces quite equable temperatures. Consequently, there are almost never sizzling hot spells or subzero weather although the winter winds can be biting. There are many gray and overcast skies and only May and June will average 7 hours of sunshine per day. The hordes of lemmings, which plunge to their death in the sea, choose this general area for their periodic migration. The sun-starved Norwegians, rushing southward to golden climes, jokingly compare themselves to these little creatures who gather in such numbers that railroad trains are, at times, compelled to halt. Trondheim is not Miami or Nice, but springtime and early summer are generally mild, quite bright, and very pleasant.

If ever there was a place that was greatly loved in spite of an almost continuously leaking sky, it must surely be Bergen, the second largest city in Norway and "capital of the fjord country." Imagine—79 inches of precipitation each year! With so much moisture, it is surprising that May and June (the bright season) manage about 6 hours of sunshine per day. That figure, of course, tapers off to less than 1 hour daily in December. There are certainly more gray than golden days. The usual score is about 145 gray days a year as against only 70 bright ones. That, however, does not keep Bergen from being almost as great a tourist attraction as Oslo itself. The familiar waterfront architecture of this once-important Hansiatic city forms a fitting backdrop for the popular marketplace. Little need to ask if fishing is of major importance. On a visit many years ago, my host amazed me by remarking that we would have fish for dinner if there were any fresh ones at the market. What he meant was, if there were any swimming in the tanks. Even that day's haul, just taken off the boats in baskets, was not designated as "fresh fish." Like Rome, Bergen is a city of seven hills, with a funicular railway that travels to the top of one hill and a cable car to another. There is so great a variety of interests within a small area, that it is easy to spend a week or more in the city and environs. We often refer to miniclimates in local areas which, sometimes, because of the topography, experience unusual conditions. Just to the east of Bergen, is Haukeland on the Masfjord. The wet air currents are funneled up the gorgelike space dropping an amazing 130 inches (almost 11 feet of water) a year. That is real "webfoot" country.

The busy community of Stavanger is both an important seaport and the hub of the Norwegian sardine industry. The recent discovery of huge gas and petroleum deposits in the North Sea between Norway and Scotland promises to transform Stavanger into a large industrial complex. The first step was the construction of refineries. It is to be hoped that the technically advanced but also very common-sensed Norwegians will control this major development and avoid the pollution conditions which all too often accompany such events.

Except for a lower precipitation (43 inches a year) and a bit more sunshine, the climate down on this southwestern tip is quite similar to that of Bergen. Even so, there will be some rainfall every third or fourth day during spring and summer. Overseas visitors quickly learn to disregard sprinkles and drizzle just as do the local residents. The relative scarcity of fog in fall and winter makes Sola, Stavanger's airport, an excellent gateway. With the probable industrial development, this area may become less attractive to tourists but, in the meantime, it is a very pleasant place to visit. Incidentally, just to keep the record straight, the word "sardine" has become pretty much a generic term. In Norway, they are called *bristlings* but the English call them *sprats*. Their finny cousins down toward Portugal way are *pilchards*. Almost anything that can squeeze into that little tin, except the tasty anchovy, can be labeled a sardine.

Every visitor to Stavanger makes the short excursion to stand on the "Pulpit." This is a sheer, 2,000-foot-high rock formation jutting out over the blue waters of Lyse Fjord. Those who can spare the time also journey from either Bergen or Stavanger to see the Jostedalsbreen Glacier. It covers 340 square miles and is the largest in Europe. Many remark on the deep-blue color common to glaciers. This is because this ice is very dense, due to having been formed under tremendous pressure.

If Norway could boast a Riviera, Kristiansand would be it. Although it averages a whopping 52 inches of precipitation, this section of coast manages to also find time for 1,750 hours of sunshine a year. May, June, July, and August normally average 7 or more hours a day. While there are plenty of winter nights when the mercury slumps below 32°, subzero temperatures are a rarity. Summers are cool by American standards, but the weather is generally pleasant. There are constant breezes which, in summer, are invigorating, but in winter, particularly when they reach about 25-mile-an-hour velocity, are more than nippy. There are fewer foggy days and overcast skies than on most parts of the coast. The usual equable temperatures are due primarily to a small, warm branch of the North Atlantic Current that finds its way into the Skagerrak, which separates Norway from Denmark.

This Kristiansand coast is not a Costa del Sol but it is a very popular summer resort area. There are small sand beaches, interspersed with rocky headlands and deep inlets. The water is especially clear along this stretch. This is a quiet, peaceful place dotted with small summer homes and agreeable hotels.

The capital city, Oslo, is the first goal of most visitors. If one were to see only the fantastically beautiful city hall, Vigeland's world-famous sculptures in the Frogner Park, and the original *Kon-Tiki* in its perfect setting, the trip from America would indeed be one to long remember.

Oslo is rather fortunate in its weather, being quite a bright spot with only moderate precipitation. Its 27-inch rainfall per year compares well with the 43-inch total of New York City. Note on the City Weather Table that, during spring and early summer, there will be more than 1/10-inch of rainfall only ½ day in 5 or 6. Actually, that much precipitation will not occur more than 75 days a year. Since there are about a dozen thunderstorms during summer, the showers may be heavy but brief, with rapid clearing.

The months of May, June, July, and August normally average about 7 hours of sunshine per day. That is also the season of long days, with June 21 getting almost 19 hours of daylight. There is plenty of time for sightseeing and other outdoor activities.

Oslo experiences a wider range of temperatures than the lower Atlantic Coast. There are, however, very few 90°- or subzero readings. About 50 days a year will get a snowfall, and there can be considerable fog in winter. Summers are about as cool as Quebec City, Canada.

Although summer is, by far, the most popular tourist season, visitors arrive at all times of the year. Unfortunately, practically none allow enough time in their itineraries to see all that Oslo has to offer.

The Interior of Norway

Mountains may be just mountains in some places, but not in Norway. The long, narrow ranges form the sheer, towering walls of the innumerable fjords—the likes of which are to be found nowhere else in the world.

Spectacular scenery is difficult to avoid in Norway but one of the most satisfying experiences is the several-day drive across the hump between Oslo and Bergen. A shorter, more direct, railroad trip can be made in about 8 hours. This is an excellent way for those on limited schedules to see some of inland Norway.

The car route heads north from Oslo along Mjösa, the largest lake in the country, and through some of the finest agricultural lands. Lillehammer at the upper end of the lake (740 feet above sea level) is at the beginning of the high mountain chains. The summer temperatures range between 45 and 75°. This is one of Norway's drier areas. Lillehammer averages only 25 inches of precipitation while Röros just to the north gets a meager 18-inch total.

The road follows the Guddrands River, which is bordered with 4,000- to 5,000-foot mountains to Otta. These are much more impressive than the heights might suggest since the snow line this far north can be below 4,000 feet and trees generally do not grow above about 3,500 feet altitude. The route then turns southwesterly toward Jostedalsbreen. Off to the left, the lofty summit of Galdhöpiggen (8,400 feet) stands out against the sky. There are many 6,000-foot peaks in this region. The road terminates at Karpanger on the mighty Sognefjord. This is a 109-mile body of beautiful blue water about 4,000 feet deep. A steamer across the Sognefjord and up the lesser Aurlandsfjord unloads at Sudvangen. This is only a stone's throw straight up to Stalheim, where every visitor should stay at least overnight. This is a spot to just stand and look. The next morning, the drive along the Hardangerfjord and through the Tokagjel Ravine leads into Bergen. There are other routes between Norway's two largest cities. This is perhaps the prize, since it includes the country's largest lake, the longest fjord, and the highest mountain—also Europe's largest glacier and scenic views galore.

Spitsbergen (Svalbard)

Anyone with a hankering for fermented trout and preferably inoculated against chilblains and goose pimples, can visit this little air-conditioned Norwegian outpost, 700 miles north of the Arctic Circle. Here, one looks southward toward Iceland to see the "Northern

Lights." The Norwegian name is Svalbard, but this tiny archipelago is more commonly called Spitsbergen. This group of five major and many small islands, seldom found in guidebooks, is almost unknown to travel agencies. It is, however, one of the few places in the free world that had more visitors at the beginning of the century than during this past year. The reason? Well, as the hopping-off spot for Arctic explorers, it was headline news and attracted the curious. But Spitsbergen was of so little world concern that it was just a spot on some maps and actually belonged to no one up until 1920. After World War I, the League of Nations presented it to Norway.

This strange, fascinating land is not a recent discovery. Coal mining started about 1900 and is still the principal occupation of the 1,000 Norwegians and 2,000 Russian residents. In this bleak, stark landscape, with 6 months of night, the people go outdoors on full-moonlit winter nights—just to enjoy the light. This long period of darkness raises other problems—at least for the mining commissar in Moscow, who felt compelled to cable: "More production of coal—and fewer babies, please."

In spite of the long period of darkness, this region gets more hours of daylight per year than is experienced in the Tropics. About 4,487 as against 4,420 hours per year at the Equator.

Most of the people live in three settlements—Longyearbyen, the little capital with several hundred; Barentsburg, the center of the Russian operations which is much more impressive; and Ny-Alesund, the most northerly settlement (not city) in the world.

Although Spitsbergen is far above the top of Alaska and Russia, it is possible to sail about 100 miles farther north. At that point, a towering wall of glacial ice, covered with pure white snow, forms an effective barrier against continuing to the North Pole just 600 miles beyond. In returning again to Spitsbergen from the north, a lucky traveler may be treated to a spectacle seen perhaps nowhere else on earth. If the sea is free of fog, seven glaciers will be spread in grand panorama before his eyes.

At times, this might be called the land of red snow. That happens when the surface is covered with red algae—perhaps akin to Florida's red tide. Although the ground is said to be frozen to a depth of 2,000 feet with a 3-foot deep thawed strip only along the southern edges, there is plant life in great variety. There are about 130 species, many of which cannot be found 800 miles south in Norway and some nowhere else in the world.

"America's finest export"—the Gulf Stream, makes all of that possible. Where this warm stream encounters the cold Polar Current at Bear

Island, the whole area is fogbound 340 days of the year. But with it all, Spitsbergen is a healthy place, quite free from bacteria, save only during the few summer months when visitors arrive with their colds and other ailments of civilization.

The Weather Table of Isfjorden, which is on the western coast, will give an idea of the climate in this region. Tourists will, of course, be concerned only with the weather conditions from June through August. Most tours arrange to have heavy parkas and other protective clothing for visitors. Precipitation is not usually a problem as the ½-inch per month occurs as light showers about once a week. Our remarks concerning foods of Norway do not necessarily apply in far-off Spitsbergen. A common joke told by the islanders is that their style of rakorret is made by filling a wooden tub with trout or other fish, leaving it under the bed until it ferments, and becoming so potent that either the householder or fish must leave. At that stage, it is considered a prime delicacy.

The Best Seasons to Visit Norway

By far, the greatest number of visitors arrive during summer. That pattern has been slowly changing. Europeans, in particular, are finding many pleasant reasons for going to Norway at all seasons. The imaginative SAS Airlines and the Norwegian National Travel Offices at 505 Fifth Avenue in New York City and 612 South Flower Street in Los Angeles have developed many tempting tours, special events, and combination package arrangements. Since marine passenger travel has almost disappeared in the United States, overseas visitors often overlook the ferry and ship services in Europe. Norwegian ports have service to England, Germany, Denmark, and The Netherlands as well as throughout Norway. Norwegian travel offices will have information about boat travel.

Norway is summer country; every section of the land can be enjoyed during that all too short period. While the days are generally bright and pleasant, many overseas visitors are surprised to learn that summer is not quite as sunny as May and June. The Oslo area and the coast around to Stravanger average 6 to 8 hours of sunshine per day at this season. The northern coastal segment will get somewhat less.

It will rain about 1 day in 4, but since there are three to six thunderstorms per month, many of the showers will be heavy but brief. The 65 to 85% relative humidity is not a problem as really hot weather is virtually unknown and there is an almost constant light breeze. Normally, the thermometer will reach a daytime high of about 75°, which, at night, drops to the mid-50's. This makes very comfortable sightseeing weather. Since the sea water even at Kristiansand does not get up to 65° until August, few will be tempted to try the surf.

This is also a prime time for seeing the fjords. The farther north, the less sunshine and the lower the temperatures. Precipitation also decreases between Bergen and North Cape. Tourists who like to crowd in a lot of sightseeing get a big bonus. The daylight hours increase from about 18 in the south up to Spitsbergen where the sun does not drop below the horizon for about 3 months. This is rather misty country— or, as an English friend remarked, "Isn't this lovely gray weather?"

The long Sulitjelma Mountain interior is very pleasant in summer. This is wonderful hiking, fishing, or just plain loafing country. The air is clear and invigorating, but there is generally somewhat less sunshine than toward the coast.

Almost all of coastal Norway gets heavier rainfall in September and October than during spring and summer. There is also less sunshine and more cloudiness. Occasionally, there will be an unusually fine autumn season with bright days and foliage that can rival the most flamboyant display of New England or Canada.

The year 1972 was such a one when almost all of Western Europe experienced most disagreeable, cold, wet weather during spring and early summer. But a glorious autumn season more than evened the score. This climatic happening was attributed to a great flow of small icebergs—which sometimes separate from the Arctic icefields and drift slowly southward. Although coastal temperatures do not appear at all severe, the fall air is chilly and, when coupled with a brisk ocean wind, can be mighty uncomfortable.

The interior mountain country is more agreeable at this time of year. The air is inclined to be still, clear, and perhaps a bit brisk. Note on the City Weather Tables of Röros that precipitation usually tapers off somewhat from the summer figures. This is also true of Oslo.

The long winter period generally suggests two things—snow and skiing. These hardy Norsemen were not born with skis anymore than Londoners arrive with webbed feet. Anyone who has witnessed "Ski Day" during early March in Norway, however, might well wonder. Over 900 ski clubs throughout the country participate in the outdoor activities which include everything from highly skilled precision tests to the more usual ski races, competitions, and family events.

A recent trend that might appeal to many overseas outdoor people is the emphasis being put on cross-country skiing as against the Alpine type. This has become a very popular club, group, and family pastime and will, no doubt, be taken up in many parts of the world. After finding out how the Norwegians do it, why not enjoy a little one-upmanship by being the first to organize a cross-country jaunt at home? Oslo, too, is becoming a more popular winter spot. Some visitors do nothing but ski, others everything but ski—a combination can be a perfect holiday.

April, May, and June is the time of greatest sunshine and least rainfall. And remember, June 21, the longest day of the year, never seems to end. The most avid sightseer must get leg-weary before sun finally sets after being above the horizon from 18 to 24 hours. The 45 to 65°-temperatures call for wool suits and a top coat, but the air is clear and conditions fine for walking and touring. The lower foothills are generally very pleasant; the higher sections may be blocked with snow or disagreeable underfoot. Flowers seem to be everywhere—in the fields, the marketplaces, and at least a few sticking out of every housewife's shopping bag or basket.

There is a slogan: "Come to Norway any season, but vacation in springtime." Another thing not to overlook: Vacation schedules in Norway are very generous and mostly taken in summer—particularly July and August. That means many people hurrying to the same places at the same time. Except during special occasions, springtime is almost never as crowded.

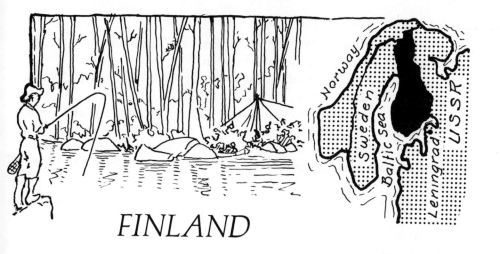

FINLAND

HISTORIANS tell us that centuries ago tribes from east of the Ural Mountains trekked to mid-Europe where they split into two groups. One veered southward to Hungary while the other ended up in Finland. Perhaps so. At any rate, the Finnish language is also said to be quite similar to the Estonian, just across the Gulf of Finland, but only remotely related to Hungarian.

In the world of current tourism, Finland is grouped with the Scandinavian countries and indeed displays many of the common traits that delight every visitor. Being gracious, honest, and industrious, with mania for cleanliness, the Finns are also a very intelligent and artistic people. Couple all of that with a wonderful northland setting, and the result is a combination which well justifies a detour off the customary beaten paths.

Many tourists are getting away from the practice of trying to cover Europe in a week. They find that spending the entire vacation in one country or area is more enjoyable, less tiring, and, in most cases, easier on the purse. The latter is particularly true if the time spent in the larger cities is carefully limited. It also helps if one can find areas where the off-season weather is as good or better than during the more popular tourists' periods.

Topography

From the viewpoint of the potential visitor, the 30,000 islands, 60,000 lakes, and 1,450 miles of coast line must sound like a pretty good start for a place to vacation. Those, of course, are not all the tourist attractions that Finland has to offer. It is possible to squeeze quite a variety of scenery, places, and conditions into a country 340 miles wide and 750 from north to south. While it is just about the size of the British Isles, it has a population of less than London.

About one-third of Finland lies above the Arctic Circle, which qualifies it as a "Land of the Midnight Sun."

The country can be divided into four large topographical zones: the low coastal belts, the large lake region in the southeast which is a plateau, northern Finland which is densely forested uplands and heathered moors, and a high portion along the Norwegian border. Mt. Haltia (4,344 feet) is in this area.

About 9% of the country, some 130,000 square miles, is occupied by inland lakes and waterways, while only about an equal area is cultivated farms and pasturelands. More than half of Finland, mostly in the north, is covered by pine and various evergreens, along with a sprinkling of willow, birch, and other deciduous trees.

Those who enjoy boat trips will be pleased to know that they can sail for 2,700 miles without retracing a single foot in Finland's lake region, which is the most extensive in Europe.

Climate

There is a greater difference in temperatures than precipitation throughout the year. In the northern portion of the country, the annual precipitation is very low, being in the 18-inch range. Most of this falls as snow (and it should be remembered that 1 inch of precipitation equals about 10 inches of snow), so there is no lack of good skiing. The snow remains on the ground continuously from the end of October through the middle of May in Ivalo and the northern portion of the country.

Precipitation increases toward the south but rarely averages more than 24 inches. Over the many years that statistics have been compiled, the maximum recorded was a bit over 29 inches which is just two-thirds as much as New York City gets any normal year. During the driest year on record, Finland had only 16½ inches of precipitation. The annual precipitation does not vary nearly as widely as is the case in most other European countries.

There is a wider spread of temperatures, particularly during winter. In the far north, at that time, the mercury will range between 22 and 39°, which is about 10° colder than in the south. Because there are no pronounced topographical variations in the terrain, the differences in weather conditions are graduated quite uniformly from top to bottom of the country. For the same reasons, there are relatively few localized miniclimates such as we have noted so often throughout the book. As might be expected, there is a much greater spread between summer and winter temperatures in such inland places as Sodankylä, Kuusamo, and

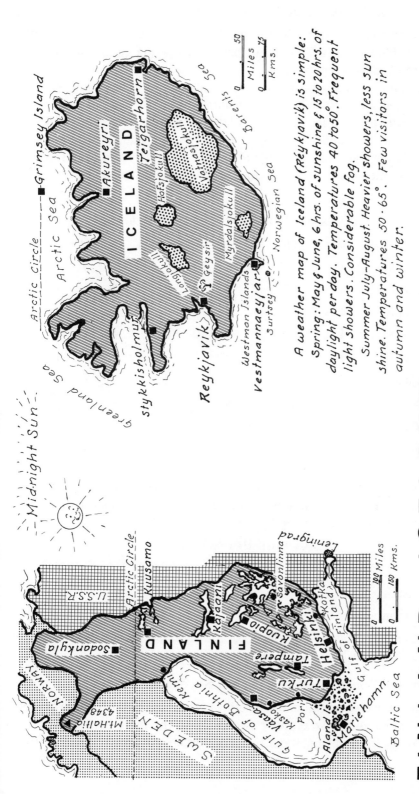

A weather map of Iceland (Réykjavík) is simple: Spring: May & June, 6 hrs. of sunshine & 15 to 20 hrs. of daylight per day. Temperatures 40 to 50°. Frequent light showers. Considerable fog.

Summer July-August. Heavier showers, less sunshine. Temperatures 50 - 65°. Few visitors in autumn and winter.

Chart 40

FINLAND ICELAND

Kuopio than in cities directly on the coast. The prevailing westerly winds blowing in off the ocean have a tempering effect and produce more equable temperatures.

Because so many people associate sunshine with vacations, we have indicated the average hours of sunshine during the various seasons on the vacation weather maps. Finland is often pictured as one of those gray, northern lands. Although there are normally about twice as many cloudy days as clear ones, the abundant sunshine from May through August may come as a pleasant surprise to many planning a visit to Finland.

AVERAGE HOURS OF SUNSHINE PER DAY

	J	F	M	A	M	J	J	A	S	O	N	D
Helsinki	1	2	5	7	9	10	10	8	5	3	1	1
Turku	1	3	6	7	8	10	9	7	5	3	1	1
Jokioinen	1	2	5	6	8	10	8	7	5	2	1	1
Jyväskvlä	1	2	5	6	7	10	9	7	4	2	1	—
Vaasa	1	3	6	7	9	11	10	7	5	3	1	1
Kajaani	1	2	5	7	8	10	9	7	4	2	1	—
Kemi	1	2	5	7	8	9	10	7	4	2	1	—
Sodankýlä	—	2	5	7	8	10	9	6	4	2	1	—
Ivalo	—	1	5	6	7	8	8	5	3	2	—	—

The Finnish chambers of commerce often explain that during some periods of the year, Helsinki will average more sunshine than London, and present the following figures:

AVERAGE HOURS OF SUNSHINE

	June	July	August
Helsinki	330	250	230
London	210	200	190

Quite understandably, they do not highlight the October through February sunshine record in Finland.

Average Temperatures of Lake and Sea Water

Most overseas visitors will be quite content to use the bathtub or even the sauna and leave outdoor swimming to the more hardy Finns.

While the following water temperatures are not greatly different from those along the North Sea and around the British Isles, they are not too tempting to most visitors.

AVERAGE WATER TEMPERATURES

	June	*July*	*August*
South Seacoast	55°	62°	63°
Southern Inland Lakes	59°	66°	63°

HELSINKI

To say that the capital of Finland is a seaport is an understatement. There are miles of waterfront and dozens of islands connected by bridges, causeways, and ferries. Almost 30% of this exciting capital city is landscaped. There are 240 parks in addition to the Garden City of Tapiola, a planned suburb of lakes, fountains, and manicured green belts only 6 miles from the heart of Helsinki. There really is a feeling that, "spring is bursting out all over"—when all of the blossoms, flowering shrubs, and budding trees wake up in May and June. It takes a second thought to remember that, next only after Iceland and Norway, this is the most northerly of all countries in Europe.

Perhaps June is the prize weather month of the year. Here are some statistics for those prime 30 days:

JUNE WEATHER IN FINLAND

Sunshine	Average 10 hours per day
Daylight	About 17 hours per day
Temperatures	Day, 64°, night, 50°
Rain	A low 2 inches for the month
Showers	Only one per week
Thunderstorms	About two this month
Fog	About 2 days this month

It should also be remembered that June 21 is the longest day of the year. The sun rises at about 3 A.M. and does not set until almost 10 P.M. May and June are generally off-season for both Finns and foreign tourists. Accommodations are more readily available and the city is less crowded. A note of caution: Helsinki has spring and autumn festivals featuring opera, ballet, and classical and jazz concerts. It is always well to check with your travel agent regarding such events since the most desirable hotel rooms may be rather scarce for the limited period.

Many visitors from America look forward to roaming the old, and often even medieval, sections of the European cities. That is less of an attraction in Helsinki which, however, is a very great treat for those interested in distinguished, and sometimes quite daring, examples of contemporary architecture.

By and large, the older buildings have rather drab, gray façades although, in many cases, they house some of the most attractive and smartest shops in town. Part of the reason for this may be that Helsinki was relocated on its present site in 1640 but did not attain prestige and importance until it became the nation's capital in 1812. Most of the growth has taken place since Finland gained its independence from Russia in 1917. One colorful old part of the city, which attracts almost everyone, is the Market Square. The fish stalls and flower and vegetable stands make this a lively meeting place of brilliant colors. Every visitor should also set aside a few hours to enjoy the Finnish Design Center, where almost all of Finland's remarkable arts and crafts are on display. No one is in Helsinki very long before he gets to know Stockman's, the Macy's or Selfridge's of this city. It is easy to find anything that you forgot to pack and many more that you did not.

Visitors are not so apt to make Helsinki their headquarters while visiting nearby places of interest as so many do in Copenhagen. Costs in Helsinki, just as in most major Scandinavian cities, are rather high but well worthwhile for several days of local sightseeing. Many find it more satisfactory to spend most of the stay in the fantastic lake region, the seashore and coastal islands, and the smaller and perhaps more colorful towns.

Tour to Turku, Vaasa, and Southwest Coast

A pleasant drive in the spring is along Route No. 8 to Turku (150,000 population), the oldest town in Finland with many ancient structures such as the fortress castle built in 1260 and the cathedral of about the same vintage. The Aura River runs through the middle of this busy seaport making it an interesting stopover while seeing the area. Vaasa, farther up the coast, with some of the old city walls, is also worth a visit. The City Weather Tables of these two cities show that there is very little difference between springtime weather here and Helsinki. The day and night temperatures will be about 60 and 50°, although there will usually be a half-dozen nights in Turku when the mercury will touch the freezing point and, perhaps, twice as many in Vaasa.

It may also be a bit breezy, but note that Turku averages 8 and 10 hours of sunshine a day in May and June, while Vaasa enjoys a whopping 9 and 11. That leaves very little time for rain and sometimes this area even experiences a spring drought. This whole district gets only 1½ to 2 inches of rainfall per month in May and June. New York City and Chicago normally accumulate almost exactly twice that rainfall during this season. The trip makes a pleasant 5- or 6-day jaunt and can be completed back to Helsinki via Tampere. This is the site of the unique and beautiful Pyynikki summer theater. It is an open-air auditorium in which the entire seated audience revolves, so that they successively face the various stages; three are usually used. In many cases, the natural settings, along with some props and furniture, serve as the stage set.

Tour Along Baltic Coast to USSR Border

Another fine tour is in the opposite direction along the Baltic shore to Borgå (Porvoo) with its many rows of attractive, old wooden homes and quaint streets that make it seem like a world quite apart from modern Helsinki, only a short distance to the west. Kotka is a Baltic port quite close to the Russian border and has much the flavor of its giant neighbor. One of the tourist sights is the fishing cottage used by Czar Alexander III—in a delightful natural setting. There are also several onion-topped Orthodox churches in the area. Actually this is about as close to Leningrad as Helsinki. Unlike the other Scandinavian countries, *vodka* is a more common drink than aquavit.

The Southeastern Lake Region of Finland

Just to the north is the largest lake region in Europe. Lappeenranta, called the capital or gateway into this fabulous country of 60,000 blue mirrors, is a fine stopping place. This low plateau area is generally about 300 feet above sea level. During May and June, there will be a light shower not more than once a week.

There are numerous boat trips through the various waterway chains but that to Savonlinna is one of the most attractive. This town is in the center of an area where there are lakes to every point of the compass. Here is a gloriously peaceful marine land where foot-weary tourists can escape the usual chore of cathedral and museum collecting. Those addicted to fishing have to be deported or shanghaied out of this trout-infested region.

There are many other land and water trips out of Helsinki. A popular one is northeastward up Highway No. 5 to the scenic lake and forest country around Kuopio. A quick glance at the City Weather Table shows that the familiar pleasant May-June weather patterns prevail. This 50,000-population city is also a center from which roads and waterways radiate in all directions to more peaceful nature spots.

Lappi Province

Only about 1,500 Lapps now live in Finland. They are no longer true nomads but do follow their herds of reindeer, numbering some 100,000, as dictated by grazing conditions.

Being well above the Arctic Circle, there is practically continuous daylight during spring and early summer in Lappi. Because of the almost round-the-clock activity, it is easy to believe the common saying, "We save our sleep for winter." Note on the Sodankýlä City Weather Table that weather conditions are generally very agreeable. Even though August usually gets about twice the rainfall of May and the skies are often cloudy, the days are not all gray. July averages a very respectable 9 hours of sunshine per day, but, of course, August is lucky to get about 6. This is among the last of the large, unspoiled, natural wilderness regions remaining in Europe. It is marvelous outdoor country: a fisherman's delight, a naturalist's dream, and a darned good place for any vacationist to loaf away a most enjoyable several weeks. Incidentally, costs are considerably lower than in Helsinki and other large European cities.

Winter can provide some unusual attractions in Lappi. The adventurous tourist can live in a Lapp cabin and go to the reindeer school. A brief training period qualifies the novice to use one of these versatile animals as the power tool for cross-country skiing. There is a long skiing season in this part of Finland. The snow comes to stay at about the end of October and does not leave until the middle of May. The best skiing, however, is in the March-April period.

The lucky visitor may also see the annual reindeer roundup and branding which is an exciting event. The Lapp herders are often dressed in brilliantly colored native costumes, and the whole family participates in this important activity.

Another important Lapp pursuit is fishing through the ice, a few hours of which makes one understand why the fiery aquavit and vodka originated in these northern climes. Years ago, fishing through a hole on frozen Lake Erie was a popular sport but that was before pollution. So far, Finland is almost free of serious man-made air and water contamination, but there have been some signs that cement plants and other heavy industrial developments may soon dent the fine record. The progressive Finnish government, however, will no doubt do everything possible to avoid hazardous conditions.

The Åland Archipelago

Summer is the heyday of the 30,000 islands clustered along the coasts of Finland. The little Åland archipelago off Turku alone comprises 6,500 of these land specks, 80 of which are inhabited. Mariehamn is probably the most popular summer place in this island group. If the normal 55 to 70°-day-temperatures and 40 to 50°-night-figures are not a deterrent, this could be a delightful vacation spot. All of the other weather conditions are almost ideal: ample sunshine, lack of rain, and gloriously long days.

Rent an Island in Finland

A first, for even the most experienced traveler, is the opportunity to rent a whole island complete with a nicely furnished cottage by the week just for your own use. Imagine—king and ruler of an entire body of land, even if only for a short reign! Ask your travel agent about the Finnish "Rent an Island Plan." The more gregarious, but also those with an eye to costs, might like to try one of the 70 Holiday Villages. Incidentally, they range from very simple to almost luxurious. This, of course, is a very pleasant way to really get to know and enjoy the Finnish people. Those familiar with the language might even like to try a vacation at a farmhouse. The Finnish government has also helped to sponsor a dozen such tourist programs.

Boat Trips Around Finland

There are all kinds of water trips to various parts of Finland as well as neighboring countries, including the U.S.S.R. A word of caution before booking sleep-aboard boat trips. Almost without exception, they are clean and safe, but while many are bright, spanking new and most comfortable, there are other ancient veterans that leave something to be desired.

The Sauna

Although indulged in throughout the year, summer is the most prudent time of the year to try that national Finnish pastime, the *sauna.* Like the huge, often communal hot tubs in Japan, no visitor to Finland would dare omit this ritual. Finns explain that when taken in midwinter, one can hop out of the 200°-steamchamber and roll in the snow, producing a very satisfying effect. We prefer to enjoy that experience vicariously—from a travel folder.

It is not necessary to go looking for a sauna; they are as common as bathtubs. Some of the municipally run saunas, however, close for a period during the summer.

Best Seasons to Visit Finland

Anyone who has once been to Finland would jump at the offer of a return visit at any season of the year. It would be even more enjoyable, however, if there were some choice regarding time and weather. While Finland is a fairly low, uniformly flat, country, there are most favorable times for visiting the various segments.

It seems repetitious to continue pointing out the glories of the spring season throughout many parts of Western Europe. The City Weather Tables, however, show that there is almost always a dramatic jump in average temperatures between April and May in practically all of Finland. That is the time when the chill of winter gives way to the warming spring sunshine. The increase in temperature usually amounts to about 10° but in some sections is even greater; there will be a substantial rise of 13° or more in the Helsinki district.

Spring is generally a delightful time to tour the southern half of the country. (Many will find the northern parts of Finland still too cool for comfort.) There will also be considerable patches of residual snow especially in the more elevated country along the Norweigian border.

Finns love outdoor life and the annual summer vacation, most often taken in July, is usually quite rigorous and healthful. Since many foreign visitors follow the same time pattern, July and August are busy months in almost all parts of the country. It is, however, the best time to visit the far northland of pines, heather, and clean fresh air. Many Finns agree that Lappi Province is the place to enjoy the summer vacation and, despite its popularity, will be far less crowded than the southern parts of the country.

Autumn is the shady time of year. September will average about 5 hours of sunshine per day in the southern part of Finland, but only 2 or 3 hours in the north. Precipitation is apt to be a trifle lower than during July and August, but there is also a marked drop in temperatures. On the whole, normal Finnish October would just about fit the American visitor's description of winter. Finland is occasionally blessed with a glorious autumn with beautifully colored foliage and more than usual sunshine, but those are odds that you would hardly bet on at the poker table. Helsinki is a place where on can keep occupied at any time of the year, but it is certainly more comfortable in spring and summer than this late in the year.

Winter is a long, gray season and, except for the unusual events in Lappi Province, not the most inviting time to visit Finland. Temperatures in Helsinki are about the same as northern New England—they do not often drop below about 15°—but the days are very short and the sunshine is hardly worth talking about. The ground is covered with snow by the middle of December and patches will still be there well into April. The cultural activities and restaurants are at a peak, but even the chamber of commerce can not make much of a case for this season in Helsinki.

Americans almost invariably enjoy their visits to Finland. Even those who have never been nor ever expect to go have kind feelings toward the Finnish people. Many know very little about the country itself but remember or have heard of its Olympic records and the name Paavo Nurmi. Some recall with respect and admiration that, in spite of a strained postwar budget, this brave little nation paid back every last *markkaa* of its international loans right on schedule. Of all the Finnish characteristics the three which seem stand out are cleanliness, honesty, and physical fitness.

This is a country of surprising contrasts. Finland gave women equal status in the polling booths before others even thought about it. It is also rated as having the highest literacy rate in Europe and perhaps the world. On the other hand, one sees windmills, water wheels, and eighteenth-century rural life patterns in parts of the country. Few nations have had more impact on contemporary architecture or the decorative arts. All of these are bits and pieces that, collectively, make this one of the most enjoyable and interesting countries in Europe in which to spend a vacation.

ICELAND

MANY Americans, who previously had only a vague idea of Iceland's whereabouts, now often associate it with that far northern region where the championship chess games garnered so much free publicity. Of course, the word "Iceland" is only about 50% descriptive. The natives jest that the title is the coldest thing about the country. The ancient name for this unusual island was "Land of Fire," which is also only half accurate.

Actually, it is possible to sit on the edge of a gigantic ice field, equal to the combined areas of Rhode Island and Delaware, and almost spit into a huge natural pool of boiling spring water. But is this little island nation really that far north or that frigid? The Arctic Circle passes through Rifstangi, its topmost point, but large segments of its Scandinavian cousins, Norway, Sweden, and Finland, lie well above that imaginary cold line.

Iceland is the westernmost country of Europe, only 200 miles from Greenland. Because of its relative isolation, Iceland is often not shown on maps of Europe, which does not seem to bother these independent people much. It sits out in the mid-North Atlantic, 645 miles west of Norway and 520 northwest of Scotland. Although often blanketed in mist or fog, it is easy to find and was used by old Norsemen as a handy stepping stone to America.

Topography

Iceland is 39,702 square miles of volcanoes, hot springs, stark, crater-pocked lava beds, immense glaciers, desert areas, and vast, glistening snowfields. Much of the island is also green with wide areas of ground moss, lichens, and considerable pasture land, but only about 1% of the country is under cultivation.

254

Iceland consists mostly of plateau lands averaging 2,000 to 2,500 feet above sea level. Mt. Hvannadalshnukur, at 6,952 feet, near the southeastern coast, is the highest peak. There are almost continuous mountain chains extending from northeast to southwest across the island. They contain a few other 6,000-footers but more are in the 3,000 to 5,000-foot range.

The only lowlands are along the shores of Faxa Bay, the alluvial plains of the Hvita and Thjorsa river valleys and the sandy flats of the southeastern coast. The 125-mile Thjorsa River is the longest in the country. Many of the streams cut deep channel beds, sometimes flowing through subterranean tunnels or tumbling over waterfalls on the way to the deep outlets into the sea.

The 3,700 miles of shore line are generally high and rocky with deep fjord inlets and large bays. Much of the coast descends to sea level in sharp, terraced steps, which make spectacular high-rise apartments for a multitude of marine birdlife.

There are also some unusual features: the 107 volcanoes, several of which are active. Some of the busy ones are below the icecaps in the great snowfields, and when they occasionally "steam up," the melted ice can cause serious floods.

Major volcanic eruptions were the 4,900-foot Mt. Hekla in 1949, and, in 1963 off the southwestern coast, the ocean bottom spurted upward, creating the new island of Surtsey.

The most catastrophic volcanic disaster in modern Icelandic history occurred on Heymaey Island early in 1973. This three-mile-long island headquarters of the country's important fishing industry is about midway between Surtsey and the main shore line. Vestmannaeyjar, its capital city of 5,000 inhabitants, had to be evacuated. The flow of molten lava into the sea raised the water temperature above 110°, making fire fighting and even rescue work very difficult. Thankfully, there was no loss of life due directly to the eruption. Because the export of fish is Iceland's most important industry, this unfortunate occurrence will seriously affect the economy of the country for many years. But be assured, these brave people will take it all in stride.

Then there are the glaciers. Vatnajökull, the largest in Europe, covers 3,000 square miles. Two smaller glaciers, Langjökull and Hofsjökull— only 500 square miles each—are in the central portion of the country. The only other major ice field, Myrdalsjökull, is in the extreme south. Glaciers of one kind or another occupy 11.5% of Iceland's total surface.

We should, of course, mention the "Great Geysir" which gave that name to all others in the world. Some unkind fellow nicknamed it "Old Unfaithful," since, unlike its namesake in Yellowstone Park, the Great

Geysir is a mite tempermental and spouts only on whim. When it does, however, after a coaxing dose of yellow soap, it fountains a majestic blast of boiling water to a height of 150 to 200 feet in the air. There are other strange and interesting things that we will mention in the following little weather story of Iceland.

Climate

In some respects, the weather is almost as varied as the physical features of the country. Iceland is situated on the boundary between the Temperate and Arctic Zones in a cyclonic region. It is buffeted between the cold, sharp blasts from the north or northeast and the moist moderating winds from the south or southwest. This may sound like a battleground of climates—and indeed, at times it is just that.

These air movements are closely related to the waters surrounding the island. The frigid East Greenland Current surges in around the north and northeastern coasts. But Iceland's lifesaver is the warm Irminger Current that flows along the south and southwest and a bit around the northwestern corner. That benevolent stream, that keeps this island from being an almost perpetual ice cube, is a small branch of the North Atlantic Drift which, in turn, is a direct descendant of our own balmy Gulf Stream.

Warm and cold air masses are constantly striving for mastery. When they meet head-on, there are often strong, gale force winds. When more passive, the hot-cold contact can produce the inevitable fog, mist, or overcast skies. The southern portions of the island are usually more windy than either the north or inland areas.

Lest we alarm prospective visitors, we hasten to make mention of just a few of Iceland's many agreeable attractions. Where else these days would one dare drink freely from any spring, stream, or lake? Scenically, the beauty and grandeur of the immense glaciers and snowfields are intensified by the clear, pure air. Because contamination is still just about nonexistent, the sun is hotter and really sparkles on the glacial ice and snow. In spite of a less than perfect climate, this is an outdoor country where swimming in pools of naturally warm water is popular even in midwinter. At times, the atmosphere is so clear that the snow-peaked Snaefellsjökull can be seen from Reykjavík 75 miles away.

One advantage of being so near the top of the globe is that the days during late spring and early summer are wonderfully long. On June 21, the longest day of the year, 17 to 20 hours of the 24 will be daylight throughout Iceland.

The following table shows the average hours of sunshine per day and year in the northern and southern parts of the island.

	J	F	M	A	M	J	J	A	S	O	N	D	Year
Reykjavík	1	2	3	5	6	6	6	5	4	2	1	—	1,249
Akureyri	—	1	2	4	6	6	5	4	2	2	—	0	962

Reykjavík

Every visitor will probably start and end his visit in this most northerly capital city of the world. Almost half of the 190,000 total population of the country lives in Reykjavík. Although this is an ancient place, said to have been settled in 874 by a Norse chieftain, there is no "old city" district of narrow, winding streets. There are practically no really old structures of any kind. Neither are there slum areas or run-down and abandoned districts. It is made up mostly of single-family homes, some high-rise apartments, and modern commercial and government buildings.

The city is built around three sides of an attractive lake; the smaller of two airports borders on the fourth side. The lake is only a few hundred yards from the harbor on Faxa Bay. Most of the public buildings, embassies, hotels, and the Tourist Information Bureau are in the space between these bodies of water.

Reykjavík is certainly one of the cleanest cities in the world. There is almost no smoke, litterbugs are unknown, and (U.S.A. and others, please note) dogs are not permitted on the city streets. Perhaps the only thing an ecology purist might sniff at would be the odor of fish. But let us remember that this is truly the home of the herring and cod.

Taxes are very high, but these people, nevertheless, enjoy a very good standard of living. This is one of the most literate nations in the world; Icelanders are said to buy more books per capita than any other national group. The Finns think that they may hold that honor, but in both countries, there is a book shop on almost every other street. The Icelandic language has been zealously and jealously kept untainted by foreign additions or slang expressions. As a result, present-day Icelandic is closer to the pure, old Nordic than any of the other Scandinavian tongues.

There are several interesting sights that can be seen on short excursions out of Reykjavík. The famous Great Geysir (or Geyser), already mentioned, is only 70 miles to the east. Gullpass (Golden Falls), with a 100-foot drop, can be included in a 1-day jaunt. This is a very impressive sight. The river has cut a wide gorge and the high, luminous clouds of mist, which give the falls its name, make it a sight worth visiting.

At Thingvellir, only 35 miles from the capital, is the site of the ancient parliament called the *Althing*. It was established in 930 as the central legislative and judicial assembly for the entire country. At that time, most of Europe was made up of small tribal or local factions.

Central Iceland

Everyone will certainly want to take a peek at the unique 1,422-square-mile Odadahraun lava field in central Iceland. The Askja crater alone covers 34 square miles. Perhaps the only other place to see such a stark, pock-marked, volcanic wasteland, aside from a jaunt to the moon, is on the island of Lanzarote in the Canaries.

Most will also be greatly impressed by the scenic beauty in flying over the 3,000-square-mile Vatnajökull, the largest glacier in Europe. These trips can best be taken in July and August although May and June are also quite favorable months.

Northern Iceland

Akureyri, the second largest community in Iceland, with a population of 10,000, is the most logical headquarters while in the north. It is one of the important herring grounds in Europe and has the largest herring-processing plant in the world. Akureyri is also a popular vacation area for the outdoor Icelanders who hike, fish, and swim in the warm waters of the open-air pools. Incidentally, few visitors to Iceland display the slightest interest in ocean swimming. The 36° January surf temperature zooms to a sizzling 53° by mid-July—and that only down on the warm southern seacoast. Akureyri is also becoming an important resort and winter activities' center. The best skiing is March through May but may even extend into June. It is not the daring Alps sport but rather considerable cross-country skiing because of the more modest slopes. Because of minimal winter daylight and sunshine, this will probably never become an international attraction.

Kerlingarfjoll, farther south but at over 3,000-foot elevation, has a year-round snow cover. July and August temperatures are comfortable, while days are bright and sparkling. With suitable facilities, this could easily become a favored summertime mecca for skiing buffs from all over the world.

While this far north, some visitors hop over to the interesting little Isle of Grímsey with its 70 inhabitants. Since this earth spot is on the Arctic Circle, every vistor steps across that magic line to earn the polar certificate. The City Weather Tables of Akureyri and Grímsey show just how meager precipitation is in summer in these northern regions. Rainfall averages only 1 to 2 inches in July and August, but the equally scant 2 to 4 hours of sunshine per day illustrates how close at hand the Iceland winter is.

Summer is the time of year to explore scenic areas in the north aboard a gentle, sure-footed Viking horse. Many beauty spots can be reached in no other manner. Perhaps "pony" is the best designation since the handsome animals rarely stand more than 13 hands high. These patient little fellows, who can tote the lardiest tourist the whole day, look very much like the magnificent Norwegian pony. Because pony trekking is absolutely safe for even the complete novice, this very comfortable form of transportation can be a most enjoyable experience.

Westman Islands

There are several interesting sights down at the lower tip of Iceland which make the 70-mile drive from Reykjavik worthwhile. About 7 miles off the southern coast is the tiny archipelago of Vestmannaeyjar, more commonly called Westman Islands. This latter name may have been derived from the early Celtic settlers, called Westmen, who arrived from Ireland about 800 A.D. Now the chief inhabitants are about 5,000 Icelanders and what appear to be millions of puffins, petrel, quillemot, kittiwake, the rare gannet, and many other of their feathered friends. These 15 tiny land specks, the most southerly habitation in Iceland, remain green through much of the winter, owing to a comparatively mild climate that results from the Gulf Stream. Surtsey, a newcomer in this group, was born on the morning of November 14, 1963, when it was heaved up from the ocean floor. Heymaey Island was the scene of volcanic disaster 10 years later. Iceland is located over a great crack or fault in the ocean bottom which explains the volcanic activity on the island. The Westman Islands, only 20 minutes by Iceland Airways from Reykjavík, make a fine 1-day excursion. Tote your camera but also a raincoat. The 55 inches of precipitation is distributed throughout the year but even May and June get a goodly share with some mist or sprinkles about every other day.

Natural Hot Spring Heating in Iceland

Many visitors will be interested in visiting some of the greenhouses where flowers, tomatoes, cucumbers, and other produce are grown on a large scale. This form of agriculture is not uncommon and such places as Guernsey, in the Channel Islands, have extensive installations. The unusual feature in Iceland is that the heating is done by circulating water from the natural thermal springs. The best place to see this operation is in the little town of Hveragerdi.

Practically all the homes and public buildings are also heated by this hot water, which is collected in large reservoirs and pumped through a network of distribution piping. This, along with the generation of electric energy by water power, are part, but certainly not all, the reason why Reykjavík deserves the title of "clean city."

Eating in Iceland

Foods in Iceland are generally quite simple. Most overseas visitors will love the wonderful fresh fish but not the equally popular smoked lamb and mutton. One rare delicacy, perhaps found no other place but Norway, is occasionally available for the alert epicure who spots *rjupur* on a menu. This is a grouselike bird (we know it as the ptarmigan) that turns white in winter and grows feathers on its feet to walk on top of the snow. In a cream sauce, it is a dish fit for a king. If you really want to feel like royalty, start the feast with freshly caught fish and top it off with the delicately flavored Arctic cloudberries. Because so many foodstuffs must be imported, costs are about 25% higher than on the Continent. That does not apply to commodities grown or produced on the island. Nor need potential visitors be too concerned about hotel rates. It is also a great pleasure to be able to drink the water and eat any food offered without fear of inner repercussions.

The Best Seasons to Visit Iceland

When to visit? The answer is spring, summer, and, perhaps, winter, depending upon whether the purpose is sightseeing or skiing. Our guess is that there will be a continuous increase of visitors during the first two periods but suggest that there are more attractive places to spend the cold months.

May and June are the most favorable months from the viewpoint of rainfall. Reykjavík, which gets 34 inches annually, averages only 1.6 and 1.7 inches in May and June. It is important to note, however, that there is a little precipitation—often just light mist—more than half the days of each month. That calls for a light raincoat, not hip boots, but it suggests rather gray skies. Actually, there are usually not more than 3 or 4 completely clear days in this 2-month period. The remainder are divided about equally between cloudy and partly so. There will also be a few foggy days.

With all that mist and overcast weather, it is surprising that Reykjavík will enjoy an average of at least 6 hours of sunshine per day all through May and June.

The day to night temperatures of 55° to 40° are fine for active sightseeing but not ideal for sitting at an open cafe. But it is really quite mild, particularly when one remembers that Reykjavík is almost as far north as Yellowknife in Canada's Northwest Territories. Most tourists find a sweater and raincoat more suitable than a heavy overcoat.

The northern part of Iceland may also be agreeable during these 2 months but if there is a choice, July and August usually experience somewhat more inviting weather. All of this is quite remarkable when we remember that Iceland is closer to the North Pole than almost all of mainland Canada and two-thirds of Alaska.

It should be noted that precipitation in northern Iceland is extremely low and there is seldom more than a scant ½ to 1 inch per month in either May or June. Moreover, both of these months will average at least 6 hours of sunshine a day. If the possibility of some sharp, cold winds is not a deterrent, this period can be just right for many people. At any rate, one would certainly not want to forego a trip to the exciting north if one were in Iceland during May or June.

As shown on the City Weather Table of Reykjavík, rainfall in July and August increases over the previous 2 months. In the north, however, it still averages only 1½ inches each for July and August. Days begin getting shorter and the hours of sunshine decrease; it is somewhat milder with less frequent stiff winds. This may all sound more like saying which season is the least bad than which is best. Actually, neither of these two periods is really too uncomfortable if the visitor is prepared for rather cool spring and summer weather.

Keep the latitude in mind when considering September and October. Days get very short and the average 2 to 4 hours of sunshine a day leave plenty of time for darkened skies. Iceland does not experience the gloriously golden autumns we can expect in New England and northern United States. Precipitation is usually higher than even August and, on the whole, this is a more satisfactory season to sit at home and read about the wonders of spring and summertime in Iceland.

Usually, the only detailed weather information available to most people are the temperature and precipitation figures of a single city. On that basis, winter Iceland might look quite attractive. Below is a comparison between Reykjavík and New York City which, incidentally, is some 1,200 miles farther south.

		December	*January*	*February*
Reykjavík	*Precipitation*	3.7 in.	4 in.	3.1 in.
	Temperature	30–38°	28–36°	28–37°
New York City	*Precipitation*	3.6 in.	3.7 in.	3.8 in.
	Temperature	29–41°	24–37°	24–38°

It is easy to understand how very misleading such limited information can be. Adding only one more set of figures changes the weather picture completely. New York City enjoys more than twice the 1,249 hours of sunshine that Reykjavík is limited to each year. The capital of Iceland gets an average of only 8, 21, and 57 hours of sunshine, respectively, during the months of December, January, and February. New York City's comparable figures are 155, 154, and 157 hours, so even a little weather research can be quite profitable.

Some are surprised to learn that there is so little snow, especially along the coastal lands of Iceland. It is also true that the harbor of Reykjavík and most often Akureyri are free from ice freeze. The great floes of icebergs, large and small, that appear to float south from Iceland, generally originate along the eastern shores of Greenland.

CITY
WEATHER
TABLES

AARHUS — DENMARK

Elevation 82 Feet
Latitude 56° 18' N   Longitude 10° 37' E

Month	TEMPERATURES—IN °F Daily Average Max.	Min.	Extreme Max.	Min.	No. of Days Over 90°	Under 32°	0°	RELATIVE HUMIDITY A.M. 7:30	P.M. 1:30	PRECIPITATION Inches Total Prec.	Snow Only	No. of Days Total Prec. 0.04"	0.1"	Snow Only 1.5"	Thun. Stms.	Max. Inches in 24 Hrs.	WIND VELOCITY 18 M.P.H.	30 M.P.H.	VISIBILITY ½ MILE
Jan.	35	27	49	-12	0	20	0	90	86	1.5	—	11	5	—	—	1.1	30	5	7
Feb.	36	27	56	-4	0	21	—	91	81	1.2	—	8	4	—	0	0.6	27	6	6
March	41	30	65	0	0	19	0	89	74	1.5	—	7	5	—	—	0.8	25	3	7
April	51	37	73	17	0	6	0	84	65	1.6	—	8	5	—	—	0.9	17	1	3
May	60	43	82	28	0	—	0	75	59	1.8	—	6	6	—	1	0.7	11	—	2
June	67	50	87	35	0	0	0	77	63	1.7	0	9	5	0	2	1.1	10	1	1
July	70	54	87	43	0	0	0	82	66	2.4	0	11	6	0	4	1.8	11	1	1
Aug.	69	54	86	42	0	0	0	86	68	3.0	0	11	7	0	2	3.3	14	1	1
Sept.	63	49	81	34	0	0	0	90	73	2.1	0	12	6	0	2	2.0	16	—	1
Oct.	53	42	67	22	0	—	0	92	78	2.4	—	11	7	—	—	1.2	22	2	2
Nov.	45	37	56	18	0	4	0	93	86	2.1	—	12	6	—	0	0.9	31	4	2
Dec.	39	32	55	9	0	12	0	93	89	2.6	—	11	8	—	1	0.9	30	5	5
Year	52	40	87	-12	0	81	—	87	74	24.0	—	117	69	—	12	3.3	20	3	38

Total precipitation means the total of rain, snow, hail, etc. expressed as inches of rain.

Ten inches of snow equals approx. 1 inch of rain.

Temperatures are as °F.

Wind velocity is the percentage of observations greater than 18 and 30 M.P.H. Number of observations varies from one per day to several per hour.

Visibility means number of days with less than ½-mile visibility.

Note: "—" indicates slight trace or possibility blank indicates no statistics available

ABERDEEN   (United Kingdom) SCOTLAND

Month	Daily Average Max.	Daily Average Min.	Extreme Max.	Extreme Min.	Over 90°	Under 32°	Under 0°	R.H. A.M. 7:00	R.H. P.M. 1:00	Total Prec.	Snow Only	Total Prec. 0.01"	Total Prec. 0.1"	Snow Only 1.5"	Thun. Stms.	Max. Inches in 24 Hrs.	Wind 18 M.P.H.	Wind 30 M.P.H.	Visibility ½ Mile
Jan.	43	35	57	14	0	9	0	84	80	3.0	—	18	8	—	0.0	1.1			
Feb.	43	35	62	16	0	9	0	83	75	2.2	—	15	7	—	0.0	0.8			
March	46	36	66	12	0	8	0	84	73	2.0	—	16	6	—	0.0	0.8			
April	49	39	69	26	0	3	0	83	72	2.3	—	17	6	—	0.3	2.2			
May	53	43	73	27	0	1	0	81	73	2.7	—	15	7	—	1.0	1.0			
June	60	48	83	33	0	0	0	78	70	2.1	0	15	6	0	1.0	2.0			
July	63	52	88	39	0	0	0	83	73	3.3	0	17	8	0	2.0	2.4			
Aug.	62	51	83	36	0	0	0	85	73	2.8	0	15	7	0	2.0	1.3			
Sept.	59	48	79	33	0	0	0	86	73	2.7	0	16	7	0	0.3	2.4			
Oct.	53	43	70	26	0	2	0	85	74	3.5	—	18	8	—	0.3	2.5			
Nov.	47	39	64	13	0	5	0	84	79	3.4	—	19	8	—	0.3	1.6			
Dec.	44	36	61	14	0	6	0	84	80	3.1	—	18	9	—	0.3	1.7			
Year	52	42	88	12	0	43	0	83	75	33.0	0⁺	199	87	0⁺	7.5	2.5			

Elevation 234 Feet
Latitude 51° 10' N   Longitude 2° 12' W

Total precipitation means the total of rain, snow, hail, etc. expressed as inches of rain.

Ten inches of snow equals approx. 1 inch of rain.

Temperatures are as ° F.

Wind velocity is the percentage of observations greater than 18 and 30 M.P.H. Number of observations varies from one per day to several per hour.

Visibility means number of days with less than ½-mile visibility.

Note: "—" indicates slight trace or possibility blank indicates no statistics available

AJACCIO   (Corsica—Mediterranean Sea)   FRANCE

Month	TEMPERATURES—IN °F Daily Average Max.	Min.	Extreme Max.	Min.	No. of Days Over 90°	Under 32°	0°	RELATIVE HUMIDITY A.M. 7:30	P.M. 1:30	PRECIPITATION Inches Total Prec.	Snow Only	No. of Days Total Prec. 0.004"	0.1"	Snow Only 1.5"	Thun. Stms.	Max. Inches in 24 Hrs.	WIND VELOCITY 18 M.P.H.	30 M.P.H.	VISIBILITY ½ MILE
Jan.	56	40	72	25	0	—	0	83	64	3.0	—	10	8	—	1	1.1			
Feb.	58	42	72	23	0	—	0	82	61	2.6	—	9	8	—	1	2.2			
March	62	44	82	28	0	—	0	79	61	2.1	—	10	6	—	—	2.6			
April	66	48	88	36	0	0	0	77	60	2.0	0	8	6	0	2	1.4			
May	72	53	99	38	—	0	0	77	60	1.9	0	6	6	0	1	1.8			
June	79	60	97	45	—	0	0	71	55	1.0	0	3	3	0	2	1.2			
July	85	64	101	47	8	0	0	68	52	0.4	0	1	1	0	1	2.0			
Aug.	85	64	103	50	8	0	0	69	52	0.6	0	2	2	0	1	1.4			
Sept.	81	61	97	45	4	0	0	73	57	1.9	0	6	6	0	2	2.3			
Oct.	72	55	88	39	0	0	0	78	60	3.5	0	8	8	0	2	2.3			
Nov.	64	49	82	31	0	—	0	84	66	3.8	0	11	8	0	3	2.7			
Dec.	59	44	73	26	0	—	0	83	67	3.8	—	10	10	—	1	1.7			
Year	70	52	103	23	20+	0+	0	77	60	27.0	0+	84	72	0+	17	2.7			

Elevation 20 Feet

Latitude 41° 55' N   Longitude 8° 48' E

Total precipitation means the total of rain, snow, hail, etc. expressed as inches of rain.

Ten inches of snow equals approx. 1 inch of rain.

Temperatures are as ° F.

Wind velocity is the percentage of observations greater than 18 and 30 M.P.H. Number of observations varies from one per day to several per hour.

Visibility means number of days with less than ½-mile visibility.

Note: "—" indicates slight trace or possibility blank indicates no statistics ava lable

AKUREYRI

ICELAND

| Month | TEMPERATURES—IN °F | | | | | | RELATIVE HUMIDITY | | PRECIPITATION | | | | | | | WIND VELOCITY | | VISIBILITY ½ MILE |
| | Daily Average | | Extreme | | No. of Days | | | | Inches | | No. of Days | | | Max. Inches in 24 Hrs. | | | | |
	Max.	Min.	Max.	Min.	Over 90°	Under 32°	A.M. —	P.M. —	Total Prec.	Snow Only	Total Prec. 0.004"	0.1" Snow Only 1.5"	Thun. Stms.		18 M.P.H.	30 M.P.H.	
Jan.	34	26	57	−4					1.7		12			0.8			
Feb.	34	25	56	−4					1.5		12			0.8			
March	36	26	61	−8					1.7		11			1.1			
April	40	30	61	2					1.3		11			0.6			
May	49	37	71	16					0.6		7			0.9			
June	55	43	83	25					0.9		7			0.8			
July	57	47	74	33					1.3		10			1.1			
Aug.	56	45	77	30					1.6		11			2.1			
Sept.	51	41	72	17					1.9		13			3.6			
Oct.	43	34	64	11					2.3		14			1.2			
Nov.	37	29	60	−1					1.9		12			1.1			
Dec.	36	27	56	−5					1.9		13			1.1			
Year	44	34	83	−8					19.0		133			3.6			

Elevation 16 Feet
Latitude 65° 41' N   Longitude 18° 5' W

Total precipitation means the total of rain, snow, hail, etc. expressed as inches of rain.

Ten inches of snow equals approx. 1 inch of rain.

Temperatures are as ° F.

Wind velocity is the percentage of observations greater than 18 and 30 M.P.H. Number of observations varies from one per day to several per hour.

Visibility means number of days with less than ½-mile visibility.

Note: "—" indicates slight trace or possibility blank indicates no statistics available

AMSTERDAM

THE NETHERLANDS

Month	TEMPERATURES—IN °F Daily Average Max.	Min.	Extreme Max.	Min.	No. of Days Over 90°	Under 32°	0°	RELATIVE HUMIDITY A.M. 8:00	P.M. 2:00	PRECIPITATION Inches Total Prec.	Snow Only	No. of Days Total Prec. 0.004"	0.1"	Snow Only 1.5"	Thun. Stms.	Max. Inches in 24 Hrs.	WIND VELOCITY 18 M.P.H.	30 M.P.H.	VISI-BIL-ITY ½ MILE
Jan.	40	34	55	3	0	15	0	90	84	2.0	—	19	6	—	0.3	0.8	27	4.9	6
Feb.	41	34	56	4	0	14	—	88	79	1.4	—	15	5	—	0.3	1.0	22	2.8	7
March	46	37	65	15	0	11	0	85	71	1.3	—	13	4	—	1.0	0.9	20	3.5	6
April	52	43	74	29	0	4	0	80	66	1.6	—	14	5	—	2.0	0.9	20	2.1	2
May	60	50	88	37	0	0	0	75	63	1.8	0	12	6	0	3.0	1.1	15	1.1	1
June	65	55	95	44	—	0	0	74	63	1.8	0	12	5	0	5.0	1.2	10	—	1
July	69	59	91	50	—	0	0	76	65	2.6	0	14	7	0	5.0	1.9	15	0.7	1
Aug.	68	59	90	47	—	0	0	78	66	2.7	0	14	7	0	5.0	2.3	12	0.6	2
Sept.	64	56	89	44	0	0	0	82	69	2.8	0	15	7	0	4.0	1.2	14	1.2	3
Oct.	56	48	73	31	0	1	0	86	75	2.8	—	18	7	—	2.0	1.8	15	1.1	5
Nov.	47	41	63	19	0	5	0	89	82	2.6	—	19	7	—	1.0	1.1	17	1.6	6
Dec.	41	35	55	6	0	9	0	90	86	2.2	—	19	7	—	1.0	0.9	23	3.4	7
Year	54	46	95	3	0+	59	0+	83	72	26.0	0+	184.	73	0+	30.0	2.3	18	2.0	47

Elevation 5 Feet

Latitude 52° 23' N   Longitude 4° 55' E

Total precipitation means the total of rain, snow, hail, etc. expressed as inches of rain.

Ten inches of snow equals approx. 1 inch of rain.

Temperatures are as ° F.

Wind velocity is the percentage of observations greater than 18 and 30 M.P.H. Number of observations varies from one per day to several per hour.

Visibility means number of days with less than ½-mile visibility.

Note: "—" indicates slight trace or possibility blank indicates no statistics available

ANTWERP   BELGIUM

Month	TEMPERATURES—IN °F							RELATIVE HUMIDITY		PRECIPITATION								
	Daily Average		Extreme		No. of Days					Inches		No. of Days			Max. Inches in 24 Hrs.	WIND VELOCITY		VISIBILITY ½ MILE
	Max.	Min.	Max.	Min.	Over 90°	Under 32°	0°	A.M.	P.M.	Total Prec.	Snow Only	Total Prec. 0.1"	Snow Only 1.5"	Thun. Stms.		18 M.P.H.	30 M.P.H.	
Jan.	41	32	57	10	0.0	15	0.0	—	86	2.4	—		—	1.0				
Feb.	41	31	63	0	0.0	14	0.4	—	82	2.1	—		—	0.3				
March	50	35	70	14	0.0	11	0.0	—	79	3.8	—		—	1.0				
April	57	40	81	25	0.0	4	0.0	—	72	1.8	—		—	1.0				
May	64	46	88	34	0.0	0	0.0	—	74	2.5	0		0	4.0				
June	69	51	90	37	0.1	0	0.0	—	75	2.7	0		0	3.0				
July	72	55	93	43	0.4	0	0.0	—	76	2.9	0		0	4.0				
Aug.	72	55	88	41	0.0	0	0.0	—	79	3.7	0		0	4.0				
Sept.	67	51	95	34	0.3	3	0.0	—	81	3.1	0		0	2.0				
Oct.	59	44	73	21	0.0	6	0.0	—	83	2.0	—		—	1.0				
Nov.	49	39	68	19	0.0	12	0.0	—	85	2.9	—		—	1.0				
Dec.	43	35	63	3	0.0	12	0.0	—	87	2.6	—		—	0.3				
Year	57	43	95	0	0.8	66	0.4	—	80	33.0	0+		0+	23.0				

Elevation 39 Feet

Latitude 51° 11' N   Longitude 4° 27' E

Total precipitation means the total of rain, snow, hail, etc. expressed as inches of rain.

Ten inches of snow equals approx. 1 inch of rain.

Temperatures are as °F.

Wind velocity is the percentage of observations greater than 18 and 30 M.P.H. Number of observations varies from one per day to several per hour.

Visibility means number of days with less than ½-mile visibility.

Note: "—" indicates slight trace or possibility blank indicates no statistics available

ATHENS

GREECE

Month	TEMPERATURES—IN °F								RELATIVE HUMIDITY		PRECIPITATION						WIND VELOCITY		VISI-BIL-ITY ½ MILE
	Daily Average		Extreme		No. of Days				A.M. 7:30	P.M. 1:30	Inches		No. of Days				18 M.P.H.	30 M.P.H.	
	Max.	Min.	Max.	Min.	Over 90°	Under 32°	0°				Total Prec.	Snow Only	Total Prec. 0.1"	Snow Only 1.5"	Thun. Stms.	Max. Inches in 24 Hrs.			
Jan.	54	42	72	20	0	2	0	77	62		2.2		7		1	2	7	0	0
Feb.	55	43	73	21	0	1	0	77	61		1.6		5		1	2	14	0	0
March	60	46	83	20	0	—	0	74	54		1.4		5		1	2	4	0	0
April	67	52	91	35	0	0	0	70	47		0.8		3		1	2	6	—	0
May	77	60	104	42	1	0	0	66	44		0.8		3		2	2	4	0	0
June	85	67	109	54	8	0	0	61	40		0.6		2		2	2	5	0	0
July	90	72	106	58	21	0	0	53	32		0.2		—		1	2	8	0	0
Aug.	90	72	111	59	23	0	0	53	33		0.4		1		1	2	11	—	0
Sept.	83	66	103	48	6	0	0	60	38		0.6		3		1	2	4	0	0
Oct.	74	60	95	44	0	0	0	73	52		1.7		5		2	3	1	0	0
Nov.	64	52	87	30	0	—	0	77	60		2.8		7		2	6	2	0	0
Dec.	57	46	72	24	0	1	0	79	63		2.8		8		1	3	6	—	0
Year	71	57	111	20	59	4	0	68	49		16.0		49		16	6	6	—	0

Elevation 90 Feet

Latitude 37° 53' N   **Longitude 23° 43' E**

Total precipitation means the total of rain, snow, hail, etc. expressed as inches of rain.

Ten inches of snow equals approx. 1 inch of rain.

Temperatures are as ° F.

Wind velocity is the percentage of observations greater than 18 and 30 M.P.H. Number of observations varies from one per day to several per hour.

Visibility means number of days with less than ½-mile visibility.

Note: "—" indicates slight trace or possibility
blank indicates no statistics available

BALLYKELLY   (United Kingdom) NORTHERN IRELAND

Month	Daily Average Max.	Daily Average Min.	Extreme Max.	Extreme Min.	Over 90°	Under 32°	0°	R.H. A.M. —	R.H. P.M. —	Total Prec.	Snow Only	Total Prec. 0.1"	Snow Only 1.5"	Thun. Stms.	Max. in 24 Hrs.	Wind 18 M.P.H.	Wind 30 M.P.H.	Visibility ½ Mile
Jan.	43	35	56	9	0	8.5	0		90	3.2	—	9	—					
Feb.	44	35	57	11	0	7.6	0		85	2.1	—	7	—					
March	49	37	67	10	0	7.3	0		81	1.9	—	6	—					
April	53	39	69	24	0	4.9	0		76	1.9	—	6	—					
May	59	43	77	26	0	0.9	0		73	2.1	—	6	—					
June	63	48	83	30	0	0.1	0		76	2.7	0	7	0					
July	65	52	85	36	0	0.0	0		79	3.6	0	8	0					
Aug.	65	51	82	34	0	0.0	0		82	3.2	0	8	0					
Sept.	61	49	76	28	0	0.2	0		83	3.2	—	8	—					
Oct.	55	44	69	24	0	1.6	0		86	3.3	—	8	—					
Nov.	48	39	61	21	0	4.4	0		89	2.9	—	7	—					
Dec.	45	37	58	13	0	6.8	0		90	3.6	—	9	—					
Year	54	42	85	9	0	42.0	0		83	33.7	0+	89	0+					

Elevation 13 Feet

Latitude 55° 3' N   Longitude 7° 0' W

Total precipitation means the total of rain, snow, hail, etc. expressed as inches of rain.

Ten inches of snow equals approx. 1 inch of rain.

Temperatures are as ° F.

Wind velocity is the percentage of observations greater than 18 and 30 M.P.H. Number of observations varies from one per day to several per hour.

Visibility means number of days with less than ½-mile visibility.

Note: "—" indicates slight trace or possibility blank indicates no statistics available

BARCELONA

SPAIN

Month	Daily Average Max.	Min.	Extreme Max.	Min.	Over 90°	Under 32°	0°	A.M. 7:00	P.M. 1:00	Total Prec.	Snow Only	Total Prec. 0.004"	0.1"	Snow Only 1.5"	Thun. Stms.	Max. Inches in 24 Hrs.	18 M.P.H.	30 M.P.H.	½ MILE
Jan.	56	42	71	24	0	1	0	74	60	1.2	—	5	4	—	2	2.5	3.0	—	
Feb.	57	44	77	21	0	2	0	72	59	2.1	—	7	7	—	—	5.6	3.0	—	
March	61	47	79	32	0	—	0	75	61	1.9	0	7	6	0	2	3.6	2.0	—	
April	64	51	82	30	0	—	0	72	61	1.8	0	8	6	0	1	1.9	2.0	—	
May	71	57	90	32	—	—	0	70	61	1.8	0	8	6	0	3	1.5	1.0	—	
June	77	63	95	49	—	0	0	67	61	1.3	0	5	4	0	3	3.8	—	0	
July	81	69	96	54	3	0	0	68	61	1.2	0	4	4	0	2	3.4	—	0	
Aug.	82	69	98	56	1	0	0	73	63	1.7	0	5	5	0	5	3.3	—	0	
Sept.	78	65	91	49	—	0	0	74	64	2.6	0	7	7	0	4	3.3	—	—	
Oct.	71	58	82	39	0	0	0	77	66	3.4	0	8	8	0	4	5.3	1.0	—	
Nov.	62	50	80	31	0	—	0	75	63	2.7	0	7	7	0	1	2.7	1.0	—	
Dec.	57	44	73	25	0	—	0	73	60	1.8	—	6	6	—	2	3.6	3.0	—	
Year	68	55	98	21	4	3	0	73	62	24.0	—	77	67	—	29	5.6	1.4	—	

TEMPERATURES—IN ° F / RELATIVE HUMIDITY / PRECIPITATION (Inches / No. of Days) / WIND VELOCITY / VISIBILITY

Elevation 13 Feet
Latitude 41° 17' N   Longitude 2° 5' E
Total precipitation means the total of rain, snow, hail, etc. expressed as inches of rain.
Ten inches of snow equals approx. 1 inch of rain.
Temperatures are as ° F.

Wind velocity is the percentage of observations greater than 18 and 30 M.P.H. Number of observations varies from one per day to several per hour.
Visibility means number of days with less than ½-mile visibility.
Note: "—" indicates slight trace or possibility
blank indicates no statistics available

BASTIA (Corsica—Mediterranean Sea) FRANCE

Month	TEMPERATURES—IN °F Daily Average Max.	Min.	Extreme Max.	Min.	No. of Days Over 90°	Under 32°	0°	RELATIVE HUMIDITY A.M.	P.M.	PRECIPITATION Inches Total Prec.	Snow Only	No. of Days Total Prec. 0.1"	Snow Only 1.5"	Thun. Stms.	Max. Inches in 24 Hrs.	WIND VELOCITY 18 M.P.H.	30 M.P.H.	VISIBILITY ½ MILE
Jan.	52	42	60	31	0	—	0		69	2.9	0	8	0	1.5		2.0	0.0	0
Feb.	49	42	58	34	0	0	0		59	2.6	0	8	0	2.9		5.0	—	1
March	58	45	69	35	0	0	0		66	2.4	0	7	0	1.0		3.0	—	0
April	65	52	75	46	0	0	0		73	2.5	0	7	0	0.5		2.0	0.0	2
May	71	57	78	49	0	0	0		70	1.9	0	6	0	0.7		3.0	0.0	0
June	79	63	90	51	1	0	0		64	0.8	0	2	0	1.0		3.0	—	0
July	82	66	92	54	3	0	0		65	—	0	1	0	1.5		5.0	1.0	1
Aug.	84	67	93	56	1	0	0		65	1.0	0	3	0	2.0		2.0	0.0	0
Sept.	78	67	88	61	0	0	0		70	2.6	0	7	0	1.3		2.0	—	0
Oct.	68	52	78	46	0	0	0		72	4.3	0	9	0	2.8		2.0	—	1
Nov.	63	49	74	41	0	0	0		70	3.7	0	8	0	1.6		4.0	—	0
Dec.	55	46	68	34	0	0	0		77	3.7	0	10	0	1.6		1.0	—	0
Year	67	54	93	31	5	0+	0		68	29.0	0	76	0	18.0		2.7	0.2	5

Elevation 26 Feet

Latitude 42° 33' N   Longitude 9° 29' E

Total precipitation means the total of rain, snow, hail, etc. expressed as inches of rain.

Ten inches of snow equals approx. 1 inch of rain.

Temperatures are as ° F.

Wind velocity is the percentage of observations greater than 18 and 30 M.P.H. Number of observations varies from one per day to several per hour.

Visibility means number of days with less than ½-mile visibility.

Note: "—" indicates slight trace or possibility blank indicates no statistics available

BELFAST (United Kingdom) NORTHERN IRELAND

Month	Daily Average Max.	Daily Average Min.	Extreme Max.	Extreme Min.	Over 90°	Under 32°	R.H. A.M.	R.H. P.M.	Total Prec.	Snow Only	Days Total Prec. 0.1"	Snow Only 1.5"	Thun. Stms.	Max. Inches in 24 Hrs.	Wind 18 M.P.H.	Wind 30 M.P.H.	Visibility ½ Mile
Jan.	43	35	56	9	0	8.5	—	90	4.1	—	10	—					
Feb.	44	35	57	11	0	7.6		85	2.8	—	8	—					
March	49	37	67	10	0	7.3		81	2.4	—	7	—					
April	53	39	69	24	0	4.9		76	2.4	—	7	—					
May	59	43	77	26	0	0.9		73	2.5	—	7	—					
June	63	48	83	30	0	0.1		76	2.4	0	6	0					
July	65	52	85	36	0	0.0		79	3.4	0	8	0					
Aug.	65	51	82	34	0	0.0		82	3.5	0	8	0					
Sept.	61	49	76	28	0	0.2		83	3.3	—	8	—					
Oct.	55	44	69	24	0	1.6		86	3.9	—	9	—					
Nov.	48	39	61	21	0	4.4		89	3.6	—	8	—					
Dec.	45	37	58	13	0	6.8		90	3.9	—	10	—					
Year	54	42	85	9	0	42.3	—	83	38.3	0+	96	0+					

Elevation 24 Feet

Latitude 54° 36' N   Longitude 5° 52' W

Total precipitation means the total of rain, snow, hail, etc. expressed as inches of rain.

Ten inches of snow equals approx. 1 inch of rain.

Temperatures are as ° F.

Wind velocity is the percentage of observations greater than 18 and 30 M.P.H. Number of observations varies from one per day to several per hour.

Visibility means number of days with less than ½-mile visibility.

Note: "—" indicates slight trace or possibility blank indicates no statistics available

BERGEN

NORWAY

| Month | Temperatures—in °F | | | | | | | Relative Humidity | | Precipitation | | | | | | | Wind Velocity | | Visibility |
| | Daily Average | | Extreme | | No. of Days | | | A.M. 7:30 | P.M. 1:30 | Inches | | No. of Days | | | | Max. Inches in 24 Hrs. | 18 M.P.H. | 30 M.P.H. | ½ MILE |
	Max.	Min.	Max.	Min.	Over 90°	Under 32°	0°			Total Prec.	Snow Only	Total Prec. 0.04"	0.1"	Snow Only 1.5"	Thun. Stms.				
Jan.	43	27	56	7	0	—	0	80	77	7.9	—	18	11	—	—	4			
Feb.	44	26	54	3	0	—	0	80	74	6.0	—	14	11	—	1	4			
March	47	28	68	10	0	—	0	80	68	5.4	—	13	8	—	1	5			
April	55	34	77	15	0	—	0	77	66	4.4	—	13	8	—	—	4			
May	64	41	81	25	0	—	0	77	65	3.9	—	11	7	—	1	2			
June	70	46	89	35	0	0	0	81	70	4.2	0	13	9	0	1	3			
July	72	51	89	39	0	0	0	85	73	5.2	0	13	10	0	2	3			
Aug.	70	50	85	39	0	0	0	87	74	7.3	0	16		0	2	4			
Sept.	64	45	79	28	0	—	0	87	75	9.2	—	17		—	1	5			
Oct.	57	38	68	22	0	—	0	84	75	9.2	—	18		—	1	4			
Nov.	49	33	59	21	0	—	0	82	77	8.0	—	16		—	1	5			
Dec.	45	28	62	6	0	—	0	81	79	8.1	—	18	11	—	1	5			
Year	57	37	89	3	0	0+	0	82	73	79.0	—	180		—	12	5			

Elevation 141 Feet
Latitude 60° 24' N   **Longitude 5° 19' E**

Total precipitation means the total of rain, snow, hail, etc. expressed as inches of rain.

Ten inches of snow equals approx. 1 inch of rain.

Temperatures are as ° F.

Wind velocity is the percentage of observations greater than 18 and 30 M.P.H. Number of observations varies from one per day to several per hour.

Visibility means number of days with less than ½-mile visibility.

Note: "—" indicates slight trace or possibility blank indicates no statistics available

BERLIN (East And West) GERMANY

Month	TEMPERATURES—IN °F Daily Average Max.	Min.	Extreme Max.	Min.	No. of Days Over 90°	Under 32°	0°	RELATIVE HUMIDITY A.M. 7:00	P.M. 2:00	PRECIPITATION Inches Total Prec.	Snow Only	No. of Days Total Prec. 0.04"	0.1"	Snow Only 1.5"	Thun. Stms.	Max. Inches in 24 Hrs.	WIND VELOCITY 18 M.P.H.	30 M.P.H.	VISIBILITY ½ MILE
Jan.	35	26	55	—1	0	20	—	89	81	1.9	5	10	6	2	0	1	6	—	4
Feb.	38	27	62	—15	0	18	—	87	73	1.3	6	8	4	1	0	1	4	—	4
March	46	32	73	7	0	15	0	87	63	1.5	1	9	5	—	1	1	6	—	2
April	55	38	85	20	0	3	0	83	56	1.7	—	9	5	0	1	1	3	0	1
May	65	46	92	28	—	—	0	79	50	1.9	0	8	6	0	4	1	2	0	—
June	70	51	94	35	—	0	0	78	53	2.3	0	9	6	0	5	2	2	0	—
July	74	55	101	43	1	0	0	79	55	3.1	0	10	8	0	5	3	2	0	0
Aug.	72	54	94	44	—	0	0	86	58	2.2	0	10	6	0	4	2	2	0	1
Sept.	66	48	94	31	—	—	0	90	60	1.9	0	8	6	0	1	2	2	0	1
Oct.	55	41	77	15	0	—	0	92	68	1.7	0	8	5	0	—	1	1	0	4
Nov.	43	33	64	8	0	7	0	91	79	1.7	2	8	5	—	—	2	3	0	6
Dec.	37	29	60	—3	0	14	—	90	84	1.9	1	11	6	—	—	1	6	—	6
Year	55	40	101	—15	1	77	0+	86	65	23.0	15	108	68	3	21	3	3	—	29

Elevation 163 Feet
Latitude 52° 28' N   Longitude 13° 24' E
Total precipitation means the total of rain, snow, hail, etc. expressed as inches of rain.
Ten inches of snow equals approx. 1 inch of rain.
Temperatures are as ° F.

Wind velocity is the percentage of observations greater than 18 and 30 M.P.H. Number of observations varies from one per day to several per hour.
Visibility means number of days with less than ½-mile visibility.
Note: "—" indicates slight trace or possibility
blank indicates no statistics available

BERN

SWITZERLAND

| Month | TEMPERATURES—IN °F | | | | | | | RELATIVE HUMIDITY | | PRECIPITATION | | | | | | | WIND VELOCITY | | VISI-BILITY |
| | Daily Average | | Extreme | | No. of Days | | | | | Inches | | No. of Days | | | | Max. | | | ½ MILE |
	Max.	Min.	Max.	Min.	Over 90°	Under 32°	0°	A.M. 7:00	P.M. 1:00	Total Prec.	Snow Only	Total Prec. 0.04"	0.1"	Snow Only 1.5"	Thun. Stms.	Inches in 24 Hrs.	18 M.P.H.	30 M.P.H.	
Jan.	35	26	59	−3	0	24	—	89	79	1.9	—	10	6	—	0	1.2			
Feb.	40	27	59	−9	0	20	—	89	70	2.0	—	8	6	—	—	1.4			
March	48	33	69	7	0	15	0	88	63	2.6	—	9	7	—	—	1.5			
April	56	39	79	21	0	3	0	85	56	3.0	—	9	7	—	1	1.5			
May	64	46	85	28	0	1	0	83	56	3.7	—	8	7	—	3	2.0			
June	70	52	93	38	—	0	0	81	56	4.4	0	9	9	0	5	2.0			
July	74	56	96	38	—	0	0	83	56	4.4	0	9	9	0	6	2.4			
Aug.	73	55	97	40	—	0	0	87	59	4.3	0	10	9	0	4	1.8			
Sept.	66	50	87	30	0	—	0	90	64	3.5	0	8	8	0	2	2.3			
Oct.	55	42	75	25	0	2	0	91	71	3.5	—	8	8	—	—	1.9			
Nov.	44	34	68	15	0	11	0	91	78	2.7	—	8	7	—	0	2.7			
Dec.	36	27	56	3	0	24	0	90	81	2.5	—	11	8	—	0	1.9			
Year	55	41	97	−9	0+	100	0+	87	66	39.0	—	108	91	—	22	2.7			

Elevation 1,876 Feet

Latitude 46° 57' N   Longitude 7° 29' E

Total precipitation means the total of rain, snow, hail, etc. expressed as inches of rain.

Ten inches of snow equals approx. 1 inch of rain.

Temperatures are as ° F.

Wind velocity is the percentage of observations greater than 18 and 30 M.P.H. Number of observations varies from one per day to several per hour.

Visibility means number of days with less than ½-mile visibility.

Note: "—" indicates slight trace or possibility
blank indicates no statistics available

BIARRITZ—BAYONNE   FRANCE

Month	Daily Average Max.	Daily Average Min.	Extreme Max.	Extreme Min.	Over 90°	Under 32°	0°	R.H. A.M.	R.H. P.M.	Total Prec.	Snow Only	Total Prec. 0.004"	Total Prec. 0.1"	Snow Only 1.5"	Thun. Stms.	Max. Inches in 24 Hrs.	Wind 18 M.P.H.	Wind 30 M.P.H.	Visibility ½ Mile
Jan.	52	39	73	14	0.0	4.2	0	—	82	4.1	—	10	10	—	1	1.4	10	1.6	
Feb.	53	40	79	19	0.0	3.0	0		81	3.8	—	11	10	—	1	1.5	10	1.9	
March	59	44	84	21	0.0	1.0	0		79	4.2	—	11	8	—	1	1.4	15	2.4	
April	60	47	86	28	0.0	0.8	0		82	4.0	—	11	8	—	1	1.7	9	0.7	
May	65	52	99	32	0.1	0.1	0		81	3.4	0	11	7	0	3	1.2	11	0.3	
June	71	57	99	41	0.3	0.0	0		81	3.2	0	10	8	0	5	2.2	9	0.1	
July	74	61	99	46	0.9	0.0	0		81	2.6	0	7	7	0	4	1.4	6	0.7	
Aug.	75	61	103	41	0.7	0.0	0		81	3.2	0	8	8	0	5	1.9	5	0.0	
Sept.	72	58	98	41	0.8	0.0	0		80	4.3	0	9	9	0	4	1.6	6	0.4	
Oct.	66	51	93	30	0.2	0.2	0		82	6.4	0	11	10	0	2	2.7	16	5.1	
Nov.	58	45	80	20	0.0	0.2	0		81	5.7	—	12	10	—	1	2.0	15	2.1	
Dec.	53	41	73	16	0.0	5.7	0		83	4.7	—	14	10	—	2	1.6	15	2.8	
Year	63	50	103	14	3.0	15.0	0		81	49.0	0+	125	105	0+	30	2.7	11	1.5	

Elevation 246 Feet

Latitude 43° 28' N   **Longitude 1° 31' W**

Total precipitation means the total of rain, snow, hail, etc. expressed as inches of rain.

Ten inches of snow equals approx. 1 inch of rain.

Temperatures are as ° F.

Wind velocity is the percentage of observations greater than 18 and 30 M.P.H. Number of observations varies from one per day to several per hour.

Visibility means number of days with less than ½-mile visibility.

Note: "—" indicates slight trace or possibility blank indicates no statistics available

BLACKSOD POINT

IRELAND

| Month | TEMPERATURES—IN °F | | | | | | | RELATIVE HUMIDITY | | PRECIPITATION | | | | | | | WIND VELOCITY | | VISI-BIL-ITY |
| | Daily Average | | Extreme | | No. of Days | | | | | Inches | | No. of Days | | | | Max. Inches in 24 Hrs. | | | ½ MILE |
	Max.	Min.	Max.	Min.	Over 90°	Under 32°	0°	A.M. 6:30	P.M. 5:30	Total Prec.	Snow Only	Total Prec. 0.01"	0.1"	Snow Only 1.5"	Thun. Stms.		18 M.P.H.	30 M.P.H.	
Jan.	48	39	58	20	0	—	0	92	90	5.1	—	24	11.0	—		2.1			
Feb.	48	40	57	28	0	—	0	92	85	4.1	—	18	9.8	—		1.5			
March	49	41	60	29	0	—	0	93	78	4.1	0	19	7.5	0		1.1			
April	52	43	62	32	0	—	0	92	70	2.9	0	18	7.0	0		0.8			
May	56	47	73	36	0	0	0	90	68	2.8	0	18	6.9	0		1.0			
June	60	51	78	39	0	0	0	90	70	2.8	0	17	7.1	0		1.5			
July	62	54	79	40	0	0	0	92	75	3.2	0	21	7.7	0		1.5			
Aug.	63	54	77	43	0	0	0	94	75	4.6	0	21	9.2	0		2.3			
Sept.	60	52	70	41	0	—	0	95	80	3.9	0	20	8.5	0		1.7			
Oct.	55	47	66	32	0	—	0	94	86	5.0	0	24	9.4	0		1.4			
Nov.	51	43	62	34	0	0	0	93	90	5.2	0	23	9.5	0		1.4			
Dec.	48	41	59	25	0	—	0	93	91	6.1	—	24	11.0	—		1.8			
Year	54	46	79	20	0	0+	0	92	80	49.7	0+	247	104.0	0+		2.3			

Elevation 18 Feet

Latitude 54° 6′ N   Longitude 10° 4′ W

Total precipitation means the total of rain, snow, hail, etc. expressed as inches of rain.

Ten inches of snow equals approx. 1 inch of rain.
Temperatures are as ° F.

Wind velocity is the percentage of observations greater than 18 and 30 M.P.H. Number of observations varies from one per day to several per hour.

Visibility means number of days with less than ½-mile visibility.

Note: "—" indicates slight trace or possibility
blank indicates no statistics available

BIRMINGHAM

(United Kingdom) ENGLAND

Month	TEMPERATURES—IN °F Daily Average Max.	Min.	Extreme Max.	Min.	No. of Days Over 90°	Under 32°	0°	RELATIVE HUMIDITY A.M. 7:00	P.M. 1:00	PRECIPITATION Inches Total Prec.	Snow Only	No. of Days Total Prec. 0.01"	0.1"	Snow Only 1.5"	Thun. Stms.	Max. Inches in 24 Hrs.	WIND VELOCITY 18 M.P.H.	30 M.P.H.	VISIBILITY ½ MILE
Jan.	42	35	59	11	0.0	8	0	90	82	2.0	—	18	6	—	0.3	1.3			
Feb.	43	35	59	13	0.0	9	0	88	77	1.7	—	14	6	—	0.3	1.0			
March	49	37	70	19	0.0	7	0	88	67	1.9	—	13	6	—	0.3	0.7			
April	53	40	75	28	0.0	2	0	85	66	1.7	—	15	5	—	1.0	1.2			
May	60	45	85	30	0.0	—	0	82	63	2.1	0	14	6	0	2.0	1.5			
June	66	50	87	37	0.0	0	0	82	63	2.3	0	13	6	0	2.0	3.5			
July	69	54	92	44	—	0	0	84	65	2.3	0	15	6	0	3.0	1.7			
Aug.	68	53	91	42	0.1	0	0	87	65	2.7	0	15	7	0	2.0	2.2			
Sept.	63	50	83	37	0.0	0	0	89	68	1.8	0	14	5	0	1.0	1.6			
Oct.	55	45	79	28	0.0	1	0	90	73	2.8	—	16	7	—	1.0	1.6			
Nov.	47	39	67	25	0.0	2	0	90	80	2.4	—	17	6	—	3.0	1.5			
Dec.	43	37	58	20	0.0	8	0	89	84	2.7	—	17	8	—	0.0	1.9			
Year	55	43	92	11	0.1	37	0	87	71	27.0	0+	181	74	0+	16.0	3.5			

Elevation 325 Feet

Latitude 52° 27' N   Longitude 1° 44' W

Total precipitation means the total of rain, snow, hail, etc. expressed as inches of rain.

Ten inches of snow equals approx. 1 inch of rain.

Temperatures are as ° F.

Wind velocity is the percentage of observations greater than 18 and 30 M.P.H. Number of observations varies from one per day to several per hour.

Visibility means number of days with less than ½-mile visibility.

Note: "—" indicates slight trace or possibility

blank indicates no statistics available

BODO

NORWAY

TEMPERATURES—IN ° F

Month	Daily Average Max.	Daily Average Min.	Extreme Max.	Extreme Min.	No. of Days Over 90°	No. of Days Under 32°	No. of Days 0°	RELATIVE HUMIDITY A.M. 8:00	RELATIVE HUMIDITY P.M. 2:00	PRECIPITATION Inches Total Prec.	PRECIPITATION Inches Snow Only	No. of Days Total Prec. 0.04"	No. of Days Total Prec. 0.1"	No. of Days Snow Only 1.5"	No. of Days Thun. Stms.	Max. Inches in 24 Hrs.	WIND VELOCITY 18 M.P.H.	WIND VELOCITY 30 M.P.H.	VISIBILITY ½ MILE
Jan.	33	25	49	0	0	28	0	72	72	3.9	0	15	10	0	1	2	30	4	4
Feb.	33	23	49	−2	0	26	—	73	72	3.4	0	13	9	0	0	2	45	7	2
March	35	25	53	3	0	24	0	73	71	2.7	0	12	7	0	0	3	35	7	2
April	42	31	64	11	0	18	0	73	68	2.3	0	11	6	0	0	2	23	1	1
May	45	38	74	20	0	6	0	71	66	2.1	0	10	6	0	0	2	12	—	1
June	55	44	82	30	0	—	0	75	70	2.6	—	11	7	—	—	2	9	—	2
July	61	50	85	35	0	0	0	77	72	2.5	—	10	7	—	1	2	6	—	1
Aug.	60	49	82	33	0	0	0	79	73	3.6	—	14	8	—	0	3	9	—	1
Sept.	53	43	73	25	0	—	0	80	73	5.2	0	17	10	0	—	2	13	1	—
Oct.	43	36	61	10	0	7	0	78	74	5.1	0	17	10	0	—	3	26	5	—
Nov.	37	30	52	5	0	19	0	75	75	4.7	0	15	9	0	1	3	30	5	1
Dec.	33	26	51	−4	0	21	—	73	72	3.4	0	13	9	0	1	2	38	7	1
Year	45	35	85	−4	0	148	0⁺	75	71	42.0	—	158	97	0	4	3	23	3	16

Elevation 42 Feet

Latitude 67° 16' N   Longitude 14° 22' E

Total precipitation means the total of rain, snow, hail, etc. expressed as inches of rain.

Ten inches of snow equals approx. 1 inch of rain.

Temperatures are as ° F.

Wind velocity is the percentage of observations greater than 18 and 30 M.P.H. Number of observations varies from one per day to several per hour.

Visibility means number of days with less than ½-mile visibility.

Note: "—" indicates slight trace or possibility blank indicates no statistics available

BOLOGNA

ITALY

Elevation 131 Feet
Latitude 44° 3' N   Longitude 11° 17' E

Month	Temperatures—in °F Daily Average Max.	Min.	Extreme Max.	Min.	No. of Days Over 90°	Under 32°	0°	Relative Humidity A.M. 6:00	P.M. 1:00	Precipitation Inches Total Prec.	Snow Only	No. of Days Total Prec. 0.04"	0.1"	Snow Only 1.5"	Thun. Stms.	Max. Inches in 24 Hrs.	Wind Velocity 18 M.P.H.	30 M.P.H.	Visibility ½ Mile
Jan.	41	29	64	10	0	—	0	90	79	2.1	—	7	7	—					
Feb.	48	31	72	9	0	—	0	88	71	1.6	—	6	5	—					
March	58	39	79	22	0	—	0	87	61	2.4	—	7	7	—					
April	68	46	85	32	0	—	0	89	55	3.2	0	7	7	0					
May	82	54	87	41	0	0	0	87	57	2.9	0	7	7	0					
June	82	60	97	45	5	0	0	81	48	2.4	0	7	7	0					
July	87	64	101	51	11	0	0	82	47	1.7	0	5	5	0					
Aug.	86	64	103	49	10	0	0	83	49	1.7	0	5	5	0					
Sept.	80	60	95	44	8	0	0	89	57	2.5	0	7	7	0					
Oct.	65	50	82	29	0	—	0	96	73	4.3	0	9	9	0					
Nov.	53	41	71	26	0	—	0	94	80	3.1	—	8	8	—					
Dec.	44	33	65	12	0	—	0	94	85	1.9	—	7	6	—					
Year	66	48	103	9	34	0+	0	88	64	30.0	0+	82	80	0+					

Total precipitation means the total of rain, snow, hail, etc. expressed as inches of rain.

Ten inches of snow equals approx. 1 inch of rain.
Temperatures are as ° F.

Wind velocity is the percentage of observations greater than 18 and 30 M.P.H. Number of observations varies from one per day to several per hour.

Visibility means number of days with less than ½-mile visibility.

Note: "—" indicates slight trace or possibility blank indicates no statistics available

BOLZANO

ITALY

Month	TEMPERATURES—IN °F							RELATIVE HUMIDITY		PRECIPITATION							WIND VELOCITY		VISIBILITY ½ MILE
	Daily Average		Extreme		No. of Days					Inches		No. of Days							
								A.M. 6:00	P.M. 1:00	Total Prec.	Snow Only	Total Prec.		Snow Only 1.5"	Thun. Stms.	Max. Inches in 24 Hrs.	18 M.P.H.	30 M.P.H.	
	Max.	Min.	Max.	Min.	Over 90°	Under 32°	0°					0.04"	0.1"						
Jan.	41	23	60	5	0	—	0	84	63	0.9	—	4	3	—					
Feb.	48	27	71	10	0	—	0	84	57	1.4	—	5	5	—					
March	59	35	78	15	0	—	0	82	50	1.4	—	5	5	—					
April	69	43	90	25	0	—	0	82	49	1.7	—	7	5	—					
May	74	49	95	28	—	—	0	83	50	2.3	—	8	6	—					
June	80	55	94	33	—	0	0	83	53	3.5	0	10	8	0					
July	84	58	100	43	7	0	0	83	53	3.1	0	9	8	0					
Aug.	83	57	98	42	6	0	0	87	54	3.5	0	9	8	0					
Sept.	79	52	92	34	—	0	0	88	55	2.4	0	6	6	0					
Oct.	66	42	81	24	0	—	0	91	55	2.0	—	6	6	—					
Nov.	53	29	70	14	0	—	0	88	59	2.5	—	7	7	—					
Dec.	44	25	62	6	0	—	0	89	69	1.2	—	5	4	—					
Year	65	41	100	5	13+	0+	0	85	56	26.0	0+	81	71	0+					

Elevation 778 Feet
Latitude 46° 28' N   Longitude 11° 19' E

Total precipitation means the total of rain, snow, hail, etc. expressed as inches of rain.

Ten inches of snow equals approx. 1 inch of rain.

Temperatures are as ° F.

Wind velocity is the percentage of observations greater than 18 and 30 M.P.H. Number of observations varies from one per day to several per hour.

Visibility means number of days with less than ½-mile visibility.

Note: "—" indicates slight trace or possibility blank indicates no statistics available

BORDEAUX

FRANCE

Elevation 161 Feet
Latitude 44° 49' N   **Longitude 0° 42' W**

| Month | TEMPERATURES—IN °F | | | | | | | RELATIVE HUMIDITY | | PRECIPITATION | | | | | | | WIND VELOCITY | | VISIBILITY ½ MILE |
| | Daily Average | | Extreme | | No. of Days | | | | | Inches | | No. of Days | | | Thun. Stms. | Max. Inches in 24 Hrs. | | | |
	Max.	Min.	Max.	Min.	Over 90°	Under 32°	0°	A.M. 7:00	P.M. 1:00	Total Prec.	Snow Only	Total Prec. 0.004"	0.1"	Snow Only 1.5"			18 M.P.H.	30 M.P.H.	
Jan.	48	35	66	10	0.0	14	0	93	76	2.7	—	16	8	—	0.3	1.4	4.0	0.3	6
Feb.	52	36	79	9	0.0	9	0	91	68	2.8	—	14	8	—	0.8	1.5	5.0	0.2	4
March	58	40	84	21	0.0	6	0	91	66	2.9	—	12	7	—	1.0	1.6	4.0	0.1	3
April	63	44	88	22	0.0	1	0	89	57	2.6	—	15	7	—	2.0	1.9	5.0	0.2	2
May	69	49	95	32	0.7	—	0	87	59	2.5	0	15	7	0	5.0	1.3	4.0	0.0	2
June	75	54	101	37	1.2	0	0	86	58	2.3	0	13	6	0	5.0	2.1	2.0	0.0	1
July	80	58	101	41	3.7	0	0	87	57	2.0	0	11	6	0	4.0	1.6	2.0	0.1	2
Aug.	80	57	102	33	2.4	0	0	90	54	1.9	0	10	5	0	3.0	1.9	2.0	0.0	2
Sept.	75	54	100	29	0.6	—	0	94	59	2.2	0	12	6	0	3.0	1.5	2.0	0.3	4
Oct.	66	47	88	22	0.0	1	0	95	67	3.0	—	13	7	—	1.0	1.5	2.0	0.1	7
Nov.	55	41	77	18	0.0	6	0	95	77	3.9	—	16	9	—	0.8	1.9	4.0	0.2	8
Dec.	49	37	70	9	0.0	9	0	94	81	3.9	—	17	10	—	0.9	1.1	4.0	0.2	7
Year	64	46	102	9	9.0	46	0	90	65	33.0	0+	164	86	0+	27.0	2.1	3.3	0.1	48

Total precipitation means the total of rain, snow, hail, etc. expressed as inches of rain.

Ten inches of snow equals approx. 1 inch of rain.

Temperatures are as ° F.

Wind velocity is the percentage of observations greater than 18 and 30 M.P.H. Number of observations varies from one per day to several per hour.

Visibility means number of days with ess than ½-mile visibility.

Note: "—" indicates slight trace or possibility blank indicates no statistics available

BRAGANCA

PORTUGAL

| Month | TEMPERATURES—IN °F | | | | | | | RELATIVE HUMIDITY | | PRECIPITATION | | | | | | | WIND VELOCITY | | VISI-BIL-ITY |
| | Daily Average | | Extreme | | No. of Days | | | | | Inches | | No. of Days | | | | Max. Inches in 24 Hrs. | | | ½ MILE |
	Max.	Min.	Max.	Min.	Over 90°	Under 32°	0°	A.M. 9:00	P.M. 1:00	Total Prec.	Snow Only	Total Prec. 0.04"	0.1"	Snow Only 1.5"	Thun. Stms.		18 M.P.H.	30 M.P.H.	
Jan.	46	31	66	10	0	12	0	89	83	12.0	—	12		—		9.4	8	1.0	
Feb.	50	31	66	16	0	9	0	85	73	6.9	—	13	11	—		4.6	9	2.0	
March	53	37	77	23	0	3	0	79	70	7.7	—	7		—		3.9	7	—	
April	59	39	83	27	0	3	0	70	66	3.7	—	9	7	—		1.6	5	—	
May	65	44	88	28	0	—	0	68	61	3.0	—	9	7	—		1.1	4	0.0	
June	74	50	96	37	2	0	0	66	57	1.6	0	5	5	0		1.1	4	—	
July	80	54	103	37	5	0	0	58	54	0.5	0	2	1	0		0.7	3	—	
Aug.	81	54	99	33	6	0	0	58	49	0.6	0	2	2	0		1.1	5	—	
Sept.	74	49	95	33	2	0	0	69	57	1.5	0	5	5	0		1.9	3	0.0	
Oct.	62	42	82	26	0	—	0	78	69	3.0	—	7	7	—		2.4	2	—	
Nov.	52	36	72	22	0	5	0	85	76	6.3	—	11	10	—		2.6	6	—	
Dec.	46	31	63	16	0	11	0	89	74	7.1	—	11	11	—		4.3	10	—	
Year	62	42	103	10	14	43	0	74	66	54.0	—	93		—		9.4	6	0.4	

Elevation 2,356 Feet

Latitude 41° 49' N   Longitude 6° 40' W

Total precipitation means the total of rain, snow, hail, etc. expressed as inches of rain.

Ten inches of snow equals approx. 1 inch of rain.

Temperatures are as ° F.

Wind velocity is the percentage of observations greater than 18 and 30 M.P.H. Number of observations varies from one per day to several per hour.

Visibility means number of days with less than ½-mile visibility.

Note: "—" indicates slight trace or possibility blank indicates no statistics available

BREMEN (West) FED. REP. GERMANY

| Month | TEMPERATURES—IN °F | | | | | | | RELATIVE HUMIDITY | | PRECIPITATION | | | | | | | WIND VELOCITY | | VISIBILITY ½ MILE |
| | Daily Average | | Extreme | | No. of Days | | | A.M. 6:30 | P.M. 1:30 | Inches | | No. of Days | | | | Max. Inches in 24 Hrs. | 18 M.P.H. | 30 M.P.H. | |
	Max.	Min.	Max.	Min.	Over 90°	Under 32°	0°			Total Prec.	Snow Only	Total Prec. 0.04"	0.1"	Snow Only 1.5"	Thun. Stms.				
Jan.	37	30	55	−7	0	17	—	92	85	1.9		12	6		—	2	17	3	6
Feb.	40	30	62	−7	0	16	—	91	79	1.6		9	5		—	1	16	2	6
March	45	34	72	0	0	15	—	91	72	1.8		10	6		—	1	13	1	8
April	53	38	81	20	0	4	0	87	62	1.5	0	10	5		2	1	15	1	3
May	63	46	94	28	—	—	0	84	59	2.1		10	6		4	1	3	0	4
June	68	51	90	35	—	0	0	82	62	2.6	0	10	7	0	5	1	4	—	1
July	71	55	93	41	—	0	0	84	65	3.2	0	11	8	0	6	2	7	—	3
Aug.	69	55	91	38	1	0	0	87	66	2.8	0	12	7	0	4	1	8	0	4
Sept.	64	50	90	33	—	0	0	91	68	2.1	0	9	6	0	2	2	12	—	5
Oct.	54	43	76	22	0	4	0	93	76	2.2		11	6		1	1	9	1	8
Nov.	45	36	65	11	0	8	0	92	83	2.0		10	6		—	1	15	1	7
Dec.	39	32	59	3	0	14	0	92	87	2.2		12	7		—	1	21	2	8
Year	54	42	94	−7	1+	78	0+	89	72	26.0		126	75		24	2	12	1	63

Elevation 11 Feet

Latitude 53° 2' N   Longitude 8° 47' E

Total precipitation means the total of rain, snow, hail, etc. expressed as inches of rain.

Ten inches of snow equals approx. 1 inch of rain.

Temperatures are as °F.

Wind velocity is the percentage of observations greater than 18 and 30 M.P.H. Number of observations varies from one per day to several per hour.

Visibility means number of days with less than ½-mile visibility.

Note: "—" indicates slight trace or possibility

blank indicates no statistics available

BREST

FRANCE

Month	TEMPERATURES—IN °F Daily Average Max.	Min.	Extreme Max.	Min.	No. of Days Over 90°	Under 32°	0°	RELATIVE HUMIDITY A.M. 6:30	P.M. 1:00	PRECIPITATION Inches Total Prec.	Snow Only	No. of Days Total Prec. 0.004"	0.1"	Snow Only 1.5"	Thun. Stms.	Max. Inches in 24 Hrs.	WIND VELOCITY 18 M.P.H.	30 M.P.H.	VISIBILITY ½ MILE
Jan.	48	39	68	12	0	1.5	0	84	83	5.2	—	18	11	—	1.0	2.8	25	1.0	
Feb.	48	38	69	12	0	2.6	0	84	81	3.8	—	15	10	—	1.0	4.1	24	0.9	
March	53	41	75	23	0	1.3	0	84	77	3.3	—	15	7	—	1.0	1.9	27	0.9	
April	55	42	83	27	0	0.5	0	84	76	2.7	—	14	7	—	0.2	3.4	27	0.0	
May	60	46	90	30	—	0.1	0	85	77	2.7	0	13	7	0	1.0	4.0	25	0.4	
June	65	51	92	34	—	0.0	0	85	77	2.2	0	10	6	0	2.0	5.1	17	0.0	
July	67	54	100	43	—	0.0	0	85	76	2.4	0	12	7	0	1.0	2.7	21	0.0	
Aug.	68	54	95	41	—	0.0	0	86	77	3.2	0	13	8	0	1.0	2.1	17	0.0	
Sept.	65	52	95	38	—	0.0	0	86	77	3.5	0	12	8	0	2.0	1.8	18	0.4	
Oct.	59	48	86	30	0	—	0	85	79	4.1	0	17	9	0	1.0	2.7	15	0.3	
Nov.	53	43	72	18	0	0.4	0	84	82	5.4	—	18	10	—	1.0	2.4	19	0.7	
Dec.	49	40	63	19	0	2.5	0	84	83	5.9	—	20	11	—	2.0	3.8	22	1.5	
Year	58	46	100	12	0+	9.0	0	85	79	45.0	0+	177	101	0+	14.0	5.0	21	0.5	

Elevation 338 Feet

Latitude 48° 27' N   Longitude 4° 25' W

Total precipitation means the total of rain, snow, hail, etc. expressed as inches of rain.

Ten inches of snow equals approx. 1 inch of rain.
Temperatures are as ° F.

Wind velocity is the percentage of observations greater than 18 and 30 M.P.H. Number of observations varies from one per day to several per hour.

Visibility means number of days with less than ½-mile visibility.

Note: "—" indicates slight trace or possibility blank indicates no statistics available

BRIANÇON

FRANCE

Month	TEMPERATURES—IN °F Daily Average Max.	Min.	Extreme Max.	Min.	No. of Days Over 90°	Under 32°	0°	RELATIVE HUMIDITY A.M. —	P.M. —	PRECIPITATION Inches Total Prec.	Snow Only	No. of Days Total Prec. 0.004" / 0.1"	Snow Only 1.5"	Thun. Stms.	Max. Inches in 24 Hrs.	WIND VELOCITY 18 M.P.H.	30 M.P.H.	VISIBILITY ½ MILE
Jan.	35	18	50	7						1.9		7			3.3			
Feb.	38	18	63	8						1.9		7			1.1			
March	42	25	58	9						2.0		7			1.6			
April	50	31	73	20						2.1		8			1.5			
May	57	38	75	28						2.4		9			0.8			
June	65	45	75	33						2.3		7			1.2			
July	73	50	83	38						1.7		6			1.9			
Aug.	72	49	89	39						1.7		5			1.4			
Sept.	64	46	79	34						1.9		7			1.3			
Oct.	54	37	71	21						3.9		8			1.5			
Nov.	44	29	60	13						3.0		8			1.1			
Dec.	36	22	53	9						2.6		7			3.6			
Year	52	34	89	7						27.0		86			3.6			

Elevation 4,613 Feet

Latitude 44° 53' N   **Longitude 6° 39' E**

Total precipitation means the total of rain, snow, hail, etc. expressed as inches of rain.

Ten inches of snow equals approx. 1 inch of rain.

Temperatures are as ° F.

Wind velocity is the percentage of observations greater than 18 and 30 M.P.H. Number of observations varies from one per day to several per hour.

Visibility means number of days with less than ½-mile visibility.

Note: "—" indicates slight trace or possibility

blank indicates no statistics available

BRINDISI

ITALY

| Month | TEMPERATURES—IN °F | | | | | | | RELATIVE HUMIDITY | | PRECIPITATION | | | | | | WIND VELOCITY | | VISI- BIL- ITY |
| | Daily Average | | Extreme | | No. of Days | | | | | Inches | | No. of Days | | | Max. Inches | | | ½ |
	Max.	Min.	Max.	Min.	Over 90°	Under 32°	0°	A.M. 6:00	P.M. 1:00	Total Prec.	Snow Only	Total Prec. 0.1"	Snow Only 1.5"	Thun. Stms.	in 24 Hrs.	18 M.P.H.	30 M.P.H.	MILE
Jan.	55	43	68	26	0	1	0	82	73	3.4	—	9	—	2	4	12	—	3
Feb.	57	43	73	29	0	1	0	81	68	1.7	0	6	0	2	3	20	1.0	1
March	60	45	80	24	0	1	0	82	67	1.6	—	5	—	1	3	14	—	2
April	65	50	83	36	0	0	0	81	67	1.0	0	3	0	2	4	12	—	2
May	73	57	95	44	—	0	0	80	67	1.3	0	4	0	3	2	13	0.0	1
June	80	64	99	50	1	0	0	75	65	0.3	0	1	0	2	3	9	—	1
July	84	68	100	54	4	0	0	75	64	0.8	0	2	0	3	3	12	—	1
Aug.	84	69	103	51	4	0	0	75	62	1.1	0	3	0	2	2	13	—	—
Sept.	80	65	103	51	—	0	0	80	66	1.6	0	5	0	3	2	7	—	2
Oct.	70	58	84	45	0	0	0	83	70	3.5	0	8	0	4	3	10	1.0	2
Nov.	64	52	78	37	0	0	0	83	72	3.1	0	8	0	1	4	14	1.0	2
Dec.	58	46	73	33	0	0	0	84	73	3.3	0	9	0	1	4	15	—	2
Year	69	55	103	24	9	3	0	80	68	23.0	—	63	—	26	4	12	0.3	19

Elevation 69 Feet

Latitude 40° 39' N   Longitude 17° 57' E

Total precipitation means the total of rain, snow, hail, etc. expressed as inches of rain.

Ten inches of snow equals approx. 1 inch of rain.

Temperatures are as ° F.

Wind velocity is the percentage of observations greater than 18 and 30 M.P.H. Number of observations varies from one per day to several per hour.

Visibility means number of days with less than ½-mile visibility.

Note: "—" indicates slight trace or possibility blank indicates no statistics available

BRUSSELS   BELGIUM

| Month | Temperatures—in °F | | | | | | | Relative Humidity | | Precipitation | | | | | | | Wind Velocity | | Visibility ½ Mile |
| | Daily Average | | Extreme | | No. of Days | | | A.M. 3:30 | P.M. 1:30 | Inches | | No. of Days | | | | Max. Inches in 24 Hrs. | 18 M.P.H. | 30 M.P.H. | |
	Max.	Min.	Max.	Min.	Over 90°	Under 32°	0°			Total Prec.	Snow Only	Total Prec. 0.04"	0.1"	Snow Only 1.5"	Thun. Stms.				
Jan.	40	33	57	10	0	—	0	94	82	2.7	—	12	8	—	0.0	1.5			
Feb.	41	31	64	1	0	—	0	95	79	1.9	—	10	6	—	0.3	1.0			
March	49	36	70	14	0	—	0	93	71	2.0	—	11	6	—	1.0	0.9			
April	57	40	86	28	0	—	0	92	62	2.5	—	12	7	—	1.0	1.5			
May	64	47	88	32	0	—	0	93	62	2.7	0	10	7	0	4.0	1.4			
June	69	52	88	37	0	0	0	94	62	1.7	0	11	5	0	3.0	1.7			
July	73	56	97	43	—	0	0	95	63	3.1	0	11	8	0	4.0	1.9			
Aug.	73	56	90	43	0	0	0	94	63	2.7	0	11	7	0	3.0	1.4			
Sept.	67	52	93	37	—	0	0	96	67	2.4	0	10	6	0	3.0	1.7			
Oct.	59	45	75	23	0	—	0	95	76	1.7	—	12	5	—	1.0	1.2			
Nov.	48	40	68	23	0	—	0	96	82	3.2	—	12	8	—	0.3	1.5			
Dec.	43	36	63	3	0	—	0	96	87	3.1	—	13	9	—	0.3	1.3			
Year	57	44	97	1	0+	0+	0	94	71	30.0	0+	135	82	0+	21.0	1.9			

Elevation 180 Feet

Latitude 50° 54' N   Longitude 4° 29' E

Total precipitation means the total of rain, snow, hail, etc. expressed as inches of rain.

Ten inches of snow equals approx. 1 inch of rain.

Temperatures are as ° F.

Wind velocity is the percentage of observations greater than 18 and 30 M.P.H. Number of observations varies from one per day to several per hour.

Visibility means number of days with less than ½-mile visibility.

Note: "—" indicates slight trace or possibility

         blank indicates no statistics available

CAGLIARI

SARDINIA, ITALY

| Month | TEMPERATURES—IN °F | | | | | | | RELATIVE HUMIDITY | | PRECIPITATION | | | | | | | WIND VELOCITY | | VISI-BIL-ITY |
| | Daily Average | | Extreme | | No. of Days | | | | | Inches | | No. of Days | | | | Max. Inches in 24 Hrs. | | | ½ |
	Max.	Min.	Max.	Min.	Over 90°	Under 32°	Under 0°	A.M. 5:30	P.M. 12:30	Total Prec.	Snow Only	Total Prec. 0.04"	0.1"	Snow Only 1.5"	Thun. Stms.		18 M.P.H.	30 M.P.H.	MILE
Jan.	57	42	68	26	0	3	0	87	73	1.7	—	9	5	—	1	1	6	—	1
Feb.	58	43	69	26	0	2	0	87	69	1.4	—	7	5	—	2	1	12	1	1
March	62	45	81	28	0	1	0	88	66	2.1	—	6	6	—	1	1	8	1	2
April	65	49	82	35	0	0	0	86	65	1.6	0	6	5	0	2	1	8	—	1
May	72	55	85	44	0	0	0	85	65	1.0	0	5	3	0	2	1	7	0	2
June	80	56	100	47	2	0	0	82	58	0.9	0	3	3	0	2	2	9	—	1
July	86	56	100	55	6	0	0	81	58	0.1	0	1	0	0	1	—	9	—	1
Aug.	85	67	101	54	5	0	0	82	61	0.4	0	2	1	0	2	1	8	—	—
Sept.	80	64	102	51	—	0	0	86	61	1.1	0	4	4	0	3	2	5	0	1
Oct.	74	57	88	43	0	0	0	89	64	2.0	0	7	6	0	2	2	3	0	1
Nov.	66	50	77	33	0	0	0	87	67	4.0	0	9	9	0	2	7	7	1	1
Dec.	62	45	71	32	0	—	0	87	73	2.7	0	10	8	0	1	3	6	1	2
Year	71	52	102	26	13	6	0	86	65	19.0	0+	69	55	0+	21	7	7	0+	0+

Elevation 12 Feet
Latitude 39° 14' N   Longitude 9° 3' E

Total precipitation means the total of rain, snow, hail, etc. expressed as inches of rain.

Ten inches of snow equals approx. 1 inch of rain.

Temperatures are as ° F.

Wind velocity is the percentage of observations greater than 18 and 30 M.P.H. Number of observations varies from one per day to several per hour.

Visibility means number of days with less than ½-mile visibility.

Note: "—" indicates slight trace or possibility blank indicates no statistics available

CARDIFF (United Kingdom) WALES

Month	Daily Average Max.	Daily Average Min.	Extreme Max.	Extreme Min.	No. of Days Over 90°	No. of Days Under 32°	No. of Days 0°	Rel. Hum. A.M. 9:00	Rel. Hum. P.M. 9:00	Inches Total Prec.	Inches Snow Only	Days Total Prec. 0.01"	Days Total Prec. 0.1"	Days Snow Only 1.5"	Thun. Stms.	Max. Inches in 24 Hrs.	Wind 18 M.P.H.	Wind 30 M.P.H.	Visib. ½ Mile
Jan.	45	36	59	2	0	—	0	90	88	3.7	—	18	9	—	0.4	2.0			
Feb.	46	36	59	12	0	—	0	86	86	2.9	—	15	8	—	0.1	1.3			
March	50	37	70	18	0	—	0	81	83	3.1	—	14	7	—	0.4	1.2			
April	55	41	75	27	0	—	0	75	81	2.5	—	14	7	—	0.5	1.3			
May	61	45	84	31	0	—	0	74	82	2.5	0	14	7	0	1.0	4.2			
June	66	51	87	39	0	0	0	73	80	2.5	0	13	7	0	0.5	1.3			
July	69	54	90	45	—	0	0	76	83	3.1	0	14	8	0	2.0	2.2			
Aug.	68	54	91	41	—	0	0	78	85	4.2	0	15	9	0	1.0	2.3			
Sept.	64	51	83	36	0	0	0	81	85	3.1	0	15	8	0	0.5	2.3			
Oct.	57	45	76	26	0	—	0	85	87	4.7	—	17	9	—	0.5	2.3			
Nov.	50	40	65	23	0	—	0	87	87	4.1	—	17	9	—	0.2	2.5			
Dec.	46	37	59	19	0	—	0	88	88	5.0	—	18	11	—	0.8	1.6			
Year	56	44	91	2	0+	0+	0	81	85	41.0	0+	184	99	0+	8.0	4.2			

Elevation 220 Feet
Latitude 51° 24' N   Longitude 3° 21' W
Total precipitation means the total of rain, snow, hail, etc. expressed as inches of rain.
Ten inches of snow equals approx. 1 inch of rain.
Temperatures are as ° F.

Wind velocity is the percentage of observations greater than 18 and 30 M.P.H. Number of observations varies from one per day to several per hour.
Visibility means number of days with less than ½-mile visibility.
Note: "—" indicates slight trace or possibility
blank indicates no statistics available

CHERBOURG

FRANCE

Month	Daily Average Max.	Min.	Extreme Max.	Min.	No. of Days Over 90°	Under 32°	0°	Relative Humidity A.M. 7:00	P.M. 1:00	Precipitation Inches Total Prec.	Snow Only	No. of Days Total Prec. 0.004"	0.1"	Snow Only 1.5"	Thun. Stms.	Max. Inches in 24 Hrs.	Wind Velocity 18 M.P.H.	30 M.P.H.	Visibility ½ Mile
Jan.	47	40	59	21	0	—	0	86	80	3.3	—	19	9	—	0.3	1.5			
Feb.	48	39	65	14	0	—	0	85	78	2.9	—	14	8	—	0.0	2.8			
March	50	41	73	25	0	—	0	85	73	2.7	—	14	7	—	0.3	1.8			
April	54	43	78	33	0	0	0	84	73	2.0	0	14	6	0	1.0	2.4			
May	60	49	86	38	0	0	0	83	74	1.9	0	12	6	0	1.0	1.4			
June	64	53	89	43	0	0	0	84	75	1.8	0	9	5	0	1.0	0.9			
July	67	57	89	46	0	0	0	85	75	1.9	0	12	5	0	1.0	1.8			
Aug.	68	58	91	48	—	0	0	86	75	3.0	0	14	5	0	2.0	2.2			
Sept.	65	55	87	43	—	0	0	87	75	2.9	0	14	7	0	1.0	1.4			
Oct.	59	50	79	36	0	0	0	84	73	4.6	0	16	9	0	1.0	1.4			
Nov.	53	45	66	31	0	—	0	84	78	5.1	0	18	10	0	1.0	3.0			
Dec.	49	41	62	22	0	—	0	84	79	5.2	—	19	11	—	0.3	1.4			
Year	57	48	91	14	0+	0+	0	85	76	37.0	0+	175	90	0+	10.0	3.0			

Elevation 456 Feet

Latitude 49° 39' N   Longitude 1° 28' W

Total precipitation means the total of rain, snow, hail, etc. expressed as inches of rain.

Ten inches of snow equals approx. 1 inch of rain.

Temperatures are as ° F.

Wind velocity is the percentage of observations greater than 18 and 30 M.P.H. Number of observations varies from one per day to several per hour.

Visibility means number of days with less than ½-mile visibility.

Note: "—" indicates slight trace or possibility blank indicates no statistics available

CLERMONT—FERRAND

FRANCE

Month	TEMPERATURES—IN °F							RELATIVE HUMIDITY		PRECIPITATION						WIND VELOCITY		VISIBILITY ½ MILE
	Daily Average		Extreme		No. of Days					Inches		No. of Days						
	Max.	Min.	Max.	Min.	Over 90°	Under 32°	0°	A.M. 5:00	P.M. 2:00	Total Prec.	Snow Only	Total Prec. 0.1"	Snow Only 1.5"	Thun. Stms.	Max. Inches in 24 Hrs.	18 M.P.H.	30 M.P.H.	
Jan.	44	30	66	—4	0.0	3	—	82	67	1.0	—	3	—	0.3		13	0.6	
Feb.	48	31	71	—10	0.0	9	—	81	59	1.1	—	4	—	0.0		15	0.5	
March	54	35	78	8	0.0	8	0	80	55	1.5	—	5	—	0.3		13	0.6	
April	60	39	88	24	0.0	6	0	78	54	2.1	—	6	—	0.3		10	0.1	
May	68	45	97	29	—	—	0	82	54	2.9	0	7	0	3.0		6	0.0	
June	74	51	101	36	0.4	0	0	85	54	3.1	0	8	0	3.0		4	0.0	
July	78	54	106	39	4.6	0	0	80	51	2.3	0	6	0	3.0		4	0.0	
Aug.	78	53	107	39	2.0	0	0	84	50	2.7	0	7	0	3.0		5	0.0	
Sept.	73	49	95	28	0.9	3	0	87	56	2.4	—	6	—	2.0		5	0.1	
Oct.	63	42	85	23	0.0	12	0	88	62	2.2	—	6	—	0.3		7	0.0	
Nov.	52	36	74	10	0.0	14	0	85	69	1.8	—	5	—	0.3		11	0.8	
Dec.	46	33	70	—1	0.0	21	—	83	70	1.4	—	5	—	0.3		13	0.3	
Year	62	42	107	—10	8.0	76	0+	83	59	25.0	0+	68	0+	16.0		9	0.3	

Elevation 1,089 Feet

Latitude 45° 47' N   Longitude 3° 9' E

Total precipitation means the total of rain, snow, hail, etc. expressed as inches of rain.

Ten inches of snow equals approx. 1 inch of rain.

Temperatures are as ° F.

Wind velocity is the percentage of observations greater than 18 and 30 M.P.H. Number of observations varies from one per day to several per hour.

Visibility means number of days with less than ½-mile visibility.

Note: "—" indicates slight trace or possibility
blank indicates no statistics available

COIMBRA

PORTUGAL

| Month | TEMPERATURES—IN °F | | | | | | | RELATIVE HUMIDITY | | PRECIPITATION | | | | | | | WIND VELOCITY | | VISI- BIL- ITY ½ MILE |
| | Daily Average | | Extreme | | No. of Days | | | | | Inches | | No. of Days | | | | Max. Inches in 24 Hrs. | | | |
	Max.	Min.	Max.	Min.	Over 90°	Under 32°	0°	A.M. 8:30	P.M. 2:30	Total Prec.	Snow Only	Total Prec. 0.04"	0.1"	Snow Only 1.5"	Thun. Stms.		18 M.P.H.	30 M.P.H.	
Jan.	57	42	73	26	0	3	0	84	67	4.9	—	12	11	—	1	3.0	4	0	
Feb.	60	43	79	25	0	3	0	80	61	3.9	—	10	10	—	1	2.2	3	0	
March	64	47	87	28	0	1	0	74	60	4.7	—	13	8	—	2	2.0	3	0	
April	68	49	96	34	—	0	0	69	56	3.5	0	11	7	0	3	1.6	3	0	
May	72	52	99	40	2	0	0	70	57	2.8	0	10	7	0	3	1.2	2	—	
June	79	57	105	44	3	0	0	68	53	1.4	0	5	4	0	3	1.8	—	0	
July	83	59	114	48	7	0	0	67	48	0.6	0	3	2	0	1	0.8	—	0	
Aug.	85	59	111	48	7	0	0	66	45	0.6	0	3	2	0	1	1.4	—	—	
Sept.	81	58	104	42	5	0	0	70	50	1.9	0	6	6	0	2	4.8	1	0	
Oct.	73	54	98	36	1	0	0	74	57	3.5	0	9	8	0	2	2.5	2	0	
Nov.	63	47	83	30	0	—	0	81	66	4.4	0	10	9	0	1	2.4	6	—	
Dec.	58	43	72	27	0	1	0	83	70	5.2	—	11	11	—	1	2.8	5	—	
Year	70	51	114	25	25	8+	0	74	57	38.0	—	103	83	—	22	4.8	3	—	

Elevation 459 Feet

Latitude 40° 12' N   Longitude 8° 25' W

Total precipitation means the total of rain, snow, hail, etc. expressed as inches of rain.

Ten inches of snow equals approx. 1 inch of rain.

Temperatures are as ° F.

Wind velocity is the percentage of observations greater than 18 and 30 M.P.H. Number of observations varies from one per day to several per hour.

Visibility means number of days with less than ½-mile visibility.

Note: "—" indicates slight trace or possibility
blank indicates no statistics available

COPENHAGEN

DENMARK

| Month | TEMPERATURES—IN °F | | | | | | | RELATIVE HUMIDITY | | PRECIPITATION | | | | | | | WIND VELOCITY | | VISIBILITY ½ MILE |
| | Daily Average | | Extreme | | No. of Days | | | A.M. 8:00 | P.M. 2:00 | Inches | | No. of Days | | | | Max. Inches in 24 Hrs. | 18 M.P.H. | 30 M.P.H. | |
	Max.	Min.	Max.	Min.	Over 90°	Under 32°	0°			Total Prec.	Snow Only	Total Prec. 0.04"	0.1"	Snow Only 1.5"	Thun. Stms.				
Jan.	35	28	49	−10	0	20	—	89	86	1.5	—	9	5	—	—	0.6	38	5	5
Feb.	36	27	54	−13	0	19	—	92	86	1.2	—	7	4	—	0	0.3	36	4	5
March	40	30	62	−1	0	21	—	87	77	1.5	—	8	5	—	0	0.7	31	4	4
April	50	36	79	20	0	6	0	80	66	1.6	—	9	5	—	1	0.8	21	1	1
May	61	44	85	26	0	—	0	73	60	1.6	—	8	5	—	1	1.1	19	—	1
June	69	51	90	31	—	—	0	73	61	1.9	0	8	5	0	2	1.2	16	—	1
July	72	54	91	39	—	0	0	77	63	2.4	0	9	6	0	4	1.6	15	—	—
Aug.	69	53	89	33	0	0	0	82	67	3.0	0	12	7	0	3	1.6	18	—	1
Sept.	63	48	86	26	0	—	0	87	70	2.0	—	8	6	—	1	1.3	24	—	2
Oct.	53	42	74	19	0	—	0	90	77	2.2	—	9	6	—	1	1.1	28	1	3
Nov.	43	35	57	5	0	5	0	89	83	1.9	—	10	5	—	—	1.1	36	3	2
Dec.	38	31	54	−1	0	12	—	89	87	2.1	—	11	7	—	—	0.9	40	4	2
Year	52	40	91	−13	0+	84	0+	84	74	23.0	—	108	67	—	13	1.6	27	2	29

Elevation 16 Feet

Latitude 55° 38' N   Longitude 12° 40' E

Total precipitation means the total of rain, snow, hail, etc. expressed as inches of rain.

Ten inches of snow equals approx. 1 inch of rain.

Temperatures are as ° F.

Wind velocity is the percentage of observations greater than 18 and 30 M.P.H. Number of observations varies from one per day to several per hour.

Visibility means number of days with less than ½-mile visibility.

Note: "—" indicates slight trace or possibility
blank indicates no statistics available

CORK

IRELAND

Elevation 502 Feet
Latitude 51° 50' N   Longitude 8° 29' W
Total precipitation means the total of rain, snow, hail, etc. expressed as inches of rain.
Ten inches of snow equals approx. 1 inch of rain.
Temperatures are as ° F.

Wind velocity is the percentage of observations greater than 18 and 30 M.P.H. Number of observations varies from one per day to several per hour.
Visibility means number of days with less than ½-mile visibility.
Note: "—" indicates slight trace or possibility blank indicates no statistics available

Month	TEMPERATURES—IN ° F Daily Average Max.	Min.	Extreme Max.	Min.	No. of Days Over 90°	Under 32°	0°	RELATIVE HUMIDITY A.M. 8:00	P.M.	PRECIPITATION Inches Total Prec.	Snow Only	No. of Days Total Prec. 0.04"	0.1"	Snow Only 1.5"	Thun. Stms.	Max. Inches in 24 Hrs.	WIND VELOCITY 18 M.P.H.	30 M.P.H.	VISIBILITY ½ MILE
Jan.	48	38	60	15	0	—	0	89	—	4.9	—	16	11	—		2.1			
Feb.	49	37	58	20	0	—	0	88		3.6	—	13	9	—		2.5			
March	52	39	69	23	0	—	0	86		3.3	—	12	7	—		2.1			
April	55	41	72	27	0	—	0	81		2.6	—	11	7	—		2.5			
May	61	45	79	30	0	—	0	77		2.9	0	11	7	0		2.3			
June	66	50	84	37	0	0	0	78		2.0	0	9	6	0		1.8			
July	68	53	83	42	0	0	0	80		2.9	0	11	7	0		1.6			
Aug.	68	53	85	37	0	0	0	82		3.1	0	11	8	0		2.5			
Sept.	64	50	79	34	0	0	0	86		2.9	0	11	7	0		1.9			
Oct.	58	44	74	23	0	—	0	90		3.9	—	14	9	—		3.1			
Nov.	51	39	62	23	0	—	0	90		4.5	—	15	9	—		2.8			
Dec.	48	37	58	19	0	—	0	89		4.7	—	16	10	—		1.8			
Year	57	44	85	15	0	0+	0	85		41.3	0+	150	97	0+		3.1			

DEN HELDER

THE NETHERLANDS

Month	TEMPERATURES—IN °F							RELATIVE HUMIDITY		PRECIPITATION							WIND VELOCITY		VISI-BIL-ITY ½ MILE
	Daily Average		Extreme		No. of Days			A.M. 8:00	P.M. 2:00	Inches		No. of Days				Max. Inches in 24 Hrs.	18 M.P.H.	30 M.P.H.	
	Max.	Min.	Max.	Min.	Over 90°	Under 32°	0°			Total Prec.	Snow Only	Total Prec. 0.04"	0.1"	Snow Only 1.5"	Thun. Stms.				
Jan.	40	34	53	6	0	11	0	89	87	2.0	—	11	6	—	0.3	0.9	44	11	3.6
Feb.	40	33	59	0	0	12	—	88	84	1.5	—	9	5	—	0.3	1.4	34	6	5.3
March	44	36	68	19	0	9	0	87	79	1.7	—	10	5	—	1.0	2.0	29	5	5.4
April	49	41	76	25	0	—	0	82	75	1.4	—	8	5	—	1.0	1.2	34	6	1.6
May	57	47	83	31	0	—	0	80	74	1.5	0	8	5	0	3.0	2.2	29	2	1.5
June	62	53	87	36	—	0	0	79	74	1.7	0	7	5	0	4.0	1.9	25	1	1.1
July	66	57	93	44	0	0	0	80	74	2.2	0	9	6	0	4.0	1.9	30	3	0.9
Aug.	66	57	87	43	0	0	0	80	73	3.0	0	11	7	0	4.0	2.2	26	3	0.4
Sept.	63	54	86	38	0	0	0	80	72	2.9	0	11	7	0	3.0	2.7	32	4	0.6
Oct.	56	46	77	27	0	—	0	84	77	3.3	—	13	8	—	2.0	2.0	31	7	3.1
Nov.	47	40	64	13	0	2	0	87	83	2.7	—	13	7	—	2.0	1.4	30	5	3.9
Dec.	43	36	54	10	0	6	0	89	87	2.4	—	13	7	—	1.0	1.1	39	11	5.3
Year	53	45	93	0	0+	40	0+	84	78	26.0	0+	123		0+	26.0	2.7	32	5	33.0

Elevation 20 Feet
Latitude 52° 58' N   Longitude 4° 45' E

Total precipitation means the total of rain, snow, hail, etc. expressed as inches of rain.

Ten inches of snow equals approx. 1 inch of rain.

Temperatures are as ° F.

Wind velocity is the percentage of observations greater than 18 and 30 M.P.H. Number of observations varies from one per day to several per hour.

Visibility means number of days with less than ½-mile visibility.

Note: "—" indicates slight trace or possibility blank indicates no statistics available

DIJON

FRANCE

Month	TEMPERATURES—IN °F Daily Average Max.	Min.	Extreme Max.	Min.	No. of Days Over 90°	Under 32°	0°	RELATIVE HUMIDITY A.M. 7:30	P.M. 1:30	PRECIPITATION Inches Total Prec.	Snow Only	No. of Days Total Prec. 0.004"	0.1"	Snow Only 1.5"	Thun. Stms.	Max. Inches in 24 Hrs.	WIND VELOCITY 18 M.P.H.	30 M.P.H.	VISIBILITY ½ MILE
Jan.	41	29	62	1	0.0	20	0	90	79	1.9	—	14	6	—	0.0	1.3			
Feb.	42	31	68	−8	0.0	20	—	88	69	1.6	—	13	5	—	0.3	1.3			
March	53	36	74	11	0.0	15	0	81	58	1.9	—	12	6	—	0.3	1.3			
April	61	41	84	15	0.0	3	0	78	58	2.0	—	14	6	—	2.0	1.1			
May	68	48	94	30	—	—	0	78	58	2.3	0	14	6	0	4.0	1.9			
June	74	53	97	33	0.3	0	0	77	58	2.7	0	12	7	0	5.0	1.9			
July	78	57	107	37	3.7	0	0	74	57	2.5	0	11	7	0	6.0	2.5			
Aug.	77	56	100	34	3.1	0	0	77	53	2.5	0	12	7	0	4.0	2.5			
Sept.	71	51	92	29	—	—	0	83	59	2.1	0	11	6	0	3.0	1.3			
Oct.	60	43	83	21	0.0	2	0	89	70	2.9	—	13	7	—	1.0	1.5			
Nov.	49	36	69	13	0.0	8	0	92	76	2.8	—	16	7	—	0.3	1.7			
Dec.	42	32	62	7	0.0	19	0	91	81	2.3	—	15	7	—	0.0	1.3			
Year	60	43	107	−8	7.0	88	0+	83	65	28.0	0+	157	77	0+	26.0	2.5			

Elevation 728 Feet

Latitude 47° 16' N   Longitude 5° 5' E

Total precipitation means the total of rain, snow, hail, etc. expressed as inches of rain.

Ten inches of snow equals approx. 1 inch of rain.

Temperatures are as ° F.

Wind velocity is the percentage of observations greater than 18 and 30 M.P.H. Number of observations varies from one per day to several per hour.

Visibility means number of days with less than ½-mile visibility.

Note: "—" indicates slight trace or possibility
blank indicates no statistics available

DOUGLAS (Isle Of Man)   (United Kingdom) ENGLAND

Month	Daily Average Max.	Daily Average Min.	Extreme Max.	Extreme Min.	No. of Days Over 90°	No. of Days Under 32°	No. of Days 0°	Rel. Hum. A.M. 8:30	Rel. Hum. P.M. 4:30	Inches Total Prec.	Inches Snow Only	No. of Days Total Prec. 0.01"	No. of Days Total Prec. 0.1"	No. of Days Snow Only 1.5"	No. of Days Thun. Stms.	Max. Inches in 24 Hrs.	Wind Velocity 18 M.P.H.	Wind Velocity 30 M.P.H.	Visibility ½ Mile
Jan.	45	38	55	16	0	3	0	87	88	5.1	0	21	13	0	0.0	1.9	47	10	0.0
Feb.	44	37	56	14	0	1	0	85	85	3.2	—	17	8	—	0.0	1.4	56	9	0.3
March	47	38	65	22	0	2	0	83	80	2.9	—	16	8	—	0.0	1.6	37	3	0.2
April	51	41	69	23	0	1	0	79	79	2.6	0	15	5	0	0.0	1.1	20	2	1.0
May	56	45	75	31	0	—	0	77	79	2.7	0	15	5	0	0.0	2.1	12	—	1.1
June	61	50	80	36	0	0	0	79	76	2.6	0	14	7	0	0.0	2.4	23	1	0.8
July	63	53	82	39	0	0	0	83	78	3.2	0	16	8	0	0.0	1.9	19	—	0.6
Aug.	63	53	81	36	0	0	0	80	84	3.6	0	16	6	0	0.2	1.7	11	0	1.4
Sept.	60	51	77	32	0	—	0	83	81	4.2	0	17	10	0	0.2	2.7	22	1	0.2
Oct.	55	47	70	29	0	—	0	82	84	5.0	0	19	12	0	0.2	2.3	41	7	0.2
Nov.	49	41	61	25	0	1	0	84	85	4.9	0	19	10	0	0.5	1.9	39	8	0.2
Dec.	46	39	55	22	0	2	0	85	87	4.7	—	21	11	—	0.0	1.8	43	9	0.8
Year	53	44	82	14	0	10	0	82	82	45.0	0+	206	103	0+	1.1	2.7	31	4	7.0

Elevation 284 Feet
Latitude 54° 10' N   Longitude 4° 28' W
Total precipitation means the total of rain, snow, hail. etc. expressed as inches of rain.
Ten inches of snow equals approx. 1 inch of rain.
Temperatures are as ° F.

Wind velocity is the percentage of observations greater than 18 and 30 M.P.H. Number of observations varies from one per day to several per hour.
Visibility means number of days with less than ½-mile visibility.
Note: "—" indicates slight trace or possibility blank indicates no statistics available

DUBLIN

IRELAND

Month	TEMPERATURES—IN °F Daily Average Max.	Min.	Extreme Max.	Min.	No. of Days Over 90°	Under 32°	0°	RELATIVE HUMIDITY A.M. 9:00	P.M. 2:00	PRECIPITATION Inches Total Prec.	Snow Only	No. of Days Total Prec. 0.04"	0.1"	Snow Only 1.5"	Thun. Stms.	Max. Inches in 24 Hrs.	WIND VELOCITY 18 M.P.H.	30 M.P.H.	VISI-BIL-ITY ½ MILE
Jan.	46	35	62	4	0	—	0	87	80	2.7	—	13	8	—	0.3	1.9			
Feb.	47	34	65	8	0	—	0	85	76	2.2	—	11	7	—	0.3	1.8			
March	49	35	72	15	0	—	0	84	72	2.0	—	10	6	—	0.3	1.5			
April	53	37	72	19	0	—	0	78	67	1.9	—	11	6	—	0.3	1.1			
May	58	42	80	22	0	—	0	77	69	2.3	—	11	7	—	1.0	1.5			
June	64	47	84	31	0	—	0	76	69	2.0	0	11	6	.0	1.0	1.1			
July	66	51	86	35	0	0	0	79	70	2.8	0	13	7	0	2.0	1.8			
Aug.	65	50	85	33	0	0	0	83	72	3.0	0	13	7	0	1.0	1.5			
Sept.	62	46	82	29	0	—	0	85	73	2.8	0	12	7	0	1.0	2.7			
Oct.	55	41	73	22	0	—	0	86	75	2.7	—	12	7	—	0.3	2.2			
Nov.	50	38	67	15	0	—	0	87	79	2.7	—	12	7	—	0.3	1.5			
Dec.	47	35	63	7	0	—	0	87	81	2.6	—	13	8	—	0.0	1.9			
Year	55	41	86	4	0	0+	0	83	74	29.6	0+	142	83	0+	7.8	2.7			

Elevation 155 Feet
Latitude 53° 22' N   Longitude 6° 21' N
Total precipitation means the total of rain, snow, hail, etc. expressed as inches of rain.
Ten inches of snow equals approx. 1 inch of rain.
Temperatures are as ° F.

Wind velocity is the percentage of observations greater than 18 and 30 M.P.H. Number of observations varies from one per day to several per hour.
Visibility means number of days with less than ½-mile visibility.
Note: "—" indicates slight trace or possibility blank indicates no statistics available

ECHTERNACH

LUXEMBOURG

Month	TEMPERATURES—IN °F Daily Average Max.	Min.	Extreme Max.	Min.	No. of Days Over 90°	Under 32°	0°	RELATIVE HUMIDITY A.M. 6:30	P.M. 1:30	PRECIPITATION Inches Total Prec.	Snow Only	No. of Days Total Prec. 0.001"	0.1"	Snow Only 1.5"	Thun. Stms.	Max. Inches in 24 Hrs.	WIND VELOCITY 18 M.P.H.	30 M.P.H.	VISI-BIL-ITY ½ MILE
Jan.	39	28	55	7	0	—	0	90	82	2.2	—	13	7	—		0.7	0.8	0	0
Feb.	42	28	64	−8	0	—	—	89	77	2.3	—	11	7	—		1.0	0.2	0	0
March	52	32	75	10	0	—	0	89	63	1.7	—	11	5	—		0.8	0.3	0	0
April	59	37	88	21	0	—	0	89	55	1.9	—	10	6	—		0.8	0.1	0	0
May	67	43	91	25	—	—	0	89	56	2.1	—	10	6	—		2.2	0.4	0	0
June	72	50	95	34	—	0	0	90	59	2.8	0	12	7	0		1.6	0.0	0	0
July	77	54	99	39	—	0	0	89	59	2.1	0	12	6	0		1.1	0.3	0	0
Aug.	74	53	93	37	—	0	0	90	59	3.0	0	13	7	0		1.2	0.1	0	0
Sept.	69	49	93	34	—	0	0	93	64	2.7	0	15	7	0		1.1	0.1	0	0
Oct.	59	41	77	19	0	—	0	93	73	1.8	—	11	5	—		1.0	0.1	0	0
Nov.	47	35	63	14	0	—	0	93	80	2.6	—	14	7	—		0.7	0.0	0	0
Dec.	41	33	61	10	0	—	0	91	85	3.2	—	15	9	—		1.2	0.4	0	0
Year	58	40	99	−8	0+	0+	0+	90	68	28.0	0+	147	79	0		2.2	0.2	0	0

Elevation 535 Feet

Latitude 49° 49' N   **Longitude 6° 25' E**

Total precipitation means the total of rain, snow, hail, etc. expressed as inches of rain.

Ten inches of snow equals approx. 1 inch of rain.

Temperatures are as ° F.

Wind velocity is the percentage of observations greater than 18 and 30 M.P.H. Number of observations varies from one per day to several per hour.

Visibility means number of days with less than ½-mile visibility.

Note: "—" indicates slight trace or possibility blank indicates no statistics available

EDINBURGH   (United Kingdom)   SCOTLAND

Month	TEMPERATURES—IN °F Daily Average Max.	Daily Average Min.	Extreme Max.	Extreme Min.	No. of Days Over 90°	Under 32°	0°	RELATIVE HUMIDITY A.M. 7:00	P.M. 1:00	PRECIPITATION Inches Total Prec.	Snow Only	No. of Days Total Prec. 0.01"	0.1"	Snow Only 1.5"	Thun. Stms.	Max. Inches in 24 Hrs.	WIND VELOCITY 18 M.P.H.	30 M.P.H.	VISIBILITY ½ MILE
Jan.	44	38	58	24	0	5	0	84	83	1.7	—	18	6	—	0.1	1.0	24	4	1.4
Feb.	44	37	58	20	0	5	0	83	83	1.6	—	15	5	—	0.0	1.1	20	3	2.1
March	46	37	62	23	0	6	0	81	82	1.8	—	15	6	—	0.1	1.7	11	1	2.0
April	49	39	63	30	0	1	0	76	81	1.4	0	16	5	0	5.0	1.3	10	1	2.0
May	54	44	73	32	0	—	0	76	81	1.9	0	15	6	0	1.0	1.5	9	1	2.6
June	59	49	79	40	0	0	0	73	80	1.9	0	15	5	0	1.0	1.3	6	1	2.4
July	63	53	81	43	0	0	0	77	82	2.6	0	17	7	0	3.0	1.9	6	0	2.0
Aug.	62	53	79	43	0	0	0	79	84	3.1	0	17	8	0	2.0	2.8	8	1	2.1
Sept.	59	50	79	39	0	0	0	80	83	1.9	0	16	6	0	8.0	2.1	11	1	1.2
Oct.	53	45	69	31	0	—	0	81	84	2.6	0	18	7	0	3.0	1.5	23	4	0.5
Nov.	47	41	61	29	0	—	0	83	83	2.2	0	18	6	0	0.0	1.5	14	2	2.0
Dec.	44	39	57	26	0	2	0	84	84	2.1	—	17	7	—	0.0	1.0	17	5	2.5
Year	52	44	81	20	0	19	0	80	82	25.0	0+	197	74	0+	23.0	2.8	13	2	23.0

Elevation 135 Feet

Latitude 55° 56' N   Longitude 3° 20' W

Total precipitation means the total of rain, snow, hail, etc. expressed as inches of rain.

Ten inches of snow equals approx. 1 inch of rain.

Temperatures are as ° F.

Wind velocity is the percentage of observations greater than 18 and 30 M.P.H. Number of observations varies from one per day to several per hour.

Visibility means number of days with less than ½-mile visibility.

Note: "—" indicates slight trace or possibility blank indicates no statistics available

EVORA

PORTUGAL

| Month | TEMPERATURES—IN °F | | | | | | | RELATIVE HUMIDITY | | PRECIPITATION | | | | | | | WIND VELOCITY | | VISI-BILITY |
| | Daily Average | | Extreme | | No. of Days | | | | | Inches | | No. of Days | | | | Max. Inches | | | ½ MILE |
	Max.	Min.	Max.	Min.	Over 90°	Under 32°	0°	A.M. 8:30	P.M. 2:30	Total Prec.	Snow Only	Total Prec. 0.04"	0.1"	Snow Only 1.5"	Thun. Stms.	in 24 Hrs.	18 M.P.H.	30 M.P.H.	
Jan.	54	42	69	24	0	1	0	85	69	3.3	—	9	9	—	—	2.9	8	1	
Feb.	56	43	75	25	0	2	0	80	60	2.8	—	8	8	—	—	1.8	7	—	
March	60	46	80	31	0	—	0	78	58	3.5	0	8	7	0	1	1.9	8	—	
April	65	49	87	36	0	0	0	71	52	2.3	0	7	6	0	1	1.4	6	0	
May	70	51	92	36	1	0	0	67	47	1.8	0	6	6	0	1	1.6	6	—	
June	80	57	101	44	4	0	0	60	39	0.9	0	3	3	0	1	2.5	8	0	
July	86	61	108	50	12	0	0	55	31	0.2	0	1	—	0	1	1.3	8	0	
Aug.	87	62	104	50	13	0	0	56	30	0.1	0	1	—	0	—	0.7	6	0	
Sept.	80	59	101	46	4	0	0	63	39	1.3	0	4	4	0	1	2.1	4	0	
Oct.	71	55	92	39	—	0	0	72	51	2.5	0	7	7	0	1	4.1	7	1	
Nov.	61	48	79	33	0	0	0	82	64	2.9	0	8	7	0	—	2.0	7	1	
Dec.	55	43	79	28	0	—	0	84	69	3.2	—	9	9	—	—	2.3	12	1	
Year	69	51	108	24	35	3	0	71	51	25.0	—	72	66	—	8	4.0	7	4	

Elevation 1,053 Feet
Latitude 38° 34' N   Longitude 7° 54' W
Total precipitation means the total of rain, snow, hail, etc. expressed as inches of rain.
Ten inches of snow equals approx. 1 inch of rain.
Temperatures are as ° F.

Wind velocity is the percentage of observations greater than 18 and 30 M.P.H. Number of observations varies from one per day to several per hour.
Visibility means number of days with less than ½-mile visibility.
Note: "—" indicates slight trace or possibility
blank indicates no statistics available

FARO

PORTUGAL

Month	Temperatures—in °F							Relative Humidity		Precipitation						Wind Velocity		Visibility
	Daily Average		Extreme		No. of Days					Inches		No. of Days			Max. Inches in 24 Hrs.	18 M.P.H.	30 M.P.H.	½ Mile
	Max.	Min.	Max.	Min.	Over 90°	Under 32°	0°	A.M.	P.M.	Total Prec.	Snow Only	Total Prec. 0.1"	Snow Only 1.5"	Thun. Stms.				
Jan.	60	47	75	31	0	—	0	—	76	2.3	0	7	0	—		1.0	0	
Feb.	62	48	84	30	0	1	0	—	75	2.0	0	6	0	1		1.0	—	
March	64	51	81	36	0	0	0	—	73	2.3	0	6	0	1		—	0	
April	68	53	89	39	0	0	0	—	71	1.3	0	5	0	1		—	0	
May	72	56	91	41	—	0	0	—	71	0.7	0	2	0	1		—	0	
June	78	62	97	46	2	0	0	—	76	0.3	0	1	0	—		—	—	
July	83	66	102	52	7	0	0	—	66	0.0	0	0	0	—		1.0	—	
Aug.	83	67	100	46	4	0	0	—	66	—	0	0	0	—		—	—	
Sept.	78	64	98	47	1	0	0	—	70	1.0	0	3	0	1		—	0	
Oct.	73	57	99	41	—	0	0	—	72	2.0	0	5	0	—		—	0	
Nov.	66	53	83	36	0	0	0	—	75	3.0	0	7	0	—		1.0	0	
Dec.	61	48	75	34	0	0	0	—	75	2.0	0	7	0	1		1.0	—	
Year	71	56	102	30	14	1	0	—	72	17.0	0	50	0	6		0.5	—	

Elevation 121 Feet

Latitude 37° 1' N   Longitude 7° 55' W

Total precipitation means the total of rain, snow, hail, etc. expressed as inches of rain.

Ten inches of snow equals approx. 1 inch of rain.

Temperatures are as ° F.

Wind velocity is the percentage of observations greater than 18 and 30 M.P.H. Number of observations varies from one per day to several per hour.

Visibility means number of days with less than ½-mile visibility.

Note: "—" indicates slight trace or possibility blank indicates no statistics available

FELDKIRCH — AUSTRIA

Month	Daily Average Max.	Daily Average Min.	Extreme Max.	Extreme Min.	No. of Days Over 90°	No. of Days Under 32°	No. of Days 0°	Rel. Hum. A.M. 6:30	Rel. Hum. P.M. 1:30	Inches Total Prec.	Inches Snow Only	No. of Days Total Prec. 0.1"	No. of Days Snow Only 1.5"	Thun. Stms.	Max. Inches in 24 Hrs.	Wind Vel. 18 M.P.H.	Wind Vel. 30 M.P.H.	Visibility ½ Mile
Jan.	36	24	68	−8	0	—	—	86	73	2.7	—	8	—	0	1.9			
Feb.	39	25	68	−13	0	—	—	85	67	2.1	—	6	—	—	1.5			
March	49	32	76	5	0	—	0	81	55	2.6	—	7	—	—	1.5			
April	58	39	83	18	0	—	0	80	53	3.1	—	7	—	1	1.5			
May	65	46	89	26	0	—	0	78	55	4.3	—	8	—	3	3.1			
June	72	52	97	37	—	0	0	81	57	5.4	0	10	0	5	2.9			
July	75	55	99	41	—	0	0	81	57	6.4	0	10	0	6	3.5			
Aug.	73	55	95	37	—	0	0	84	59	6.1	0	10	0	5	2.6			
Sept.	66	50	88	32	0	—	0	89	61	4.4	0	9	0	1	2.3			
Oct.	56	41	81	22	0	—	0	90	65	3.2	—	8	—	—	2.3			
Nov.	45	34	72	3	0	—	0	87	71	2.8	—	7	—	—	3.1			
Dec.	36	26	66	−15	0	—	—	87	76	2.7	—	8	—	0	1.5			
Year	56	40	99	−15	0+	0+	0+	84	63	46.0	—	96	—	22	3.5			

Elevation 1,761 Feet

Latitude 47° 15' N   Longitude 9° 35' E

Total precipitation means the total of rain, snow, hail, etc. expressed as inches of rain.

Ten inches of snow equals approx. 1 inch of rain.

Temperatures are as ° F.

Wind velocity is the percentage of observations greater than 18 and 30 M.P.H. Number of observations varies from one per day to several per hour.

Visibility means number of days with less than ½-mile visibility.

Note: "—" indicates slight trace or possibility blank indicates no statistics available

FLORENCE

ITALY

Month	TEMPERATURES—IN °F							RELATIVE HUMIDITY		PRECIPITATION							WIND VELOCITY		VISI-BIL-ITY ½ MILE
	Daily Average		Extreme		No. of Days			A.M. 5:30	P.M. 12:30	Inches		No. of Days				Max. Inches in 24 Hrs.	18 M.P.H.	30 M.P.H.	
	Max.	Min.	Max.	Min.	Over 90°	Under 32°	0°			Total Prec.	Snow Only	Total Prec. 0.04"	0.1"	Snow Only 1.5"	Thun. Stms.				
Jan.	49	35	64	18	0	—	0	83	69	1.9	—	9	6	—	2				7
Feb.	53	36	71	18	0	—	0	84	65	2.1	—	9	7	—	—				—
March	60	40	76	11	0	—	0	85	58	2.7	—	10	10	—	0				5
April	68	46	87	28	0	—	0	87	56	2.9	—	7	7	—	1				1
May	75	53	91	37	—	0	0	87	57	3.0	0	9	7	0	0				1
June	84	58	99	44	7	0	0	85	56	2.7	0	7	7	0	1				0
July	89	63	104	46	14	0	0	82	45	1.5	0	4	4	0	1				0
Aug.	88	62	105	50	13	0	0	85	47	1.9	0	5	5	0	2				0
Sept.	81	58	98	42	4	0	0	88	54	3.3	0	8	8	0	0				2
Oct.	69	51	87	31	0	—	0	87	64	4.0	0	10	9	0	5				6
Nov.	58	42	72	22	0	—	0	89	72	3.9	—	9	9	—	4				3
Dec.	50	37	65	19	0	—	0	88	74	2.8	—	9	5	—	—				—
Year	69	48	105	11	38+	0+	0	86	60	33.0	0+	96	84	0+	16+				—

Elevation 144 Feet

Latitude 43° 48' N   Longitude 11° 12' E

Total precipitation means the total of rain, snow, hail, etc. expressed as inches of rain.

Ten inches of snow equals approx. 1 inch of rain.

Temperatures are as ° F.

Wind velocity is the percentage of observations greater than 18 and 30 M.P.H. Number of observations varies from one per day to several per hour.

Visibility means number of days with less than ½-mile visibility.

Note: "—" indicates slight trace or possibility blank indicates no statistics available

FORT WILLIAM (United Kingdom) SCOTLAND

| Month | TEMPERATURES—IN °F | | | | | | | RELATIVE HUMIDITY | | PRECIPITATION | | | | | | | WIND VELOCITY | | VISIBILITY ½ MILE |
| | Daily Average | | Extreme | | No. of Days | | | | | Inches | | No. of Days | | | | Max. Inches in 24 Hrs. | | | |
	Max.	Min.	Max.	Min.	Over 90°	Under 32°	0°	A.M. —	P.M. —	Total Prec.	Snow Only	Total Prec. 0.01"	0.1"	Snow Only 1.5"	Thun. Stms.		18 M.P.H.	30 M.P.H.	
Jan.	44	34	57	7						8.9		22				2.8			
Feb.	45	33	56	11						7.1		19				3.8			
March	48	37	69	11						5.0		17				2.2			
April	52	39	71	20						5.2		19				2.6			
May	59	43	80	27						3.2		16				2.5			
June	62	48	84	34						4.5		17				2.0			
July	64	51	86	38						5.3		21				2.4			
Aug.	64	51	81	35						5.8		19				2.5			
Sept.	60	48	77	27						7.7		21				3.6			
Oct.	54	43	68	27						9.1		22				3.1			
Nov.	49	38	64	16						7.8		21				4.0			
Dec.	45	36	61	11						9.4		23				3.8			
Year	54	42	86	7						79.0		237				4.0			

Elevation 171 Feet

Latitude 56° 49' N   Longitude 5° 7' W

Total precipitation means the total of rain, snow, hail, etc. expressed as inches of rain.

Ten inches of snow equals approx. 1 inch of rain.

Temperatures are as ° F.

Wind velocity is the percentage of observations greater than 18 and 30 M.P.H. Number of observations varies from one per day to several per hour.

Visibility means number of days with less than ½-mile visibility.

Note: "—" indicates slight trace or possibility blank indicates no statistics available

FREIBURG (West) FED. REP. GERMANY

Month	Daily Average Max	Min	Extreme Max	Min	No. of Days Over 90°	Under 32°	0°	R.H. A.M. 6:30	P.M. 1:30	Total Prec. (in)	Snow Only (in)	Total Prec. 0.04"	0.1"	Snow Only 1.5"	Thun. Stms.	Max. Inches in 24 Hrs.	Wind 18 M.P.H.	30 M.P.H.	Visibility ½ Mile
Jan.	39	29	62	−7	0	20	—	81	78	1.8		10	6		—	1	2	—	6
Feb.	43	30	70	−7	0	17	—	77	71	1.7		9	6		—	1	2	0	5
March	50	35	75	9	0	12	0	73	63	2.3		11	6		—	1	2	0	2
April	58	41	84	22	0	3	0	69	57	2.9		13	7		2	2	1	—	1
May	67	48	94	29	—	—	0	69	56	3.5		12	7		5	2	—	0	1
June	72	53	97	37	1	0	0	70	57	3.8	0	12	9	0	6	2	—	0	—
July	76	57	103	43	3	0	0	69	56	4.1	0	12	9	0	6	2	—	0	1
Aug.	74	56	98	41	2	0	0	70	57	3.7	0	12	8	0	4	2	1	—	—
Sept.	67	51	93	30	1	—	0	76	64	3.3	0	10	8	0	2	2	1	0	2
Oct.	56	44	81	23	0	2	0	81	72	3.1		10	7		—	2	1	0	6
Nov.	46	36	71	8	0	8	0	82	78	2.4		10	6		—	2	2	—	6
Dec.	41	31	67	−1	0	15	—	82	80	2.3		11	7		—	1	2	0	7
Year	57	43	103	−7	7	77	0+	76	66	35.0		132	86		25	2	1	—	37

Elevation 938 Feet
Latitude 48° 0' N   Longitude 7° 51' E

Total precipitation means the total of rain, snow, hail, etc. expressed as inches of rain.

Ten inches of snow equals approx. 1 inch of rain.

Temperatures are as ° F.

Wind velocity is the percentage of observations greater than 18 and 30 M.P.H. Number of observations varies from one per day to several per hour.

Visibility means number of days with less than ½-mile visibility.

Note: "—" indicates slight trace or possibility blank indicates no statistics available

GARMISCH—PARTENKIRCHEN (West) FED. REP. GERMANY

Month	TEMPERATURES—IN °F							RELATIVE HUMIDITY		PRECIPITATION						WIND VELOCITY		VISIBILITY
	Daily Average		Extreme		No. of Days					Inches		No. of Days			Max. Inches in 24 Hrs.	18 M.P.H.	30 M.P.H.	½ MILE
	Max.	Min.	Max.	Min.	Over 90°	Under 32°	0°	A.M.	P.M.	Total Prec.	Snow Only	Total Prec. 0.1"	Snow Only 1.5"	Thun. Stms.				
Jan.	36	20	62	−17	0	29	—	—	82	2.9		8			—	—	0	2
Feb.	40	22	70	−17	0	25	—		78	2.2		7		—		—	0	2
March	48	28	76	−3	0	19	—		74	2.7		7		—		0.0	0	1
April	55	34	83	10	0	8	0		74	3.9		7		1		—	0	1
May	63	42	90	21	—	1	0		75	4.8		8		5		1.0	0	—
June	67	47	90	32	—	—	0		79	6.9	0	10	0	8		1.0	0	—
July	71	51	93	36	1	0	0		79	7.3	0	—	0	8		—	0	0
Aug.	70	50	93	34	—	0	0		81	6.4	0	10	0	6		—	0	0
Sept.	65	46	86	25	0	—	0		82	4.8		9		3		—	0	—
Oct.	56	38	82	15	0	4	0		82	2.9		7		—		—	0	2
Nov.	45	29	75	3	0	15	0		83	2.5		7		0		—	0	2
Dec.	37	23	63	−7	0	25	—		84	3.2		9		—		—	0	3
Year	54	36	93	−17	1	126	0+		79	51.0		—		33		0.3	0	13

Elevation 2,322 Feet

Latitude 47° 30' N   Longitude 11° 6' E

Total precipitation means the total of rain, snow, hail, etc. expressed as inches of rain.

Ten inches of snow equals approx. 1 inch of rain.

Temperatures are as ° F.

Wind velocity is the percentage of observations greater than 18 and 30 M.P.H. Number of observations varies from one per day to several per hour.

Visibility means number of days with less than ½-mile visibility.

Note: "—" indicates slight trace or possibility blank indicates no statistics available

GENEVA   SWITZERLAND

Month	Daily Average Max.	Daily Average Min.	Extreme Max.	Extreme Min.	No. of Days Over 90°	No. of Days Under 32°	No. of Days 0°	R.H. A.M. 7:00	R.H. P.M. 1:00	Inches Total Prec.	Inches Snow Only	No. of Days Total Prec. 0.01"	No. of Days 0.1"	No. of Days Snow Only 1.5"	Thun. Stms.	Max. Inches in 24 Hrs.	Wind 18 M.P.H.	Wind 30 M.P.H.	Visibility ½ Mile
Jan.	39	29	62	3	0	20	0	89	78	1.9	—	10	6	—	—	2.4			
Feb.	43	30	68	−1	0	17	—	88	71	1.8	—	9	6	—	—	1.7			
March	51	35	76	10	0	10	0	84	61	2.2	—	10	6	—	—	1.7			
April	58	41	81	23	0	2	0	79	55	2.5	—	11	7	—	1	1.4			
May	66	48	89	29	0	—	0	79	57	3.0	0	12	7	0	3	1.8			
June	73	55	96	36	—	0	0	77	55	3.1	0	11	8	0	5	2.3			
July	77	58	101	42	2	0	0	76	55	2.9	0	9	7	0	5	2.0			
Aug.	76	57	97	41	1	0	0	81	57	3.6	0	10	8	0	5	2.8			
Sept.	69	52	91	32	—	—	0	88	63	3.6	0	10	8	0	3	3.1			
Oct.	58	44	79	20	0	1	0	89	69	3.8	—	11	8	—	1	2.0			
Nov.	47	37	74	17	0	7	0	90	76	3.1	—	11	8	—	—	3.0			
Dec.	40	31	64	5	0	20	0	88	78	2.4	—	10	7	—	—	1.4			
Year	58	43	101	−1	3	78	0+	84	65	34.0	—	124	87	—	23	3.1			

Elevation 1,388 Feet

Latitude 46° 14' N   Longitude 6° 5' E

Total precipitation means the total of rain, snow, hail, etc. expressed as inches of rain.

Ten inches of snow equals approx. 1 inch of rain.

Temperatures are as ° F.

Wind velocity is the percentage of observations greater than 18 and 30 M.P.H. Number of observations varies from one per day to several per hour.

Visibility means number of days with less than ½-mile visibility.

Note: "—" indicates slight trace or possibility blank indicates no statistics available

GENOA   ITALY

Month	TEMPERATURES—IN °F — Daily Average Max.	Daily Average Min.	Extreme Max.	Extreme Min.	No. of Days Over 90°	Under 32°	0°	RELATIVE HUMIDITY A.M. 7:30	P.M. 1:30	PRECIPITATION Inches Total Prec.	Snow Only	No. of Days Total Prec. 0.04"	0.1"	Snow Only 1.5"	Thun. Stms.	Max. Inches in 24 Hrs.	WIND VELOCITY 18 M.P.H.	30 M.P.H.	VISIBILITY ½ MILE
Jan.	50	41	66	21	0	3	0	68	63	4.1	—	13	10	—	0	2	17	—	—
Feb.	54	43	72	24	0	3	0	67	62	4.3	—	10	10	—	—	2	16	1	0
March	58	47	72	33	0	0	0	69	64	4.2	0	8	8	0	1	3	7	0	—
April	65	53	81	38	0	0	0	71	64	3.9	0	7	7	0	—	2	6	—	0
May	70	57	92	46	—	0	0	74	68	3.4	0	7	7	0	3	1	3	0	0
June	78	65	91	52	—	0	0	72	66	2.8	0	7	7	0	2	3	—	0	0
July	82	70	100	57	5	0	0	68	63	1.7	0	5	5	0	4	3	2	—	0
Aug.	82	70	91	58	5	0	0	66	64	2.4	0	6	6	0	4	2	1	0	0
Sept.	78	66	97	52	—	0	0	71	63	4.9	0	9	9	0	3	6	3	0	0
Oct.	73	58	82	39	0	0	0	71	64	7.9	0	9	—	0	1	3	9	0	0
Nov.	60	51	74	35	0	0	0	68	62	7.1	0	8	—	0	1	3	15	—	0
Dec.	53	44	68	30	0	—	0	66	63	5.0	0	11	11	0	1	3	15	1	0
Year	67	55	100	21	10+	6+	0	69	64	52.0	0+	100	—	0+	20	6	8	—	0+

Elevation 121 Feet
Latitude 44° 24' N   Longitude 8° 54' E

Total precipitation means the total of rain, snow, hail, etc. expressed as inches of rain.

Ten inches of snow equals approx. 1 inch of rain.

Temperatures are as ° F.

Wind velocity is the percentage of observations greater than 18 and 30 M.P.H. Number of observations varies from one per day to several per hour.

Visibility means number of days with less than ½-mile visibility.

Note: "—" indicates slight trace or possibility blank indicates no statistics available

GOTEBORG

SWEDEN

| Month | TEMPERATURES—IN °F | | | | | | | RELATIVE HUMIDITY | | PRECIPITATION | | | | | | | WIND VELOCITY | | VISIBILITY |
| | Daily Average | | Extreme | | No. of Days | | | A.M. 8:00 | P.M. 2:00 | Inches | | No. of Days | | | | Max. Inches in 24 Hrs. | 18 M.P.H. | 30 M.P.H. | ½ MILE |
	Max.	Min.	Max.	Min.	Over 90°	Under 32°	0°			Total Prec.	Snow Only	Total Prec. 0.04"	0.1"	Snow Only 1.5"	Thun. Stms.				
Jan.	35	27	48	−11	0	21	—	86	84	2.5	—	10	8	—	0	2.0			
Feb.	35	26	46	−13	0	22	2	84	78	2.0	—	8	6	—	0	1.4			
March	39	29	64	−4	0	24	—	83	70	2.0	—	9	6	—	0	1.2			
April	48	36	79	16	0	13	0	74	59	1.7	—	7	5	—	0	1.1			
May	59	44	81	28	0	9	0	67	55	1.9	—	9	6	—	1	1.6			
June	67	52	87	37	0	0	0	68	57	2.2	0	8	6	0	1	1.3			
July	69	56	88	45	0	0	0	72	60	2.8	0	10	7	0	2	1.9			
Aug.	66	54	86	40	0	0	0	78	64	3.7	0	12	8	0	3	2.9			
Sept.	60	49	79	32	0	—	0	81	66	3.1	0	10	8	0	2	2.9			
Oct.	51	42	68	21	0	5	0	83	73	3.1	—	12	8	—	1	2.3			
Nov.	43	35	56	8	0	8	0	85	80	2.7	—	10	7	—	0	2.0			
Dec.	37	30	51	−4	0	16	—	86	84	2.8	—	11	8	—	0	1.1			
Year	51	40	88	−13	0	109	2	79	69	31.0	—	116	83	—	10	2.9			

Elevation 13 Feet

Latitude 57° 43' N   Longitude 11° 47' E

Total precipitation means the total of rain, snow, hail, etc. expressed as inches of rain.

Ten inches of snow equals approx. 1 inch of rain.

Temperatures are as ° F.

Wind velocity is the percentage of observations greater than 18 and 30 M.P.H. Number of observations varies from one per day to several per hour.

Visibility means number of days with less than ½-mile visibility.

Note: "—" indicates slight trace or possibility

blank indicates no statistics available

GRANADA

SPAIN

Month	Daily Average Max.	Daily Average Min.	Extreme Max.	Extreme Min.	No. of Days Over 90°	No. of Days Under 32°	No. of Days 0°	R.H. A.M. 8:00	R.H. P.M. —	Inches Total Prec.	Inches Snow Only	No. of Days Total Prec. 0.1"	No. of Days Snow Only 1.5"	Thun. Stms.	Max. Inches in 24 Hrs.	Wind 18 M.P.H.	Wind 30 M.P.H.	Visibility ½ Mile
Jan.	53	36	73	18	0	11	0	85		1.3	—	4	—	—	1.4	4	1.0	
Feb.	56	37	82	12	0	9	0	78		1.7	—	6	—	1	0.9	6	1.0	
March	59	41	82	23	0	3	0	83		2.4	—	7	—	1	1.3	10	2.0	
April	64	45	86	32	0	—	0	78		2.0	0	6	0	5	1.3	8	—	
May	73	52	99	36	—	0	0	73		1.7	0	5	0	1	1.1	8	1.0	
June	81	58	101	42	9	0	0	67		0.8	0	2	0	3	0.9	13	—	
July	90	64	104	50	27	0	0	59		0.2	0	—	0	3	1.7	13	—	
Aug.	91	64	104	49	24	0	0	57		0.1	0	0	0	2	0.5	13	1.0	
Sept.	81	58	100	43	10	0	0	70		1.1	0	4	0	3	2.7	8	1.0	
Oct.	72	51	93	34	—	0	0	77		1.9	0	5	0	3	1.8	4	1.0	
Nov.	59	42	81	28	0	1	0	85		2.3	—	6	—	1	2.7	3	—	
Dec.	54	37	75	21	0	8	0	85		1.9	—	6	—	0	1.3	4	—	
Year	69	49	104	12	71	32	0	75		17.0	—	52	—	23	2.7	8	0.7	

Elevation 2,261 Feet

Latitude 37° 9' N   Longitude 3° 35' W

Total precipitation means the total of rain, snow, hail, etc. expressed as inches of rain.

Ten inches of snow equals approx. 1 inch of rain.

Temperatures are as ° F.

Wind velocity is the percentage of observations greater than 18 and 30 M.P.H. Number of observations varies from one per day to several per hour.

Visibility means number of days with less than ½-mile visibility.

Note: "—" indicates slight trace or possibility blank indicates no statistics available

GRAZ

AUSTRIA

| Month | TEMPERATURES—IN °F | | | | | | | RELATIVE HUMIDITY | | PRECIPITATION | | | | | | | WIND VELOCITY | | VISI-BIL-ITY ½ MILE |
| | Daily Average | | Extreme | | No. of Days | | | | | Inches | | No. of Days | | | | Max. Inches in 24 Hrs. | | | |
	Max.	Min.	Max.	Min.	Over 90°	Under 32°	0°	A.M. 7:00	P.M. 2:00	Total Prec.	Snow Only	Total Prec. 0.04"	0.1"	Snow Only 1.5"	Thun. Stms.		18 M.P.H.	30 M.P.H.	
Jan.	33	23	57	−20	0	30	5	85	72	1.2	—	6	4	—	0	1.1	1.0	0.0	12
Feb.	38	25	66	−17	0	28	3	85	64	1.2	—	5	4	—	—	1.6	—	0.0	8
March	48	32	75	−6	0	23	—	83	54	1.4	—	6	5	—	—	1.0	2.0	0.0	9
April	58	41	81	19	0	9	0	80	51	2.4	—	9	7	—	2	1.4	5.0	—	2
May	66	49	86	23	0	2	0	76	53	3.4	—	11	7	—	6	2.2	2.0	0.0	3
June	73	55	97	34	—	0	0	79	54	4.6	0	11	9	0	7	3.5	1.0	—	—
July	77	58	99	41	—	0	0	80	53	5.2	0	11	10	0	8	4.1	1.0	0.0	2
Aug.	75	56	96	39	—	0	0	84	53	4.3	0	10	9	0	6	2.3	—	0.0	4
Sept.	68	51	91	25	—	1	0	86	58	3.7	—	9	8	—	3	2.8	0.0	0.0	8
Oct.	57	42	79	19	0	7	0	91	64	3.1	—	8	8	—	1	2.2	1.0	0.0	13
Nov.	45	35	68	7	0	17	0	89	72	2.1	—	7	6	—	—	2.0	—	—	8
Dec.	36	27	61	−18	0	28	1	88	75	1.8	—	6	6	—	0	1.1	1.0	0.0	17
Year	56	41	99	−20	0+	145	9	84	60	34.0	—	99	82	—	34	4.1	1.3	0.1	84

Elevation 1,109 Feet

Latitude 46° 59' N   Longitude 15° 26' E

Total precipitation means the total of rain, snow, hail, etc. expressed as inches of rain.

Ten inches of snow equals approx. 1 inch of rain.

Temperatures are as ° F.

Wind velocity is the percentage of observations greater than 18 and 30 M.P.H. Number of observations varies from one per day to several per hour.

Visibility means number of days with less than ½-mile visibility.

Note: "—" indicates slight trace or possibility blank indicates no statistics available

GREAT YARMOUTH (United Kingdom) ENGLAND

Month	TEMPERATURES—IN °F							RELATIVE HUMIDITY		PRECIPITATION						Max. Inches in 24 Hrs.	WIND VELOCITY		VISIBILITY ½ MILE
	Daily Average		Extreme		No. of Days					Inches		No. of Days							
	Max.	Min.	Max.	Min.	Over 90°	Under 32°	0°	A.M. 7:00	P.M. 1:30	Total Prec.	Snow Only	Total Prec. 0.01"	0.1"	Snow Only 1.5"	Thun. Stms.		18 M.P.H.	30 M.P.H.	
Jan.	43	36	58	16				90	85	2.2		17			0.1	1.3			
Feb.	44	36	60	13				89	84	1.6		14			0.1	1.0			
March	47	37	72	18				89	77	1.3		12			0.4	0.8			
April	52	41	76	27				86	75	1.5		13			0.7	0.7			
May	57	46	82	31				84	75	1.6		11			1.5	0.9			
June	63	52	82	38				83	73	1.6		11			2.1	1.4			
July	67	56	85	44				85	73	2.2		11			2.7	2.3			
Aug.	68	56	88	42				87	72	1.9		11			2.7	1.6			
Sept.	64	53	82	37				89	74	2.3		13			1.2	1.8			
Oct.	57	47	74	30				88	76	2.4		15			1.5	1.3			
Nov.	49	41	68	25				88	82	2.9		17			0.3	1.2			
Dec.	44	38	59	22				88	86	2.1		17			0.1	1.0			
Year	55	45	88	13				87	76	24.0		162			12.0	2.3			

Elevation 5 Feet

Latitude 52° 35' N   Longitude 1° 43' E

Total precipitation means the total of rain, snow, hail, etc. expressed as inches of rain.

Ten inches of snow equals approx. 1 inch of rain.

Temperatures are as °F.

Wind velocity is the percentage of observations greater than 18 and 30 M.P.H. Number of observations varies from one per day to several per hour.

Visibility means number of days with less than ½-mile visibility.

Note: "—" indicates slight trace or possibility blank indicates no statistics available

GRIMSEY ISLAND

ICELAND

Month	TEMPERATURES—IN °F							RELATIVE HUMIDITY		PRECIPITATION							WIND VELOCITY		VISIBILITY ½ MILE
	Daily Average		Extreme		No. of Days			A.M. 8:00	P.M. 2:00	Inches		No. of Days				Max. Inches in 24 Hrs.	18 M.P.H.	30 M.P.H.	
	Max.	Min.	Max.	Min.	Over 90°	Under 32°	0°			Total Prec.	Snow Only	Total Prec. 0.004"	0.1"	Snow Only 1.5"	Thun. Stms.				
Jan.	33	27	50	10				79	80	0.5		9				0.5			
Feb.	34	27	50	8				80	80	1.1		10				1.8			
March	35	27	49	6				79	79	0.6		10				0.8			
April	36	29	56	11				80	80	0.7		9				0.4			
May	41	34	64	15				82	83	0.7		7				1.0			
June	49	30	79	29				83	84	1.2		7				1.0			
July	50	42	66	33				86	87	1.3		8				1.0			
Aug.	50	43	66	31				86	87	1.3		10				1.0			
Sept.	47	39	61	29				84	85	2.1		11				1.1			
Oct.	41	36	55	15				83	84	2.0		12				1.1			
Nov.	37	30	51	12				82	82	1.3		9				1.1			
Dec.	35	30	49	10				82	82	0.8		10				0.5			
Year	41	34	79	6				82	83	14.0		112				1.8			

Elevation 72 Feet

Latitude 66° 33' N   Longitude 17° 58' W

Total precipitation means the total of rain, snow, hail, etc. expressed as inches of rain.

Ten inches of snow equals approx. 1 inch of rain.

Temperatures are as ° F.

Wind velocity is the percentage of observations greater than 18 and 30 M.P.H. Number of observations varies from one per day to several per hour.

Visibility means number of days with less than ½-mile visibility.

Note: "—" indicates slight trace or possibility blank indicates no statistics available

GRONINGEN

THE NETHERLANDS

Month	TEMPERATURES—IN °F Daily Average Max.	Min.	Extreme Max.	Min.	No. of Days Over 90°	Under 32°	0°	RELATIVE HUMIDITY A.M. 8:00	P.M. 2:00	PRECIPITATION Inches Total Prec.	Snow Only	No. of Days Total Prec. 0.004"	0.1"	Snow Only 1.5"	Thun. Stms.	Max. Inches in 24 Hrs.	WIND VELOCITY 18 M.P.H.	30 M.P.H.	VISI-BILITY ½ MILE
Jan.	39	30	54	−3	0	18	—	90	83	2.4	—	19	8	—	0.5	0.8	22	2.4	6
Feb.	40	31	59	−2	0	16	—	89	77	1.8	—	17	6	—	0.5	0.7	18	1.5	8
March	47	35	73	10	0	14	0	87	68	1.5	—	16	5	—	0.8	0.7	16	1.9	8
April	54	40	79	24	0	6	0	80	62	2.0	—	15	6	—	1.7	1.0	12	0.9	3
May	62	46	93	31	—	1	0	76	59	2.1	—	14	6	—	3.5	1.1	8	0.4	3
June	67	51	94	38	—	—	0	77	61	2.2	0	14	6	0	5.2	1.0	5	0.2	3
July	71	56	95	44	—	0	0	80	63	3.3	0	16	6	0	6.4	3.0	8	0.2	2
Aug.	70	55	98	43	—	0	0	83	64	3.3	0	16	9	0	7.0	2.1	8	0.2	4
Sept.	65	51	91	34	—	0	0	87	67	2.8	0	17	8	0	3.0	2.2	10	0.3	4
Oct.	56	44	77	24	0	2	0	90	73	2.8	—	19	8	—	0.7	1.3	11	0.8	6
Nov.	46	38	64	17	0	8	0	91	82	2.8	—	20	8	—	0.9	1.5	13	0.7	7
Dec.	40	32	55	−5	0	12	—	90	86	2.4	—	20	8	—	0.8	0.9	20	2.1	8
Year	55	42	98	−5	0+	77	0+	85	70	29.0	0+	203	86	0+	31.0	3.0	12	1.0	62

Elevation 7 Feet

Latitude 53° 13' N   Longitude 6° 33' E

Total precipitation means the total of rain, snow, hail, etc. expressed as inches of rain.

Ten inches of snow equals approx. 1 inch of rain.

Temperatures are as ° F.

Wind velocity is the percentage of observations greater than 18 and 30 M.P.H. Number of observations varies from one per day to several per hour.

Visibility means number of days with less than ½-mile visibility.

Note: "—" indicates slight trace or possibility blank indicates no statistics available

HAMBURG (West) FED. REP. GERMANY

Month	TEMPERATURES—IN °F Daily Average Max.	Min.	Extreme Max.	Min.	No. of Days Over 90°	Under 32°	0°	RELATIVE HUMIDITY A.M. 7:30	P.M. 1:30	PRECIPITATION Inches Total Prec.	Snow Only	No. of Days Total Prec. 0.004"	0.1"	Snow Only 1.5"	Thun. Stms.	Max. Inches in 24 Hrs.	WIND VELOCITY 18 M.P.H.	30 M.P.H.	VISIBILITY ½ MILE
Jan.	35	28	56	−1	0	19	—	89	86	2.1		13	7		—	1	12	1.0	8
Feb.	37	30	59	−4	0	17	—	88	81	1.9		10	6		—	1	10	—	7
March	42	33	68	10	0	17	0	85	71	2.0		12	6		1	1	8	—	7
April	51	39	82	21	0	7	0	81	63	1.8		15	6		—	1	9	1.0	3
May	60	47	89	28	0	—	0	76	58	2.1		14	6		3	2	4	0.0	2
June	67	53	92	38	—	0	0	76	61	2.7	0	11	7	0	4	3	2	0.0	2
July	69	56	97	41	—	0	0	80	62	3.4	0	9	8	0	5	3	4	0.0	2
Aug.	67	55	90	38	—	0	0	85	65	3.2	0	8	8	0	5	3	4	—	4
Sept.	63	51	89	34	0	0	0	88	68	2.5	0	9	7	0	2	3	4	0.0	4
Oct.	53	44	73	25	0	2	0	90	75	2.6		11	7		1	1	4	—	7
Nov.	44	36	63	7	0	7	0	90	84	2.1		12	6		—	1	6	0.0	9
Dec.	38	31	59	−4	0	13	—	90	87	2.5		12	8		—	1	10	—	8
Year	52	42	97	−4	0+	82	0+	85	71	29.0		136	82		21	3	7	0.3	63

Elevation 53 Feet

Latitude 53° 38' N   Longitude 1° 0' E

Total precipitation means the total of rain, snow, hail, etc. expressed as inches of rain.

Ten inches of snow equals approx. 1 inch of rain.

Temperatures are as ° F.

Wind velocity is the percentage of observations greater than 18 and 30 M.P.H. Number of observations varies from one per day to several per hour.

Visibility means number of days with less than ½-mile visibility.

Note: "—" indicates slight trace or possibility blank indicates no statistics available

HANNOVER (West) FED. REP. GERMANY

Month	TEMPERATURES—IN °F Daily Average Max.	Min.	Extreme Max.	Min.	No. of Days Over 90°	Under 32°	0°	RELATIVE HUMIDITY A.M. 6:30	P.M. 1:30	PRECIPITATION Inches Total Prec.	Snow Only	No. of Days Total Prec. 0.04"	0.1"	Snow Only 1.5"	Thun. Stms.	Max. Inches in 24 Hrs.	WIND VELOCITY 18 M.P.H.	30 M.P.H.	VISI-BIL-ITY ½ MILE
Jan.	38	28	58	−7	0	18	—	88	85	1.7		11	6		—	1	12	1.0	4
Feb.	40	29	65	−13	0	17	1	89	80	1.4		9	5		—	1	8	—	4
March	46	33	73	3	0	17	0	88	73	1.6		10	5		1	1	7	—	5
April	55	37	83	17	0	7	0	86	63	1.5		10	5		1	1	5	0.0	2
May	65	44	94	28	—	1	0	80	60	2.0		10	6		4	1	3	—	3
June	70	50	94	34	—	0	0	78	62	2.7	0	10	7	0	4	2	2	0.0	1
July	72	53	98	41	—	0	0	83	63	3.0	0	12	7	0	5	2	5	0.0	2
Aug.	71	53	95	40	—	0	0	87	64	2.7	0	12	7	0	4	3	5	0.0	3
Sept.	66	48	96	33	—	0	0	90	68	1.8	0	10	5	0	1	2	7	—	4
Oct.	54	42	81	20	0	3	0	91	75	1.9		10	6		—	1	5	—	7
Nov.	45	35	68	4	0	8	0	91	82	1.7		9	5		—	2	7	0.0	7
Dec.	40	31	60	−6	0	13	—	89	85	1.9		11	6		—	1	11	1.0	5
Year	55	40	98	−13	0+	84	1+	87	72	24.0		124	70		20	3	6	0.2	47

Elevation 179 Feet
Latitude 52° 27' N   Longitude 9° 41' E
Total precipitation means the total of rain, snow, hail, etc. expressed as inches of rain.
Ten inches of snow equals approx. 1 inch of rain.
Temperatures are as ° F.

Wind velocity is the percentage of observations greater than 18 and 30 M.P.H. Number of observations varies from one per day to several per hour.
Visibility means number of days with less than ½-mile visibility.
Note: "—" indicates slight trace or possibility blank indicates no statistics available

HELSINKI

FINLAND

| Month | TEMPERATURES—IN °F | | | | | | | RELATIVE HUMIDITY | | PRECIPITATION | | | | | | | WIND VELOCITY | | VISI-BIL-ITY |
| | Daily Average | | Extreme | | No. of Days | | | | | Inches | | No. of Days | | | | Max. Inches in 24 Hrs. | | | ½ MILE |
	Max.	Min.	Max.	Min.	Over 90°	Under 32°	0°	A.M. 6:30	P.M. 2:30	Total Prec.	Snow Only	Total Prec. 0.04"	0.1"	Snow Only 1.5"	Thun. Stms.		18 M.P.H.	30 M.P.H.	
Jan.	27	17	47	−27	0	30	5	88	86	1.7	—	11	6	—	0	1.0	4	—	3
Feb.	26	14	49	−29	0	28	6	87	81	1.5	—	8	5	—	0	1.1	3	0	3
March	30	19	54	−18	0	30	5	87	73	1.4	—	8	5	—	0	0.9	4	—	3
April	41	30	70	0	0	24	—	84	69	1.4	—	8	5	—	—	1.5	4	0	2
May	54	40	81	21	0	6	0	75	62	1.8	—	8	6	—	2	1.5	2	—	2
June	64	50	88	32	0	—	0	75	60	1.8	0	9	5	0	2	1.9	3	0	—
July	67	55	89	39	0	0	0	77	63	2.2	0	8	6	0	3	1.8	2	—	1
Aug.	65	53	86	37	0	0	0	86	69	2.9	0	12	7	0	2	1.7	3	1	2
Sept.	56	45	77	25	0	3	0	89	72	2.5	—	11	7	—	1	2.1	3	—	3
Oct.	46	38	63	14	0	9	0	90	78	2.6	—	12	7	—	—	2.1	3	0	4
Nov.	36	29	51	−4	0	21	—	89	85	2.5	—	11	7	—	0	1.5	2	0	3
Dec.	35	21	47	−22	0	27	2	89	87	2.0	—	11	6	—	0	1.6	2	0	3
Year	46	34	89	−29	0	178	18	85	74	24.0	—	117	70	—	10	2.0	3	—	29

Elevation 167 Feet

Latitude 60° 19' N   Longitude 24° 57' E

Total precipitation means the total of rain, snow, hail, etc. expressed as inches of rain.

Ten inches of snow equals approx. 1 inch of rain.

Temperatures are as ° F.

Wind velocity is the percentage of observations greater than 18 and 30 M.P.H. Number of observations varies from one per day to several per hour.

Visibility means number of days with less than ½-mile visibility.

Note: "—" indicates slight trace or possibility blank indicates no statistics available

HOLYHEAD (United Kingdom) WALES

| Month | TEMPERATURES—IN °F | | | | | | | RELATIVE HUMIDITY | | PRECIPITATION | | | | | | | | WIND VELOCITY | | VISIBILITY ½ MILE |
| | Daily Average | | Extreme | | No. of Days | | | A.M. 6:30 | P.M. 1:00 | Inches | | No. of Days | | | | Max. Inches in 24 Hrs. | | 18 M.P.H. | 30 M.P.H. | |
	Max.	Min.	Max.	Min.	Over 90°	Under 32°	0°			Total Prec.	Snow Only	Total Prec. 0.01"	0.1"	Snow Only 1.5"	Thun. Stms.					
Jan.	46	40	55	20	0	1	0	89	85	2.9	—	20	8	—	0.1	1.6	48	8	—
Feb.	46	39	59	17	0	1	0	87	81	2.4	—	15	7	—	0.2	1.4	44	6	2
March	49	40	67	26	0	1	0	87	77	2.6	—	15	7	—	0.2	1.4	31	3	2
April	52	43	75	26	0	—	0	87	78	2.1	—	14	6	—	0.3	1.2	32	3	2
May	57	46	76	35	0	0	0	88	79	2.0	0	14	6	0	1.0	1.3	21	1	2
June	61	51	83	40	0	0	0	88	80	2.1	0	13	6	0	1.0	1.3	18	1	2
July	64	54	85	43	0	0	0	90	82	2.6	0	15	7	0	1.0	1.9	19	1	2
Aug.	64	55	86	45	0	0	0	90	81	3.2	0	16	8	0	1.0	1.3	18	1	2
Sept.	62	53	78	40	0	0	0	88	79	2.7	0	16	7	0	0.8	1.8	26	2	1
Oct.	57	49	76	32	0	—	0	86	79	4.0	0	17	9	0	0.9	1.6	46	9	—
Nov.	51	44	62	29	0	—	0	86	81	4.1	0	19	9	0	0.4	1.8	39	5	—
Dec.	47	42	57	26	0	1	0	88	85	4.2	—	19	10	—	0.3	1.4	44	10	1
Year	55	46	86	17	0	4+	0	88	81	35.0	0+	193	90	0+	7.0	1.9	32	4	16

Elevation 26 Feet

Latitude 53° 19' N   **Longitude 4° 37' W**

Total precipitation means the total of rain, snow, hail, etc. expressed as inches of rain.

Ten inches of snow equals approx. 1 inch of rain.

Temperatures are as ° F.

Wind velocity is the percentage of observations greater than 18 and 30 M.P.H. Number of observations varies from one per day to several per hour.

Visibility means number of days with less than ½-mile visibility.

Note: "—" indicates slight trace or possibility blank indicates no statistics available

INNSBRUCK

AUSTRIA

Month	TEMPERATURES—IN °F Daily Average Max.	Min.	Extreme Max.	Min.	No. of Days Over 90°	Under 32°	0°	RELATIVE HUMIDITY A.M. 7:00	P.M. 2:00	PRECIPITATION Inches Total Prec.	Snow Only	No. of Days Total Prec. 0.04"	0.1"	Snow Only 1.5"	Thun. Stms.	Max. Inches in 24 Hrs.	WIND VELOCITY 18 M.P.H.	30 M.P.H.	VISIBILITY ½ MILE
Jan.	34	20	61	−15	0	28	1	88	68	2.1	—	9	7	—	—	2.3	1.0	—	7
Feb.	40	24	60	−22	0	25	2	86	58	1.6	—	7	5	—	0	1.5	1.0	0	9
March	51	31	77	2	0	18	0	85	46	1.7	—	7	5	—	—	1.5	4.0	—	2
April	60	39	83	19	0	6	0	84	44	2.2	—	10	6	—	—	2.0	2.0	—	1
May	69	46	91	18	—	1	0	83	43	2.9	—	11	7	—	3	2.4	2.0	0	—
June	75	52	97	31	1	—	0	85	47	4.1	0	14	9	0	5	1.7	2.0	0	1
July	78	55	97	−11	2	—	—	87	49	4.8	—	14	9	—	7	2.6	—	0	—
Aug.	76	54	95	38	1	0	0	89	51	4.6	0	12	9	0	5	2.1	1.0	—	2
Sept.	69	49	87	30	0	—	0	91	53	3.0	0	11	7	0	2	1.9	2.0	0	3
Oct.	58	40	76	18	0	7	0	91	55	2.4	—	9	6	—	—	1.9	1.0	0	6
Nov.	46	32	73	−4	0	19	—	89	64	2.2	—	8	6	—	—	1.5	—	0	6
Dec.	36	24	66	−13	0	26	—	89	72	2.1	—	8	7	—	0	1.7	1.0	0	9
Year	58	39	97	−22	4	129	3	87	54	34.0	—	120	84	—	24	2.6	1.6	—	45

Elevation 1,906 Feet

Latitude 47° 15' N   Longitude 11° 20' E

Total precipitation means the total of rain, snow, hail, etc. expressed as inches of rain.

Ten inches of snow equals approx. 1 inch of rain.
Temperatures are as ° F.

Wind velocity is the percentage of observations greater than 18 and 30 M.P.H. Number of observations varies from one per day to several per hour.

Visibility means number of days with less than ½-mile visibility.

Note: "—" indicates slight trace or possibility blank indicates no statistics available

INVERNESS   (United Kingdom) SCOTLAND

Month	TEMPERATURES—IN °F Daily Average Max.	Min.	Extreme Max.	Min.	No. of Days Over 90°	Under 32°	0°	RELATIVE HUMIDITY A.M. 8:30	P.M. 4:30	PRECIPITATION Inches Total Prec.	Snow Only	No. of Days Total Prec. 0.01"	0.1"	Snow Only 1.5"	Thun. Stms.	Max. Inches in 24 Hrs.	WIND VELOCITY 18 M.P.H.	30 M.P.H.	VISIBILITY ½ MILE
Jan.	43	34	60	9	0	12	0	86	83	2.7	—	17	8	—	0.1	1.7			
Feb.	43	34	59	11	0	16	0	85	85	1.9	—	15	6	—	0.0	0.9			
March	47	35	65	9	0	19	0	83	79	1.5	—	14	5	—	0.1	0.7			
April	50	38	72	20	0	2	0	80	76	1.8	—	15	6	—	0.0	1.4			
May	55	42	77	29	0	—	0	78	76	2.1	0	15	6	0	0.5	1.6			
June	61	47	83	34	0	0	0	77	72	2.0	0	14	6	0	0.7	1.7			
July	63	51	82	36	0	0	0	81	79	2.9	0	17	7	0	0.4	1.6			
Aug.	63	51	79	38	0	0	0	83	78	3.1	0	16	8	0	0.1	3.3			
Sept.	59	47	85	30	0	—	0	84	79	2.6	0	17	7	0	0.3	2.1			
Oct.	53	42	75	26	0	3	0	85	83	2.9	—	18	7	—	0.0	1.8			
Nov.	47	37	60	12	0	10	0	85	85	2.5	—	17	7	—	0.0	1.7			
Dec.	44	35	59	11	0	6	0	87	85	2.3	—	17	7	—	0.0	1.1			
Year	52	41	85	9	0	68	0	83	80	28.4	0+	192	79	0+	2.7	3.3			

Elevation 31 Feet
Latitude 57° 32' N   Longitude 4° 3' W
Total precipitation means the total of rain, snow, hail, etc. expressed as inches of rain.
Ten inches of snow equals approx. 1 inch of rain.
Temperatures are as ° F.

Wind velocity is the percentage of observations greater than 18 and 30 M.P.H. Number of observations varies from one per day to several per hour.
Visibility means number of days with less than ½-mile visibility.
Note: "—" indicates slight trace or possibility blank indicates no statistics available

IRAKLION (Crete Island)　　GREECE

| Month | TEMPERATURES—IN °F | | | | | | | RELATIVE HUMIDITY | | PRECIPITATION | | | | | | WIND VELOCITY | | VISIBILITY |
| | Daily Average | | Extreme | | No. of Days | | | | | Inches | | No. of Days | | | | | | |
	Max.	Min.	Max.	Min.	Over 90°	Under 32°	0°	A.M. 7:30	P.M. 1:30	Total Prec.	Snow Only	Total Prec. 0.1"	Snow Only 1.5"	Thun. Stms.	Max. Inches in 24 Hrs.	18 M.P.H.	30 M.P.H.	½ MILE
Jan.	60	48	76	34	0	0	0	74	64	5.1	0	11	0	1	3			
Feb.	60	48	82	30	0	—	0	71	62	3.9	0	10	0	1	2			
March	64	50	91	35	—	0	0	70	61	2.6	0	7	0	1	1			
April	70	54	98	41	1	0	0	60	54	1.1	0	4	0	—	1			
May	76	60	100	48	1	0	0	61	59	0.6	0	2	0	1	2			
June	82	67	114	50	2	0	0	58	57	0.1	0	0	0	—	—			
July	85	72	106	61	3	0	0	59	55	—	0	0	0	0	1			
Aug.	85	71	104	56	4	0	0	59	57	0.1	0	—	0	0	1			
Sept.	82	68	102	54	1	0	0	62	59	1.3	0	4	0	—	3			
Oct.	77	62	96	45	—	0	0	67	61	1.5	0	5	0	1	2			
Nov.	71	56	87	37	0	0	0	72	64	4.8	0	9	0	1	3			
Dec.	64	51	80	37	0	0	0	76	67	6.7	0	11	0	1	5			
Year	73	59	114	30	11	0+	0	66	60	28.0	0	62	0	8	5			

Elevation 115 Feet
Latitude 35° 20' N　　Longitude 25° 11' E

Total precipitation means the total of rain, snow, hail, etc. expressed as inches of rain.

Ten inches of snow equals approx. 1 inch of rain.
Temperatures are as ° F.

Wind velocity is the percentage of observations greater than 18 and 30 M.P.H. Number of observations varies from one per day to several per hour.

Visibility means number of days with less than ½-mile visibility.

Note: "—" indicates slight trace or possibility blank indicates no statistics available

ISFJORDEN — Spitsbergen (Svalbard) NORWAY

| Month | TEMPERATURES—IN °F | | | | | | | RELATIVE HUMIDITY | | PRECIPITATION | | | | | | | WIND VELOCITY | | VISIBILITY |
| | Daily Average | | Extreme | | No. of Days | | | | | Inches | | No. of Days | | | | Max. Inches in 24 Hrs. | | | ½ MILE |
	Max.	Min.	Max.	Min.	Over 90°	Under 32°	0°	A.M. 8:00	P.M. 2:00	Total Prec.	Snow Only	Total Prec. 0.04"	0.1"	Snow Only 1.5"	Thun. Stms.		18 M.P.H.	30 M.P.H.	
Jan.	10	−4	38	−50				81	81	1.4		9				0.7			
Feb.	8	−7	39	−50				80	80	1.3		9				1.6			
March	5	−12	36	−57				80	78	1.1		8				0.6			
April	15	−3	39	−45				78	72	0.9		6				1.4			
May	30	15	42	−15				75	69	0.5		4				0.5			
June	40	31	55	13				79	76	0.4		4				0.4			
July	46	38	60	29				80	77	0.6		4				0.8			
Aug.	44	37	58	21				83	79	0.9		4				0.9			
Sept.	35	29	47	6				82	79	1.0		7				0.9			
Oct.	25	17	42	−20				80	79	1.2		7				0.6			
Nov.	17	6	39	−36				79	79	0.9		7				1.3			
Dec.	14	1	39	−42				81	81	1.5		9				1.3			
Year	24	12	60	−57				80	77	12.0		78				1.6			

Elevation 23 Feet
Latitude 78° 2' N   Longitude 14° 15' E

Total precipitation means the total of rain, snow, hail, etc. expressed as inches of rain.

Ten inches of snow equals approx. 1 inch of rain.

Temperatures are as ° F.

Wind velocity is the percentage of observations greater than 18 and 30 M.P.H. Number of observations varies from one per day to several per hour.

Visibility means number of days with less than ½-mile visibility.

Note: "—" indicates slight trace or possibility blank indicates no statistics available

JERSEY (Channel Islands)   (United Kingdom)   ENGLAND

Month	Daily Average Max.	Min.	Extreme Max.	Min.	Over 90°	Under 32°	0°	R.H. A.M.	P.M.	Total Prec.	Snow Only	0.01"	0.1"	Snow Only 1.5"	Thun. Stms.	Max. in 24 Hrs.	18 M.P.H.	30 M.P.H.	VISI-BILITY ½ MILE
Jan.	47	39	58	12	0	0.4	0	—	84	2.9	—	19	8	—	0.5	1.3	27	4.1	0.2
Feb.	47	39	62	18	0	0.6	0	—	83	2.4	—	15	7	—	0.3	1.1	29	4.2	0.6
March	50	40	67	21	0	0.6	0	—	82	2.5	—	13	7	—	0.5	1.1	17	0.7	2.0
April	55	44	84	28	0	—	0	—	82	1.9	—	13	6	—	0.4	1.8	21	1.0	0.5
May	60	48	81	33	0	0.0	0	—	82	1.8	0	11	6	0	2.0	1.5	12	0.6	1.6
June	65	53	87	39	0	0.0	0	—	82	2.1	0	9	6	0	1.0	1.1	8	0.1	1.4
July	68	56	90	44	0	0.0	0	—	83	2.1	0	11	6	0	1.0	1.7	12	0.6	1.7
Aug.	69	57	96	45	—	0.0	0	—	84	2.4	0	12	6	0	1.0	3.2	9	1.0	2.4
Sept.	66	55	89	40	0	0.0	0	—	85	2.4	0	14	6	0	1.0	2.1	13	2.0	1.0
Oct.	59	49	78	30	0	—	0	—	82	4.6	0	16	9	0	0.9	1.7	23	1.9	0.1
Nov.	53	44	66	16	0	0.1	0	—	83	4.1	—	19	9	—	0.5	1.5	23	1.9	0.4
Dec.	49	41	61	19	0	0.7	0	—	84	4.2	—	18	10	—	0.6	1.4	29	3.1	0.3
Year	57	47	96	12	0	2.4	0	—	83	33.0	0+	170	86	0+	10.0	3.2	18	1.6	12.0

Elevation 276 Feet
Latitude 49° 12' N   Longitude 2° 11' W

Total precipitation means the total of rain, snow, hail, etc. expressed as inches of rain.
Ten inches of snow equals approx. 1 inch of rain.
Temperatures are as ° F.

Wind velocity is the percentage of observations greater than 18 and 30 M.P.H. Number of observations varies from one per day to several per hour.
Visibility means number of days with less than ½-mile visibility.
Note: "—" indicates slight trace or possibility blank indicates no statistics available

JÖNKÖPING   SWEDEN

Month	Daily Average Max.	Daily Average Min.	Extreme Max.	Extreme Min.	Over 90°	Under 32°	0°	A.M. 8:00	P.M. 2:00	Total Prec.	Snow Only	Total Prec. 0.04"	0.1"	Snow Only 1.5"	Thun. Stms.	Max. Inches in 24 Hrs.	18 M.P.H.	30 M.P.H.	½ MILE
Jan.	33	23	48	−27	0	—	—	86	83	1.4	—	9	5	—		1.0			
Feb.	33	22	55	−28	0	—	—	86	79	1.0	—	7	3	—		0.6			
March	39	25	63	−21	0	—	—	84	69	0.9	—	6	3	—		0.8			
April	47	32	71	−1	0	—	—	79	64	1.3	—	8	4	—		1.1			
May	58	39	84	19	0	—	0	73	59	1.8	—	8	6	—		1.7			
June	65	47	92	32	—	—	0	71	59	2.2	0	9	6	0		3.3			
July	70	53	92	36	—	0	0	77	64	2.7	0	11	7	0		1.8			
Aug.	68	51	89	32	0	—	0	82	66	2.9	0	11	7	0		2.5			
Sept.	60	45	80	22	0	—	0	86	69	2.2	—	10	6	—		1.4			
Oct.	50	38	67	14	0	—	0	87	72	2.0	—	10	6	—		1.2			
Nov.	41	32	56	6	0	—	0	87	82	1.8	—	9	5	—		1.5			
Dec.	35	27	49	−11	0	—	—	87	85	1.2	—	9	4	—		0.6			
Year	50	36	92	−28	0+	0+	0+	82	71	21.0	—	107	62	—		3.3			

Elevation 325 Feet

Latitude 57° 46' N   Longitude 14° 11' E

Total precipitation means the total of rain, snow, hail, etc. expressed as inches of rain.

Ten inches of snow equals approx. 1 inch of rain.

Temperatures are as ° F.

Wind velocity is the percentage of observations greater than 18 and 30 M.P.H. Number of observations varies from one per day to several per hour.

Visibility means number of days with less than ½-mile visibility.

Note: "—" indicates slight trace or possibility blank indicates no statistics available

KARESUANDO

SWEDEN

Month	Daily Average Max.	Daily Average Min.	Extreme Max.	Extreme Min.	Over 90°	Under 32°	0°	A.M. 8:30	P.M. 2:30	Total Prec.	Snow Only	0.04"	0.1"	1.5"	Thun. Stms.	Max. Inches in 24 Hrs.	18 M.P.H.	30 M.P.H.	½ MILE
Jan.	18	2	41	−40	0	31	14	88	87	0.6	—	6	2	—	0	0.5			
Feb.	16	−1	43	−42	0	28	14	88	85	0.5	—	6	1	—	0	0.3			
March	26	4	50	−31	0	31	11	87	79	0.4	—	6	1	—	0	0.5			
April	29	13	60	−20	0	29	6	83	70	0.5	—	4	1	—	0	0.6			
May	46	29	76	−6	0	22	—	70	61	0.7	—	4	2	—	—	0.6			
June	59	40	86	25	0	2	0	63	53	1.6	—	7	5	—	1	0.7			
July	67	48	90	35	—	0	0	71	58	2.2	0	9	6	0	2	1.1			
Aug.	62	44	84	23	0	2	0	79	64	2.2	—	11	6	—	1	1.2			
Sept.	49	34	73	12	0	10	0	86	68	1.6	—	7	5	—	0	1.2			
Oct.	34	22	54	−11	0	28	1	90	81	0.9	—	6	3	—	0	1.1			
Nov.	25	11	45	−27	0	29	6	90	88	0.9	—	7	3	—	0	0.7			
Dec.	23	8	42	−31	0	31	15	89	87	0.6	—	7	2	—	0	0.3			
Year	38	21	90	−42	0+	243	66	82	73	13.0	—	80	38	—	4	1.2			

Elevation 1,073 Feet
Latitude 68° 27' N   Longitude 22° 30' E

Total precipitation means the total of rain, snow, hail, etc. expressed as inches of rain.

Ten inches of snow equals approx. 1 inch of rain.
Temperatures are as ° F.

Wind velocity is the percentage of observations greater than 18 and 30 M.P.H. Number of observations varies from one per day to several per hour.

Visibility means number of days with less than ½-mile visibility.

Note: "—" indicates slight trace or possibility blank indicates no statistics available

KASSEL (West) FED. REP. GERMANY

Month	Temperatures—in °F Daily Average Max.	Min.	Extreme Max.	Min.	No. of Days Over 90°	Under 32°	0°	Relative Humidity A.M. 6:30	P.M. 1:30	Precipitation Inches Total Prec.	Snow Only	No. of Days Total Prec. 0.1"	Snow Only 1.5"	Thun. Stms.	Max. Inches in 24 Hrs.	Wind Velocity 18 M.P.H.	30 M.P.H.	Visibility ½ Mile
Jan.	36	28	55	−16	0	18	—	89	81	1.7		6		—	1	3	0	3
Feb.	39	29	68	−10	0	14	—	89	74	1.4		5		—	1	2	—	2
March	46	33	73	0	0	13	0	88	65	1.4		5		1	1	3	—	4
April	55	38	84	21	0	4	0	84	57	1.7		5		1	1	2	0	1
May	65	46	97	28	—	—	0	84	56	1.9		6		4	1	2	—	2
June	70	51	95	36	—	0	0	85	58	2.2	0	6	0	5	2	1	0	1
July	73	54	99	39	—	0	0	88	62	2.9	0	7	0	5	2	—	0	1
Aug.	71	53	98	40	1	0	0	90	62	2.5	0	7	0	5	2	1	0	2
Sept.	65	48	93	30	—	—	0	92	66	2.0	0	6	0	1	1	1	0	4
Oct.	54	42	83	22	0	3	0	93	73	2.0		6		—	1	—	0	7
Nov.	43	35	67	6	0	7	0	91	80	2.0		5		—	2	3	0	3
Dec.	37	30	59	−4	0	14	—	90	83	2.0		6		—	1	2	0	5
Year	55	41	99	−16	1+	73	0+	89	68	23.0		70		22	2	2	—	35

Elevation 646 Feet
Latitude 51° 19′ N   Longitude 9° 27′ E

Total precipitation means the total of rain, snow, hail, etc. expressed as inches of rain.

Ten inches of snow equals approx. 1 inch of rain.
Temperatures are as ° F.

Wind velocity is the percentage of observations greater than 18 and 30 M.P.H. Number of observations varies from one per day to several per hour.

Visibility means number of days with less than ½-mile visibility.

Note: "—" indicates slight trace or possibility
blank indicates no statistics available

KASTRON (Límnos Island)   GREECE

Month	TEMPERATURES—IN °F Daily Average Max.	Daily Average Min.	Extreme Max.	Extreme Min.	No. of Days Over 90°	No. of Days Under 32°	0°	RELATIVE HUMIDITY A.M. 7:30	P.M. 1:30	PRECIPITATION Inches Total Prec.	Snow Only	No. of Days Total Prec. 0.04"	0.1"	Snow Only 1.5"	Thun. Stms.	Max. Inches in 24 Hrs.	WIND VELOCITY 18 M.P.H.	30 M.P.H.	VISIBILITY ½ MILE
Jan.	52	43	66	22				79	71	4.8		9				6			
Feb.	53	42	66	21				76	67	2.5		7				2			
March	56	44	74	27				74	66	3.0		6				3			
April	66	51	80	32				72	65	1.1		4				2			
May	74	57	89	36				70	63	1.0		3				1			
June	81	64	94	51				64	55	0.3		1				1			
July	88	70	98	52				60	52	0.7		—				3			
Aug.	87	70	98	62				62	50	0.3		1				1			
Sept.	80	65	93	51				69	55	0.8		1				1			
Oct.	70	56	84	45				77	63	2.5		5				2			
Nov.	63	51	75	33				81	71	2.8		7				2			
Dec.	56	46	71	32				80	74	4.7		9				2			
Year	69	55	98	21				72	63	24.0		53				6			

Elevation 23 Feet
Latitude 39° 53' N   Longitude 25° 4' E
Total precipitation means the total of rain, snow, hail, etc. expressed as inches of rain.
Ten inches of snow equals approx. 1 inch of rain.
Temperatures are as ° F.

Wind velocity is the percentage of observations greater than 18 and 30 M.P.H. Number of observations varies from one per day to several per hour.
Visibility means number of days with less than ½-mile visibility.
Note: "—" indicates slight trace or possibility blank indicates no statistics available

KERKIRA (Corfu) GREECE

Month	TEMPERATURES—IN °F Daily Average Max.	Min.	Extreme Max.	Min.	No. of Days Over 90°	Under 32°	0°	RELATIVE HUMIDITY A.M. 7:30	P.M. 1:30	PRECIPITATION Inches Total Prec.	Snow Only	No. of Days Total Prec. 0.1"	Snow Only 1.5"	Thun. Stms.	Max. Inches in 24 Hrs.	WIND VELOCITY 18 M.P.H.	30 M.P.H.	VISIBILITY ½ MILE
Jan.	56	44	68	25	0	4	0	81	70	6.3	—	11	—	1	6			
Feb.	57	44	73	23	0	3	0	79	69	5.5	—	11	—	1	8			
March	61	47	76	27	0	3	0	80	69	3.7	—	7	—	1	3			
April	67	51	83	37	0	0	0	79	66	3.1	0	7	0	1	4			
May	75	58	94	44	—	0	0	78	64	1.9	0	6	0	1	3			
June	82	64	97	50	5	0	0	76	61	1.0	0	3	0	1	3			
July	87	68	101	46	14	0	0	73	55	0.3	0	1	0	1	1			
Aug.	88	69	106	54	17	0	0	73	64	0.7	0	2	0	1	3			
Sept.	82	65	100	41	3	0	0	76	60	2.5	0	7	0	2	5			
Oct.	74	59	87	39	0	0	0	81	68	6.9	0	10	0	3	7			
Nov.	65	53	82	27	0	1	0	80	70	6.3	—	10	—	2	5			
Dec.	59	47	70	27	0	3	0	72	72	7.9	—	11	—	2	7			
Year	71	56	106	23	39	14	0	78	65	46.0	—	86	—	17	8			

Elevation 7 Feet
Latitude 39° 37' N   Longitude 19° 55' E
Total precipitation means the total of rain, snow, hail, etc. expressed as inches of rain.
Ten inches of snow equals approx. 1 inch of rain.
Temperatures are as ° F.

Wind velocity is the percentage of observations greater than 18 and 30 M.P.H. Number of observations varies from one per day to several per hour.
Visibility means number of days with less than ½-mile visibility.
Note: "—" indicates slight trace or possiblity
blank indicates no statistics available

KESWICK (United Kingdom) ENGLAND

Month	TEMPERATURES—IN °F Daily Average Max.	Min.	Extreme Max.	Min.	No. of Days Over 90°	Under 32°	0°	RELATIVE HUMIDITY	PRECIPITATION Inches Total Prec.	Snow Only	No. of Days Total Prec. 0.01"	0.1"	Snow Only 1.5"	Thun. Stms.	Max. Inches in 24 Hrs.	WIND VELOCITY 18 M.P.H.	30 M.P.H.	VISI-BIL-ITY ½ MILE
Jan.	45	34	59	0					7.0		20				2.8			
Feb.	45	34	59	13					4.2		16				1.9			
March	49	36	72	6					3.3		15				2.3			
April	53	39	74	24					3.3		15				2.9			
May	59	43	80	26					3.1		15				2.0			
June	64	49	85	34					3.4		14				2.5			
July	66	53	91	39					4.5		18				2.6			
Aug.	66	52	86	38					5.1		18				3.0			
Sept.	61	48	78	28					5.6		17				2.9			
Oct.	55	43	79	18					6.7		19				3.0			
Nov.	49	38	65	18					6.2		18				4.9			
Dec.	46	35	59	11					6.1		19				3.5			
Year	55	42	91	0					59.0		204				4.9			

Elevation 254 Feet

Latitude 54° 36' N   Longitude 3° 9' W

Total precipitation means the total of rain, snow, hail, etc. expressed as inches of rain.

Ten inches of snow equals approx. 1 inch of rain.

Temperatures are as °F.

Wind velocity is the percentage of observations greater than 18 and 30 M.P.H. Number of observations varies from one per day to several per hour.

Visibility means number of days with less than ½-mile visibility.

Note: "—" indicates slight trace or possibility blank indicates no statistics available

KIRKWALL (Orkney Islands)   (United Kingdom) SCOTLAND

Month	TEMPERATURES—IN °F Daily Average Max.	Min.	Extreme Max.	Min.	No. of Days Over 90°	Under 32°	0°	RELATIVE HUMIDITY A.M.	P.M.	PRECIPITATION Inches Total Prec.	Snow Only	No. of Days Total Prec. 0.1"	Snow Only 1.5"	Thun. Stms.	Max. Inches in 24 Hrs.	WIND VELOCITY 18 M.P.H.	30 M.P.H.	VISIBILITY ½ MILE
Jan.	43	36	54	8	0	7	0	—	87	3.5	—	9	—	0.5		32	12	0.3
Feb.	43	36	56	12	0	7	0	—	87	3.0	—	8	—	0.7		31	7	0.3
March	44	36	60	13	0	7	0	—	86	2.8	—	7	—	0.4		32	7	0.5
April	47	38	61	22	0	3	0	—	84	2.1	—	6	—	0.3		25	6	0.8
May	51	42	66	29	0	1	0	—	83	2.0	0	6	0	0.8		12	2	2.2
June	55	46	74	34	0	—	0	—	83	1.8	0	5	0	0.7		8	—	3.1
July	58	49	76	36	0	0	0	—	86	2.6	0	7	0	0.9		6	0	3.5
Aug.	58	50	71	37	0	0	0	—	87	2.8	0	7	0	0.7		7	1	3.9
Sept.	55	47	71	32	0	—	0	—	86	2.9	0	7	0	0.3		15	1	2.5
Oct.	49	42	66	29	0	1	0	—	86	3.8	0	8	0	0.6		24	6	0.0
Nov.	46	40	58	23	0	3	0	—	86	3.9	—	9	—	0.7		27	8	0.0
Dec.	44	38	56	20	0	7	0	—	87	4.2	—	10	—	0.5		32	13	—
Year	49	42	76	8	0	36	0	—	86	36.0	0+	89	0+	7.0		21	5	17.0

Elevation 69 Feet

Latitude 58° 57' N   **Longitude 2° 54' W**

Total precipitation means the total of rain, snow, hail, etc. expressed as inches of rain.

Ten inches of snow equals approx. 1 inch of rain.

Temperatures are as ° F.

Wind velocity is the percentage of observations greater than 18 and 30 M.P.H. Number of observations varies from one per day to several per hour.

Visibility means number of days with less than ½-mile visibility.

Note: "—" indicates slight trace or possibility blank indicates no statistics available

KÍTHIRA

GREECE

Elevation 535 Feet
Latitude 36° 9' N   Longitude 23° 0' E

Month	Daily Average Max.	Daily Average Min.	Extreme Max.	Extreme Min.	No. of Days Over 90°	No. of Days Under 32°	No. of Days 0°	Rel. Humidity A.M. 7:30	Rel. Humidity P.M. 1:30	Inches Total Prec.	Inches Snow Only	No. of Days Total Prec. 0.04"	No. of Days 0.1"	No. of Days Snow Only 1.5"	Thun. Stms.	Max. Inches in 24 Hrs.	Wind Velocity 18 M.P.H.	Wind Velocity 30 M.P.H.	Visibility ½ Mile
Jan.	55	46	66	30				70	68	5.0		15				3			
Feb.	56	46	68	28				72	69	3.4		11				2			
March	59	49	75	33				70	67	2.1		6				3			
April	64	54	83	39				65	63	0.8		2				1			
May	72	60	91	46				61	58	0.5		2				1			
June	79	67	98	54				58	56	0.2		6				1			
July	85	72	102	58				55	53	—		—				1			
Aug.	86	72	101	59				55	50	0.1		—				1			
Sept.	81	68	99	53				60	56	0.6		2				1			
Oct.	73	62	91	50				69	66	2.2		6				2			
Nov.	65	55	79	39				72	69	4.1		9				5			
Dec.	59	50	70	32				70	69	5.8		12				3			
Year	69	58	102	28				65	62	25.0		67				5			

Total precipitation means the total of rain, snow, hail, etc. expressed as inches of rain.

Ten inches of snow equals approx. 1 inch of rain.

Temperatures are as ° F.

Wind velocity is the percentage of observations greater than 18 and 30 M.P.H. Number of observations varies from one per day to several per hour.

Visibility means number of days with less than ½-mile visibility.

Note: "—" indicates slight trace or possibility blank indicates no statistics available

KOLN—BONN (West) FED. REP. GERMANY

Month	TEMPERATURES—IN °F Daily Average Max.	Min.	Extreme Max.	Min.	No. of Days Over 90°	Under 32°	0°	RELATIVE HUMIDITY A.M. 6:30	P.M. 1:30	PRECIPITATION Inches Total Prec.	Snow Only	No. of Days Total Prec. 0.04"	0.1"	Snow Only 1.5"	Thun. Stms.	Max. Inches in 24 Hrs.	WIND VELOCITY 18 M.P.H.	30 M.P.H.	VISIBILITY ½ MILE
Jan.	40	32	59	−3	0	8	—	86	78	2.1		11	6		—	1			
Feb.	43	33	66	0	0	12	0	86	73	1.8		10	6		—	1			
March	49	37	72	13	0	10	0	81	65	1.8		10	6		1	—			
April	57	42	82	25	0	6	0	81	57	1.9		10	6		1	1			
May	66	49	93	28	—	1	0	77	54	2.1		10	6		4	2			
June	71	54	93	37	1	0	0	77	57	2.6		10	7	0	4	2			
July	73	58	96	45	—	0	0	77	57	3.2		11	8	0	4	1			
Aug.	72	57	95	43	2	0	0	79	58	2.8		11	7	0	3	2			
Sept.	66	52	88	32	0	—	0	83	62	2.1		9	6	0	1	1			
Oct.	57	45	80	27	0	3	0	85	70	2.5		11	7		—	1			
Nov.	47	39	69	10	0	7	0	85	75	2.2		10	6		—	2			
Dec.	42	34	63	−4	0	15	—	85	79	2.5		12	7		—	1			
Year	57	44	96	−4	3	62	0+	82	65	27.0		125	78		18	2			

Elevation 300 Feet

Latitude 50° 51' N   **Longitude 7° 8' E**

Total precipitation means the total of rain, snow, hail, etc. expressed as inches of rain.

Ten inches of snow equals approx. 1 inch of rain.

Temperatures are as ° F.

Wind velocity is the percentage of observations greater than 18 and 30 M.P.H. Number of observations varies from one per day to several per hour.

Visibility means number of days with less than ½-mile visibility.

Note: "—" indicates slight trace or possibility blank indicates no statistics available

KRISTIANSAND

NORWAY

| Month | TEMPERATURES—IN °F | | | | | | | RELATIVE HUMIDITY | | PRECIPITATION | | | | | | | WIND VELOCITY | | VISIBILITY ½ MILE |
| | Daily Average | | Extreme | | No. of Days | | | | | Inches | | No. of Days | | | | Max. Inches in 24 Hrs. | | | |
	Max.	Min.	Max.	Min.	Over 90°	Under 32°	0°	A.M. 7:30	P.M. 1:30	Total Prec.	Snow Only	Total Prec. 0.04"	0.1"	Snow Only 1.5"	Thun. Stms.		18 M.P.H.	30 M.P.H.	
Jan.	32	25	56	−14	0	24	—	85	82	5.0	0	14	11	0	—	4	9	1	3
Feb.	36	25	54	−5	0	24	—	85	80	3.6	0	10	9	0	0	3	15	2	5
March	42	28	65	−7	0	26	—	83	72	3.6	0	10	7	0	0	3	11	1	4
April	50	35	72	5	0	15	0	77	67	2.7	0	9	7	0	—	2	7	0	3
May	61	42	82	25	0	1	0	72	64	2.5	0	9	7	0	—	3	8	—	2
June	67	48	86	32	0	—	0	74	67	2.8	—	9	7	—	1	3	4	0	2
July	71	53	90	40	—	0	0	77	67	3.5	—	10	8	—	2	4	3	0	2
Aug.	68	52	89	37	0	0	0	82	71	5.3	—	12	10	—	3	4	2	—	2
Sept.	61	46	82	28	0	1	0	85	74	4.7	0	11	9	0	2	3	5	—	2
Oct.	53	39	72	19	0	5	0	87	77	6.2	0	14	10	0	1	3	5	—	4
Nov.	43	33	56	10	0	11	0	87	83	5.7	0	14	10	0	—	5	15	2	2
Dec.	39	28	54	−8	0	17	1	86	84	6.4	0	16	11	0	0	3	11	1	4
Year	52	38	90	−14	0+	124	1	82	74	52.0	—	138	106	—	9	5	8	1	35

Elevation 54 Feet

Latitude 58° 12′ N   Longitude 8° 5′ E

Total precipitation means the total of rain, snow, hail, etc. expressed as inches of rain.

Ten inches of snow equals approx. 1 inch of rain.

Temperatures are as ° F.

Wind velocity is the percentage of observations greater than 18 and 30 M.P.H. Number of observations varies from one per day to several per hour.

Visibility means number of days with less than ½-mile visibility.

Note: "—" indicates slight trace or possibility blank indicates no statistics available

KUOPIO — FINLAND

Month	Daily Average Max.	Daily Average Min.	Extreme Max.	Extreme Min.	No. of Days Over 90°	No. of Days Under 32°	No. of Days 0°	R.H. A.M. 7:00	R.H. P.M. 3:00	Inches Total Prec.	Inches Snow Only	No. of Days Total Prec. 0.04"	No. of Days Total Prec. 0.1"	No. of Days Snow Only 1.5"	Thun. Stms.	Max. Inches in 24 Hrs.	Wind 18 M.P.H.	Wind 30 M.P.H.	Visibility ½ Mile
Jan.	19	9	42	−35	0	31	8	87	87	1.7	—	11	6	—	0	0.9	3	—	2
Feb.	19	9	43	−36	0	28	9	86	84	1.6	—	11	5	—	—	0.7	3	—	2
March	28	15	50	−26	0	30	9	82	70	1.4	—	8	5	—	0	0.6	2	—	1
April	40	27	67	−8	0	23	1	77	62	1.6	—	8	5	—	—	0.7	3	0	2
May	54	38	82	18	0	6	0	70	57	1.8	—	7	6	—	1	1.5	1	—	1
June	63	48	85	28	0	—	0	71	56	2.5	—	10	7	—	4	2.8	1	0	—
July	71	55	87	41	0	0	0	76	58	2.1	0	9	6	0	5	1.2	1	0	—
Aug.	65	52	86	37	0	0	0	84	65	3.3	0	12	8	0	3	1.2	1	0	1
Sept.	53	42	79	26	0	1	0	90	73	2.9	—	12	7	—	—	1.1	1	—	3
Oct.	40	32	63	8	0	11	0	89	82	2.7	—	13	7	—	—	1.0	2	0	2
Nov.	31	25	48	−7	0	23	1	89	88	2.1	—	12	6	—	0	1.0	1	0	1
Dec.	24	15	43	−26	0	30	6	88	88	1.5	—	10	5	—	0	0.7	2	—	2
Year	42	31	87	−36	0	183	34	82	73	25.0	—	123	71	—	14	2.8	2	—	17

Elevation 322 Feet

Latitude 63° 0′ N   **Longitude 27° 47′ E**

Total precipitation means the total of rain, snow, hail, etc. expressed as inches of rain.

Ten inches of snow equals approx. 1 inch of rain.

Temperatures are as ° F.

Wind velocity is the percentage of observations greater than 18 and 30 M.P.H. Number of observations varies from one per day to several per hour.

Visibility means number of days with less than ½-mile visibility.

Note: "—" indicates slight trace or possibility blank indicates no statistics available

KUUSAMO   FINLAND

| Month | TEMPERATURES—IN °F | | | | | | | RELATIVE HUMIDITY | | PRECIPITATION Inches | | No. of Days | | | | Max. Inches in 24 Hrs. | WIND VELOCITY | | VISIBILITY ½ MILE |
| | Daily Average | | Extreme | | Over 90° | Under 32° | 0° | A.M. 7:00 | P.M. 3:00 | Total Prec. | Snow Only | Total Prec. 0.04" | 0.1" | Snow Only 1.5" | Thun. Stms. | | 18 M.P.H. | 30 M.P.H. | |
	Max.	Min.	Max.	Min.															
Jan.	17	2	40	−40	0	31	13	86	86	1.5	—	9	5	—	0	0.4	1	0	1
Feb.	16	1	39	−39	0	28	11	86	84	1.2	—	9	4	—	0	0.5	2	0	4
March	24	6	44	−38	0	31	14	83	72	1.2	—	8	4	—	0	0.6	1	0	2
April	35	18	62	−24	0	28	3	82	64	1.3	—	8	5	—	—	0.3	1	0	1
May	48	31	77	7	0	16	0	73	57	1.7	—	7	5	—	1	0.8	—	0	1
June	60	43	85	25	0	4	0	71	57	2.4	—	11	6	—	2	0.9	1	0	0
July	68	50	90	31	—	—	0	76	59	3.0	0	11	7	0	3	1.1	1	0	1
Aug.	63	47	84	25	0	1	0	85	66	3.4	—	12	8	—	2	1.5	1	0	3
Sept.	50	37	76	11	0	6	0	89	71	2.6	—	11	7	—	—	1.3	1	0	4
Oct.	36	27	54	−14	0	18	—	90	84	2.4	—	11	6	—	—	1.6	1	0	3
Nov.	28	18	49	−25	0	28	2	91	90	1.7	—	11	5	—	—	0.5	2	0	3
Dec.	22	9	38	−35	0	31	12	89	89	1.6	—	9	5	—	0	0.6	2	0	2
Year	39	24	90	−40	0+	220	55	83	73	24.0	—	117	68	—	9	1.6	1	0	25

Elevation 863 Feet

Latitude 65° 58' N   Longitude 29° 10' E

Total precipitation means the total of rain, snow, hail, etc. expressed as inches of rain.

Ten inches of snow equals approx. 1 inch of rain.

Temperatures are as ° F.

Wind velocity is the percentage of observations greater than 18 and 30 M.P.H. Number of observations varies from one per day to several per hour.

Visibility means number of days with less than ½-mile visibility.

Note: "—" indicates slight trace or possibility
blank indicates no statistics available

LA CORUÑA

SPAIN

| Month | TEMPERATURES—IN °F | | | | | | | RELATIVE HUMIDITY | | PRECIPITATION | | | | | | | | WIND VELOCITY | | VISI-BILITY |
| | Daily Average | | Extreme | | No. of Days | | | | | Inches | | No. of Days | | | | Max. Inches in 24 Hrs. | | | | |
	Max.	Min.	Max.	Min.	Over 90°	Under 32°	0°	A.M. 6:30	P.M. 1:00	Total Prec.	Snow Only	Total Prec. 0.004"	0.1"	Snow Only 1.5"	Thun. Stms.		18 M.P.H.	30 M.P.H.	½ MILE
Jan.	55	44	69	28	0	—	0	81	73	4.4	—	17	10	—	—	2	21	6	2
Feb.	55	44	75	28	0	—	0	82	71	3.3	—	15	9	—	1	1	27	8	2
March	58	46	80	30	0	—	0	82	69	3.9	0	17	7	0	1	2	21	3	1
April	59	48	85	35	0	0	0	82	70	3.2	0	15	7	0	1	1	18	2	—
May	63	52	86	40	0	0	0	83	70	2.2	0	13	6	0	2	1	22	3	1
June	68	56	87	46	0	0	0	83	73	1.5	0	8	4	0	1	2	12	1	5
July	71	58	95	49	—	0	0	85	73	1.3	0	8	4	0	—	3	9	1	5
Aug.	72	59	93	48	—	0	0	86	72	1.7	0	8	5	0	1	3	12	1	2
Sept.	70	57	94	43	—	0	0	87	73	2.8	0	11	7	0	—	2	10	1	2
Oct.	65	53	88	39	0	0	0	86	73	3.9	0	14	9	0	—	3	12	1	2
Nov.	59	48	77	34	0	0	0	83	73	4.5	0	17	9	0	1	2	24	6	1
Dec.	56	45	69	30	0	—	0	83	74	5.2	0	18	11	0	1	3	19	4	1
Year	63	51	95	28	0+	0+	0	84	72	38.0	—	161	88	—	9	3	17	3	24

Elevation 187 Feet

Latitude 43° 23' N   Longitude 8° 22' W

Total precipitation means the total of rain, snow, hail, etc. expressed as inches of rain.

Ten inches of snow equals approx. 1 inch of rain.

Temperatures are as ° F.

Wind velocity is the percentage of observations greater than 18 and 30 M.P.H. Number of observations varies from one per day to several per hour.

Visibility means number of days with less than ½-mile visibility.

Note: "—" indicates slight trace or possibility
blank indicates no statistics available

LÁRISA

GREECE

| Month | Temperatures—in °F | | | | | | | Relative Humidity | | Precipitation | | | | | | Wind Velocity | | Visibility |
| | Daily Average | | Extreme | | No. of Days | | | | | Inches | | No. of Days | | | Max. Inches in 24 Hrs. | | | ½ Mile |
	Max.	Min.	Max.	Min.	Over 90°	Under 32°	0°	A.M. 7:30	P.M. 1:30	Total Prec.	Snow Only	Total Prec. 0.1"	Snow Only 1.5"	Thun. Stms.		18 M.P.H.	30 M.P.H.	
Jan.	50	33	69	9	0	—	0	89	76	1.9	—	6	—	0	2			
Feb.	54	35	79	12	0	—	0	87	71	1.7	—	6	—	—	2			
March	62	40	86	21	0	—	0	87	63	1.5	—	5	—	—	1			
April	71	47	92	30	—	—	0	80	53	1.5	0	5	0	—	2			
May	80	55	100	41	—	0	0	78	53	2.1	0	6	0	1	2			
June	86	63	109	50	9	0	0	69	45	1.3	0	4	0	1	3			
July	93	67	113	55	23	0	0	64	40	0.9	0	3	0	1	2			
Aug.	93	66	109	54	23	0	0	64	39	0.7	0	2	0	1	2			
Sept.	85	61	106	41	8	0	0	74	47	1.0	0	4	0	—	2			
Oct.	73	53	101	35	—	0	0	87	63	2.6	0	7	0	—	2			
Nov.	60	44	84	20	0	—	0	90	74	2.8	—	7	—	—	2			
Dec.	53	37	72	14	0	—	0	90	78	2.4	—	7	—	—	2			
Year	72	50	113	9	63+	0+	0	80	59	20.0	—	62	—	6	3			

Elevation 239 Feet

Latitude 39° 38' N   Longitude 22° 27' E

Total precipitation means the total of rain, snow, hail, etc. expressed as inches of rain.

Ten inches of snow equals approx. 1 inch of rain.

Temperatures are as ° F.

Wind velocity is the percentage of observations greater than 18 and 30 M.P.H. Number of observations varies from one per day to several per hour.

Visibility means number of days with less than ½-mile visibility.

Note: "—" indicates slight trace or possibility

blank indicates no statistics available

LERWICK (Shetland Islands) (United Kingdom) SCOTLAND

| Month | TEMPERATURES—IN °F | | | | | | | RELATIVE HUMIDITY | | PRECIPITATION | | | | | | | WIND VELOCITY | | VISIBILITY ½ MILE |
| | Daily Average | | Extreme | | No. of Days | | | | | Inches | | No. of Days | | | | Max. Inches in 24 Hrs. | | | |
	Max.	Min.	Max.	Min.	Over 90°	Under 32°	0°	A.M. 7:00	P.M. 1:30	Total Prec.	Snow Only	Total Prec. 0.01"	0.1"	Snow Only 1.5"	Thun. Stms.		18 M.P.H.	30 M.P.H.	½ MILE
Jan.	42	35	54	16	0	3.8	0	87	85	4.5	—	28	10	—	0.4	1.5	46	20	0.0
Feb.	42	34	53	17	0	5.2	0	88	84	3.3	—	22	9	—	0.3	1.7	41	17	0.4
March	43	35	57	18	0	4.7	0	87	83	3.1	—	21	7	—	0.1	1.2	34	10	0.4
April	46	37	62	21	0	1.8	0	81	79	2.7	—	20	7	—	0.1	1.3	29	8	1.2
May	51	41	66	27	0	0.5	0	85	79	2.2	—	16	0	_0	.20	1.60	220	30	3.3
June	54	45	71	33	0	0.0	0	86	80	2.1	0	15	6	0	0.1	1.1	21	4	3.1
July	58	49	70	37	0	0.0	0	90	84	2.5	0	16	7	0	0.4	3.4	14	1	5.8
Aug.	58	49	70	38	0	0.0	0	90	83	2.8	0	17	7	0	0.4	1.7	14	2	3.5
Sept.	55	46	69	28	0	0.1	0	89	83	3.8	—	20	8	—	0.2	1.3	23	5	2.3
Oct.	50	42	63	26	0	0.5	0	87	81	4.4	—	24	9	—	0.3	1.4	35	14	0.4
Nov.	46	38	58	21	0	0.9	0	87	84	4.6	—	25	9	—	0.2	2.6	36	15	0.2
Dec.	43	37	54	19	0	1.8	0	86	85	4.4	—	26	10	—	0.3	1.2	44	20	0.1
Year	49	41	71	16	0	19.0	0	87	83	40.5	0+	250	96	0+	3.0	3.4	30	10	21.0

Elevation 272 Feet

Latitude 60° 8' N   Longitude 1° 11' W

Total precipitation means the total of rain, snow, hail, etc. expressed as inches of rain.

Ten inches of snow equals approx. 1 inch of rain.

Temperatures are as ° F.

Wind velocity is the percentage of observations greater than 18 and 30 M.P.H. Number of observations varies from one per day to several per hour.

Visibility means number of days with less than ½-mile visibility.

Note: "—" indicates slight trace or possibility blank indicates no statistics available

LILLE

FRANCE

| Month | TEMPERATURES—IN °F | | | | | | | RELATIVE HUMIDITY | | PRECIPITATION | | | | | | | WIND VELOCITY | | VISI-BIL-ITY |
| | Daily Average | | Extreme | | No. of Days | | | | | Inches | | No. of Days | | | | Max. | | | |
	Max.	Min.	Max.	Min.	Over 90°	Under 32°	0°	A.M. 6:00	Noon	Total Prec.	Snow Only	Total Prec. 0.004"	0.1"	Snow Only 1.5"	Thun. Stms.	Inches in 24 Hrs.	18 M.P.H.	30 M.P.H.	½ MILE
Jan.	42	33	56	8	0.0	14	0	91	85	2.5	—	18	8	—	0.3	0.8			
Feb.	45	34	61	0	0.0	12	0	89	79	1.9	—	14	6	—	1.0	0.7			
March	51	37	72	19	0.0	16	0	90	71	2.5	—	12	7	—	1.0	0.7			
April	58	40	82	27	0.0	4	0	91	62	2.0	—	14	6	—	1.0	1.1			
May	66	47	87	31	0.0	2	0	89	60	2.4	0	13	7	0	2.0	1.3			
June	72	52	95	35	0.5	0	0	90	67	2.2	0	12	6	0	3.0	1.0			
July	75	55	96	35	1.8	0	0	92	65	2.8	0	15	7	0	3.0	1.2			
Aug.	75	56	96	39	1.0	0	0	93	62	2.3	0	14	6	0	3.0	1.0			
Sept.	69	51	93	34	0.7	—	0	94	65	2.6	0	13	7	0	2.0	1.9			
Oct.	59	45	76	24	0.0	2	0	94	74	3.0	—	15	7	—	1.0	0.9			
Nov.	48	39	65	18	0.0	6	0	93	82	3.0	—	17	7	—	0.3	0.9			
Dec.	43	35	60	7	0.0	13	0	92	87	3.2	—	18	9	—	0.3	0.9			
Year	59	44	96	0	4.0	68	0	91	72	30.0	0+	175	83	0+	18.0	1.9			

Elevation 157 Feet

Latitude 50° 33' N   Longitude 3° 5' E

Total precipitation means the total of rain, snow, hail, etc. expressed as inches of rain.

Ten inches of snow equals approx. 1 inch of rain.

Temperatures are as ° F.

Wind velocity is the percentage of observations greater than 18 and 30 M.P.H. Number of observations varies from one per day to several per hour.

Visibility means number of days with less than ½-mile visibility.

Note: "—" indicates slight trace or possibility blank indicates no statistics available

LINZ — AUSTRIA

Month	TEMPERATURES—IN °F Daily Average Max.	Min.	Extreme Max.	Min.	No. of Days Over 90°	Under 32°	0°	RELATIVE HUMIDITY A.M. 7:00	P.M. 2:00	PRECIPITATION Inches Total Prec.	Snow Only	No. of Days Total Prec. 0.04"	0.1"	Snow Only 1.5"	Thun. Stms.	Max. Inches in 24 Hrs.	WIND VELOCITY 18 M.P.H.	30 M.P.H.	VISIBILITY ½ MILE
Jan.	34	26	60	−8	0	27	2	84	76	2.4	5.3	10	7	1	0.3	1.1	7.0	0.7	10
Feb.	38	27	53	−8	0	24	—	82	69	2.1	10.0	10	7	3	0.3	1.2	10.0	0.5	4
March	48	33	72	5	0	18	0	80	59	2.1	4.0	8	6	1	0.3	1.3	5.0	0.1	2
April	56	40	81	25	0	4	0	81	56	2.9	—	11	7	0	1.0	2.0	6.0	0.4	2
May	67	49	85	28	0	1	0	78	52	3.2	0.0	11	7	0	3.0	3.4	2.0	0.0	—
June	71	54	94	36	—	0	0	77	53	4.1	0.0	12	9	0	4.0	2.9	3.0	0.2	2
July	75	57	94	45	1	0	0	79	52	4.9	0.0	13	9	0	6.0	2.3	2.0	0.0	1
Aug.	.73	56	95	42	1	0	0	82	56	4.1	0.0	11	9	0	4.0	2.2	3.0	0.1	5
Sept.	67	50	92	30	1	—	0	84	58	2.8	0.0	10	7	0	2.0	2.0	2.0	0.0	6
Oct.	55	42	79	20	0	4	0	85	65	2.3	0.0	9	6	0	0.3	1.2	3.0	0.2	7
Nov.	44	35	66	7	0	14	0	84	73	2.4	—	9	6	—	0.3	1.5	4.0	0.1	9
Dec.	36	27	60	−17	0	25	1	84	78	2.4	6.0	10	7	1	0.3	3.4	7.0	0.5	10
Year	55	41	95	−17	3	117	3	82	62	36.0	26.0	124	88	7	22.0	3.4	4.6	0.2	59

Elevation 974 Feet
Latitude 48° 14' N   Longitude 14° 11' E
Total precipitation means the total of rain, snow, hail, etc. expressed as inches of rain.
Ten inches of snow equals approx. 1 inch of rain.
Temperatures are as ° F.

Wind velocity is the percentage of observations greater than 18 and 30 M.P.H. Number of observations varies from one per day to several per hour.
Visibility means number of days with less than ½-mile visibility.
Note: "—" indicates slight trace or possibility blank indicates no statistics available

LISBON   PORTUGAL

Month	TEMPERATURES—IN °F Daily Average Max.	Min.	Extreme Max.	Min.	No. of Days Over 90°	Under 32°	0°	RELATIVE HUMIDITY A.M. 9:00	P.M. 3:00	PRECIPITATION Inches Total Prec.	Snow Only	No. of Days Total Prec. 0.04"	0.1"	Snow Only 1.5"	Thun. Stms.	Max. Inches in 24 Hrs.	WIND VELOCITY 18 M.P.H.	30 M.P.H.	VISIBILITY ½ MILE
Jan.	56	46	68	30	0	—	0	83	72	3.3	0	9	9	0	0.4	3.2	10	1.0	
Feb.	58	47	77	28	0	1	0	80	66	3.2	—	9	9	—	0.6	2.0	8	1.0	
March	61	49	83	34	0	0	0	76	63	3.1	0	9	7	0	0.9	2.4	9	0.7	
April	64	52	87	37	0	0	0	69	58	2.4	0	7	7	0	0.9	2.1	9	0.2	
May	69	56	94	42	1	0	0	67	57	1.7	0	6	5	0	0.9	1.7	14	0.8	
June	75	60	99	49	2	0	0	64	53	0.7	0	2	2	0	0.5	1.4	17	0.9	
July	79	63	103	52	6	0	0	61	48	0.2	0	1	—	0	0.3	1.5	26	1.0	
Aug.	80	64	102	52	6	0	0	61	46	0.2	0	—	—	0	0.2	1.8	20	0.6	
Sept.	76	62	99	51	3	0	0	67	53	1.4	0	4	4	0	0.9	3.0	11	0.7	
Oct.	69	57	93	43	—	0	0	72	59	3.1	0	8	8	0	0.5	2.2	6	0.3	
Nov.	62	52	77	34	0	0	0	80	68	4.2	0	10	9	0	0.6	3.6	10	0.9	
Dec.	57	47	66	31	0	—	0	83	72	3.6	0	10	9	0	0.6	4.4	11	1.5	
Year	67	55	103	28	19	1	0	72	60	27.0	—	76	69	—	7.3	4.4	13	0.8	

Elevation 361 Feet
Latitude 38° 46' N   Longitude 9° 8' W

Total precipitation means the total of rain, snow, hail, etc. expressed as inches of rain.

Ten inches of snow equals approx. 1 inch of rain.
Temperatures are as °F.

Wind velocity is the percentage of observations greater than 18 and 30 M.P.H. Number of observations varies from one per day to several per hour.

Visibility means number of days with less than ½-mile visibility.

Note: "—" indicates slight trace or possibility blank indicates no statistics available

LIVERPOOL (United Kingdom) ENGLAND

Month	Daily Average Max.	Daily Average Min.	Extreme Max.	Extreme Min.	No. of Days Over 90°	No. of Days Under 32°	No. of Days 0°	Rel. Hum. A.M. 7:00	Rel. Hum. P.M. 1:00	Total Prec.	Snow Only	No. of Days Total Prec. 0.01"	No. of Days 0.1"	No. of Days Snow Only 1.5"	Thun. Stms.	Max. Inches in 24 Hrs.	Wind 18 M.P.H.	Wind 30 M.P.H.	Visibility ½ Mile
Jan.	44	36	57	15	0	6	0	88	82	3.2	—	18	9	—		1.1			
Feb.	44	36	59	15	0	4	0	88	78	2.5	—	13	7	—		1.3			
March	48	38	68	20	0	5	0	86	71	2.0	—	13	6	—		1.1			
April	52	41	73	27	0	—	0	84	68	2.0	—	14	6	—		0.7			
May	58	46	79	32	0	—	0	83	68	3.0	0	14	7	0		2.1			
June	63	51	87	40	0	0	0	80	68	2.3	0	13	6	0		1.7			
July	66	55	87	47	0	0	0	83	72	3.2	0	15	8	0		1.6			
Aug.	65	55	85	41	0	0	0	87	73	3.7	0	16	8	0		1.4			
Sept.	61	51	79	39	0	0	0	87	72	3.2	0	15	8	0		1.1			
Oct.	55	46	73	29	0	1	0	87	75	3.7	0	17	8	0		1.5			
Nov.	48	41	64	26	0	2	0	88	81	3.5	—	17	8	—		1.4			
Dec.	45	37	58	23	0	4	0	89	84	3.3	—	18	9	—		1.1			
Year	54	44	87	15	0	22	0	86	74	35.0	0+	183	90	0+		2.1			

Elevation 84 Feet

Latitude 53° 20′ N   **Longitude 2° 52′ W**

Total precipitation means the total of rain, snow, hail, etc. expressed as inches of rain.

Ten inches of snow equals approx. 1 inch of rain.

Temperatures are as ° F.

Wind velocity is the percentage of observations greater than 18 and 30 M.P.H. Number of observations varies from one per day to several per hour.

Visibility means number of days with less than ½-mile visibility.

Note: "—" indicates slight trace or possibility blank indicates no statistics available

LONDON   (United Kingdom) ENGLAND

| Month | TEMPERATURES—IN °F | | | | | | | RELATIVE HUMIDITY | | PRECIPITATION | | | | | | | WIND VELOCITY | | VISI-BIL-ITY |
| | Daily Average | | Extreme | | No. of Days | | | | | Inches | | No. of Days | | | | Max. Inches | | | ½ |
	Max.	Min.	Max.	Min.	Over 90°	Under 32°	0°	A.M. 9:00	P.M. 3:00	Total Prec.	Snow Only	Total Prec. 0.01"	0.1"	Snow Only 1.5"	Thun. Stms.	24 Hrs.	18 M.P.H.	30 M.P.H.	MILE
Jan.	45	36	58	9	0.0	13	0	87	80	1.8	—	17	6	—	0.2	1.6	10.0	0.3	8.7
Feb.	46	36	63	11	0.0	12	0	84	72	1.5	—	13	5	—	0.1	0.8	11.0	0.4	6.7
March	49	37	70	17	0.0	8	0	79	63	1.7	—	11	5	—	0.8	1.1	8.0	0.2	5.7
April	55	40	82	26	0.0	3	0	72	58	1.5	—	14	5	—	1.0	0.7	7.0	0.1	0.9
May	63	46	89	30	0.0	—	0	69	57	1.7	0	13	5	0	3.0	1.0	7.0	0.1	1.3
June	68	51	94	37	0.1	0	0	68	57	2.1	0	11	6	0	2.0	1.6	4.0	0.0	1.2
July	71	55	93	42	0.3	0	0	68	55	2.2	0	13	6	0	3.0	1.4	6.0	0.1	0.7
Aug.	70	54	94	41	0.1	0	0	73	58	2.2	0	13	6	0	3.0	2.2	6.0	0.0	2.2
Sept.	65	50	92	31	—	—	0	78	63	1.9	0	13	6	0	1.0	1.3	8.0	0.1	3.5
Oct.	57	45	83	25	0.0	2	0	83	70	2.7	—	14	7	—	0.4	1.6	5.0	0.0	9.2
Nov.	49	39	63	20	0.0	5	0	87	79	2.2	—	16	6	—	0.2	1.4	8.0	0.3	8.6
Dec.	46	37	60	11	0.0	8	0	87	81	2.3	—	16	7	—	0.2	1.1	11.0	0.2	11.0
Year	57	44	94	9	0.5	51	0	78	66	24.0	0+	164	70	0+	15.0	2.2	7.4	0.2	59.0

Elevation 80 Feet

Latitude 51° 28' N   Longitude 0° 27' W

Total precipitation means the total of rain, snow, hail, etc. expressed as inches of rain.

Ten inches of snow equals approx. 1 inch of rain.

Temperatures are as ° F.

Wind velocity is the percentage of observations greater than 18 and 30 M.P.H. Number of observations varies from one per day to several per hour.

Visibility means number of days with less than ½-mile visibility.

Note: "—" indicates slight trace or possibility blank indicates no statistics available

LUGANO   SWITZERLAND

Month	TEMPERATURES—IN °F							RELATIVE HUMIDITY		PRECIPITATION							WIND VELOCITY		VISI-BILITY
	Daily Average		Extreme		No. of Days					Inches		No. of Days			Thun.	Max. Inches in 24 Hrs.			½ MILE
	Max.	Min.	Max.	Min.	Over 90°	Under 32°	0°	A.M. 7:00	P.M. 1:00	Total Prec.	Snow Only	Total Prec. 0.01"	0.1"	Snow Only 1.5"	Stms.		18 M.P.H.	30 M.P.H.	MILE
Jan.	43	29	76	7	0	27	0	80	64	2.4	—	7	7	—	0	2.7			
Feb.	48	30	77	7	0	19	0	77	57	2.4	—	8	7	—	0	2.9			
March	56	36	81	20	0	8	0	76	53	4.5	—	8	8	—	—	3.1			
April	63	43	89	28	0	1	0	74	53	6.4	—	11	10	—	1	3.2			
May	70	50	90	30	—	—	0	76	57	7.6	0	14	—	0	3	3.9			
June	78	56	98	40	1	0	0	74	55	7.3	0	12	—	0	5	3.9			
July	83	60	100	46	3	0	0	74	53	6.9	0	11	10	0	5	5.1			
Aug.	82	59	98	44	1	0	0	78	55	7.4	0	10	—	0	5	6.8			
Sept.	75	54	93	36	0	—	0	83	60	6.9	0	10	10	0	3	4.4			
Oct.	63	46	83	28	0	—	0	85	64	7.8	—	10	—	—	1	6.1			
Nov.	52	38	72	23	0	6	0	83	66	5.4	—	10	10	—	—	3.4			
Dec.	45	31	72	10	0	20	0	80	65	3.1	—	9	9	—	0	2.5			
Year	63	44	100	7	5	81	0	78	59	68.0	—	119	—	—	22	6.8			

Elevation 906 Feet

Latitude 46° 0' N   **Longitude 8° 57' E**

Total precipitation means the total of rain, snow, hail, etc. expressed as inches of rain.

Ten inches of snow equals approx. 1 inch of rain.

Temperatures are as ° F.

Wind velocity is the percentage of observations greater than 18 and 30 M.P.H. Number of observations varies from one per day to several per hour.

Visibility means number of days with less than ½-mile visibility.

Note: "—" indicates slight trace or possibility blank indicates no statistics available

LUXEMBOURG CITY

LUXEMBOURG

Month	Daily Average Max.	Daily Average Min.	Extreme Max.	Extreme Min.	Over 90°	Under 32°	0°	R.H. A.M. 6:30	R.H. P.M. 1:30	Total Prec.	Snow Only	No. Days Total Prec. 0.004"	No. Days 0.1"	Snow Only 1.5"	Thun. Stms.	Max. Inches in 24 Hrs	Wind 18 M.P.H.	Wind 30 M.P.H.	Visibility ½ Mile
Jan.	36	29	57	−10	0	—	—	92	86	2.3	—	14	7	—	0.3	1.1	10	0.3	9
Feb.	40	30	63	−7	0	—	—	91	79	2.0	—	13	6	—	0.3	0.9	12	0.7	5
March	49	33	73	7	0	—	0	88	62	1.9	—	13	6	—	0.3	0.9	10	0.6	3
April	58	40	85	21	0	—	0	85	57	2.1	—	12	6	—	2.0	1.1	8	0.3	1
May	65	46	91	26	—	—	0	88	60	2.4	—	13	7	—	3.0	2.6	4	0.3	1
June	71	52	98	29	—	—	0	89	61	2.5	0	13	7	0	5.0	1.3	3	0.0	1
July	74	55	99	38	—	0	0	89	59	2.8	0	13	7	0	5.0	1.9	4	0.0	1
Aug.	73	55	97	33	—	0	0	90	60	2.6	0	12	7	0	5.0	2.0	4	0.1	2
Sept.	65	50	91	23	—	—	0	94	69	2.4	—	12	6	—	2.0	1.7	5	0.5	3
Oct.	56	43	80	19	0	—	0	95	75	2.7	—	13	7	—	0.3	1.1	5	0.4	6
Nov.	45	37	66	13	0	—	0	92	86	2.7	—	15	7	—	0.3	1.6	7	0.6	7
Dec.	39	32	58	−8	0	—	—	95	91	2.8	—	15	8	—	0.0	1.6	10	0.8	11
Year	56	42	99	−10	0+	0+	0+	91	70	29.0	0+	158	81	0+	24.0	2.6	7	0.4	50

Elevation 1,083 Feet

Latitude 49° 37' N   Longitude 6° 3' E

Total precipitation means the total of rain, snow, hail, etc. expressed as inches of rain.

Ten inches of snow equals approx. 1 inch of rain.

Temperatures are as ° F.

Wind velocity is the percentage of observations greater than 18 and 30 M.P.H. Number of observations varies from one per day to several per hour.

Visibility means number of days with less than ½-mile visibility.

Note: "—" indicates slight trace or possibility blank indicates no statistics available

FRANCE

LYON

Month	Temperatures—in °F							Relative Humidity		Precipitation								Wind Velocity		Visibility
	Daily Average		Extreme		No. of Days			A.M. 7:30	P.M. 1:30	Inches		No. of Days				Max. Inches in 24 Hrs.		18 M.P.H.	30 M.P.H.	½ Mile
	Max.	Min.	Max.	Min.	Over 90°	Under 32°	0°			Total Prec.	Snow Only	Total Prec. 0.004"	0.1"	Snow Only 1.5"	Thun. Stms.					
Jan.	41	30.	66	−13	0.0	16	—	86	74	1.4		13	5	—	0	1.5				
Feb.	46	31	71	−9	0.0	15	—	86	65	1.4		11	5	—	—	2.4				
March	54	37.	79	8	0.0	10	0	84	58	1.8		12	6	—	1	2.2				
April	61	42	85	24	0.0	4	0	79	57	2.1		13	6	—	2	2.2				
May	69	49	94	25	0.1	—	0	77	55	2.8		13	7	—	6	1.9				
June	76	55	102	26	1.7	—	0	75	53	2.9		12	7	—	7	2.8				
July	80	58	105	42	3.2	0	0	74	51	2.8		11	7	0	7	2.9				
Aug.	80	57	105	40	3.4	0	0	81	52	2.9		10	7	0	7	4.1				
Sept.	72	53	98	32	0.2	—	0	85	56	3.1		11	8	0	4	3.3				
Oct.	61	45	90	20	0.2	2	0	88	64	3.1		13	8	—	2	4.3				
Nov.	49	39	73	15	0.0	8	0	88	73	2.6		14	7	—	—	3.4				
Dec.	42	32	65	−12	0.0	16	—	88	78	1.9		15	6	—	—	1.6				
Year	61	44	105	−13	9.0	70	0⁺	83	61	29.0		148	79	0⁺	37	4.3				

Elevation 649 Feet

Latitude 45° 44′ N   Longitude 4° 55′ E

Total precipitation means the total of rain, snow, hail, etc. expressed as inches of rain.

Ten inches of snow equals approx. 1 inch of rain.

Temperatures are as °F.

Wind velocity is the percentage of observations greater than 18 and 30 M.P.H. Number of observations varies from one per day to several per hour.

Visibility means number of days with less than ½-mile visibility.

Note: "—" indicates slight trace or possibility blank indicates no statistics available

MADRID

SPAIN

Month	TEMPERATURES—IN °F Daily Average Max.	Min.	Extreme Max.	Min.	No. of Days Over 90°	Under 32°	0°	RELATIVE HUMIDITY A.M. 7:00	P.M. 1:30	PRECIPITATION Inches Total Prec.	Snow Only	No. of Days Total Prec. 0.004"	0.1"	Snow Only 1.5"	Thun. Stms.	Max. Inches in 24 Hrs.	WIND VELOCITY 18 M.P.H.	30 M.P.H.	VISI-BIL-ITY ½ MILE
Jan.	49	34	66	14	0	14	0	89	71	1.7	—	9	5	—	0	1.0	10	1.0	3
Feb.	54	35	77	14	0	11	0	87	64	1.7	—	9	4	—	0	1.5	13	1.0	1
March	62	40	82	25	0	4	0	84	58	2.0	—	11	5	—	—	2.2	8	1.0	1
April	66	43	82	30	0	1	0	78	52	1.6	0	9	5	0	1	1.0	9	—	—
May	74	50	95	32	1	—	0	75	51	2.0	0	9	5	0	2	1.6	6	—	0
June	82	57	104	37	8	0	0	69	43	1.1	0	6	3	0	2	1.2	3	—	0
July	91	63	104	48	21	0	0	63	37	0.5	0	3	2	0	3	1.2	3	0.0	0
Aug.	89	62	104	48	17	0	0	63	36	0.7	0	2	2	0	1	1.5	3	0.0	—
Sept.	82	57	99	43	7	0	0	73	47	1.2	0	6	3	0	1	2.1	3	—	—
Oct.	68	49	84	30	0	—	0	83	57	2.6	0	8	6	0	—	1.8	2	0.0	—
Nov.	58	41	75	27	0	3	0	89	68	2.1	—	10	4	—	1	2.6	6	0.0	1
Dec.	51	36	70	19	0	11	0	90	73	2.5	—	9	5	—	—	1.2	7	1.0	3
Year	69	47	104	14	53	44	0	79	55	20.0	—	91	48	—	12	2.6	6	0.3	10

Elevation 1,972 Feet

Latitude 40° 28' N   Longitude 3° 34' W

Total precipitation means the total of rain, snow, hail, etc. expressed as inches of rain.

Ten inches of snow equals approx. 1 inch of rain.

Temperatures are as ° F.

Wind velocity is the percentage of observations greater than 18 and 30 M.P.H. Number of observations varies from one per day to several per hour.

Visibility means number of days with less than ½-mile visibility.

Note: "—" indicates slight trace or possibility blank indicates no statistics available

MAHON (Island Of Menorca) (Balearic Islands) SPAIN

Month	Daily Average Max.	Daily Average Min.	Extreme Max.	Extreme Min.	No. of Days Over 90°	No. of Days Under 32°	No. of Days 0°	R.H. A.M. 7:30	R.H. P.M. 1:30	Total Prec.	Snow Only	Days Total Prec. 0.004"	Days Total Prec. 0.1"	Snow Only 1.5"	Thun. Stms.	Max. Inches in 24 Hrs.	Wind 18 M.P.H.	Wind 30 M.P.H.	Visibility ½ Mile
Jan.	56	45	71	32	0	—	0	81	70	2.4	0	9	7	0	2	2	12	2	
Feb.	56	46	73	27	0	1	0	80	69	1.7	—	8	6	—	2	3	10	1	
March	60	49	77	35	0	0	0	80	68	1.9	0	7	6	0	1	3	10	1	
April	64	52	81	36	0	0	0	78	65	1.3	0	6	5	0	2	3	12	1	
May	70	57	86	42	0	0	0	75	62	1.2	0	4	4	0	1	3	8	1	
June	76	63	95	50	—	0	0	71	59	0.8	0	3	3	0	2	4	4	0	
July	81	68	97	52	3	0	0	69	57	0.2	0	1	—	0	2	1	4	0	
Aug.	82	70	101	57	2	0	0	72	59	0.9	0	3	3	0	3	3	4	—	
Sept.	78	66	94	52	—	0	0	74	63	2.8	0	7	7	0	2	7	4	—	
Oct.	71	60	86	44	0	0	0	77	67	5.2	0	10	10	0	6	4	8	—	
Nov.	63	53	77	36	0	0	0	79	70	3.6	0	10	8	0	2	3	13	1	
Dec.	58	48	70	31	0	—	0	80	72	3.0	0	11	9	0	2	3	13	2	
Year	68	56	101	27	5	1+	0	76	65	25.0	—	79	66	—	26	7	9	1	

Elevation 197 Feet

Latitude 39° 52' N   Longitude 4° 15' E

Total precipitation means the total of rain, snow, hail, etc. expressed as inches of rain.

Ten inches of snow equals approx. 1 inch of rain.

Temperatures are as ° F.

Wind velocity is the percentage of observations greater than 18 and 30 M.P.H. Number of observations varies from one per day to several per hour.

Visibility means number of days with less than ½-mile visibility.

Note: "—" indicates slight trace or possibility; blank indicates no statistics available

MALAGA

SPAIN

Elevation 51 Feet
Latitude 36° 39' N　　**Longitude 4° 29' W**

Month	TEMPERATURES—IN °F							RELATIVE HUMIDITY		PRECIPITATION							WIND VELOCITY		VISIBILITY ½ MILE
	Daily Average		Extreme		No. of Days					Inches		No. of Days				Max. Inches in 24 Hrs.	18 M.P.H.	30 M.P.H.	
	Max.	Min.	Max.	Min.	Over 90°	Under 32°	0°	A.M. 6:30	P.M. 1:00	Total Prec.	Snow Only	Total Prec. 0.004"	0.1"	Snow Only 1.5"	Thun. Stms.				
Jan.	61	47	73	32	0	—	0	75	67	1.9	0	6	6	0	2	3.0			
Feb.	62	48	78	35	0	0	0	73	67	2.1	0	7	7	0	1	2.6			
March	64	51	81	38	0	0	0	76	68	2.9	0	8	7	0	1	2.9			
April	69	55	89	43	0	0	0	71	64	1.5	0	5	5	0	3	2.8			
May	74	60	91	46	—	0	0	66	61	0.8	0	3	3	0	2	1.1			
June	80	66	100	52	—	0	0	66	61	0.4	0	1	1	0	1	1.1			
July	84	70	105	55	7	0	0	64	61	—	0	0	0	0	—	0.9			
Aug.	85	72	105	62	8	0	0	62	59	0.0	0	1	0	0	1	0.3			
Sept.	81	68	97	52	4	0	0	69	65	1.2	0	4	4	0	3	1.5			
Oct.	74	61	87	40	0	0	0	74	69	2.9	0	7	7	0	2	4.4			
Nov.	67	53	83	40	0	0	0	73	67	3.8	0	8	8	0	1	4.2			
Dec.	62	48	75	35	0	0	0	76	67	2.5	0	8	8	0	1	3.6			
Year	72	58	105	32	19+	0+	0	70	65	20.0	0	58	55	0	18	4.4			

Total precipitation means the total of rain, snow, hail, etc. expressed as inches of rain.
Ten inches of snow equals approx. 1 inch of rain.
Temperatures are as ° F.

Wind velocity is the percentage of observations greater than 18 and 30 M.P.H. Number of observations varies from one per day to several per hour.
Visibility means number of days with less than ½-mile visibility.
Note: "—" indicates slight trace or possibility blank indicates no statistics available

354

MALIN HEAD — IRELAND

Month	Daily Average Max.	Min.	Extreme Max.	Min.	Over 90°	Under 32°	0°	R.H. A.M. 6:30	Noon	Total Prec.	Snow Only	No. Days 0.01"	0.1"	Snow Only 1.5"	Thun. Stms.	Max. in 24 Hrs	Wind 18 M.P.H.	30 M.P.H.	Vis. ½ Mile
Jan.	45	39	57	21	0	1.9	0	57	21	2.6	—	23	7.7	—	0.5	2.0	33	4.4	0.1
Feb.	45	39	59	20	0	2.0	0	59	20	2.4	—	20	7.3	—	0.2	1.3	30	4.6	0.3
March	48	40	63	23	0	1.4	0	63	23	2.3	—	20	6.4	—	0.2	1.1	22	1.1	0.8
April	50	42	69	23	0	0.2	0	69	23	1.97	—	18	5.9	—	0.2	0.9	21	1.5	0.8
May	54	45	73	31	0	—	0	73	31	1.97	—	18	5.9	0	0.5	1.2	13	0.1	1.1
June	58	49	81	38	0	0.0	0	81	38	2.13	0	17	5.9	0	0.9	1.7	11	0.4	0.8
July	61	53	80	40	0	0.0	0	80	42	2.83	0	20	7.2	0	0.8	2.8	10	0.1	0.8
Aug.	61	53	80	41	0	0.0	0	80	41	3.54	0	22	8.2	0	0.7	1.8	12	0.2	0.4
Sept.	59	51	84	36	0	0.0	0	84	36	2.64	0	18	6.8	0	0.3	1.6	17	1.5	0.9
Oct.	54	47	73	32	0	—	0	73	32	2.95	0	20	7.3	0	0.1	1.6	36	5.6	0.1
Nov.	49	42	62	27	0	0.3	0	62	27	3.27	—	22	7.7	—	0.5	1.6	29	3.8	0.0
Dec.	46	40	62	25	0	1.7	0	62	25	3.35	—	23	9.0	—	0.3	1.2	28	4.6	0.2
Year	53	45	84	20	0	7.5	0	84	20	32.0	0+	241	85.0	0+	5.2	2.8	22	2.3	6.3

Elevation 85 Feet
Latitude 55° 22' N   Longitude 7° 20' W
Total precipitation means the total of rain, snow, hail, etc. expressed as inches of rain.
Ten inches of snow equals approx. 1 inch of rain.
Temperatures are as °F.

Wind velocity is the percentage of observations greater than 18 and 30 M.P.H. Number of observations varies from one per day to several per hour.
Visibility means number of days with less than ½-mile visibility.
Note: "—" indicates slight trace or possibility blank indicates no statistics available

MARGATE (United Kingdom) ENGLAND

Month	TEMPERATURES—IN °F							RELATIVE HUMIDITY		PRECIPITATION						WIND VELOCITY		VISIBILITY ½ MILE
	Daily Average		Extreme		No. of Days			A.M.	P.M.	Inches		No. of Days			Max. Inches in 24 Hrs.	18 M.P.H.	30 M.P.H.	
	Max.	Min.	Max.	Min.	Over 90°	Under 32°	0°			Total Prec.	Snow Only	Total Prec. 0.1"	Snow Only 1.5"	Thun. Stms.				
Jan.	44	37	57	14	0	6	0	—	89	1.9	—	6	—	0.0		28	1.8	
Feb.	44	37	61	14	0	6	0	—	85	1.4	—	5	—	0.1		23	0.9	
March	48	39	66	15	0	5	0	—	84	1.2	—	4	—	0.1		13	0.6	
April	53	43	74	27	0	1	0	—	78	1.4	—	5	—	0.9		12	0.0	
May	58	48	88	31	0	1	0	—	76	1.4	0	5	0	2.0		14	0.2	
June	65	54	88	39	0	0	0	—	77	1.2	0	4	0	1.0		9	0.4	
July	69	57	94	43	—	0	0	—	74	2.0	0	6	0	2.0		11	0.4	
Aug.	69	58	92	44	—	0	0	—	75	1.9	0	5	0	3.0		5	0.0	
Sept.	65	55	91	36	0	0	0	—	78	2.0	0	6	0	1.0		12	0.0	
Oct.	58	49	77	31	0	—	0	—	83	2.5	0	7	0	0.7		17	1.0	
Nov.	51	43	68	18	0	3	0	—	87	2.5	—	7	—	0.3		17	0.8	
Dec.	46	39	59	17	0	8	0	—	89	1.9	—	6	—	0.2		21	1.4	
Year	56	47	94	14	0+	30	0	—	81	21.0	0+	66	0+	11.0		15	0.6	

Elevation 180 Feet

Latitude 51° 21' N   Longitude 1° 20' E

Total precipitation means the total of rain, snow, hail, etc. expressed as inches of rain.

Ten inches of snow equals approx. 1 inch of rain.

Temperatures are as °F.

Wind velocity is the percentage of observations greater than 18 and 30 M.P.H. Number of observations varies from one per day to several per hour.

Visibility means number of days with less than ½-mile visibility.

Note: "—" indicates slight trace or possibility blank indicates no statistics available

MARIEHAMN — FINLAND

Month	Temperatures—in °F Daily Average Max.	Min.	Extreme Max.	Min.	No. of Days Over 90°	Under 32°	0°	Relative Humidity A.M. 6:30	P.M. 2:30	Precipitation Inches Total Prec.	Snow Only	No. of Days Total Prec. 0.04"	0.1"	Snow Only 1.5"	Thun. Stms.	Max. Inches in 24 Hrs.	Wind Velocity 18 M.P.H.	30 M.P.H.	Visibility ½ Mile
Jan.	31	22	47	−20	0	27	3	88	85	1.7	—	10	6	—	—	1.2	12	1	2
Feb.	29	19	52	−18	0	26	4	87	81	1.2	—	9	4	—	0	1.0	10	2	3
March	34	22	56	−15	0	28	3	88	73	1.1	—	6	4	—	—	0.7	6	—	3
April	43	30	66	−2	0	21	—	87	68	1.3	—	6	4	—	—	1.2	7	—	4
May	54	38	79	20	0	7	0	79	64	1.6	—	7	5	—	2	1.0	3	0	3
June	62	47	83	27	0	1	0	78	64	1.5	—	7	4	—	3	1.4	2	0	1
July	69	55	88	34	0	0	0	83	67	1.9	0	7	5	0	4	1.4	3	0	2
Aug.	67	54	88	30	0	—	0	88	70	3.1	0	10	8	0	3	3.1	5	—	1
Sept.	58	46	77	21	0	1	0	88	70	2.1	—	10	6	—	1	1.6	4	—	3
Oct.	48	39	63	11	0	6	0	91	73	2.6	—	13	7	—	—	1.1	6	—	4
Nov.	41	34	51	0	0	16	—	89	80	2.2	—	11	6	—	—	0.8	4	—	2
Dec.	35	27	52	−8	0	22	1	88	87	2.1	—	10	6	—	—	0.9	11	1	2
Year	48	36	88	−20	0	156	11	86	75	22.0	—	106	65	—	15	3.0	6	—	30

Elevation 16 Feet

Latitude 60° 7' N   Longitude 19° 53' E

Total precipitation means the total of rain, snow, hail, etc. expressed as inches of rain.

Ten inches of snow equals approx. 1 inch of rain.

Temperatures are as ° F.

Wind velocity is the percentage of observations greater than 18 and 30 M.P.H. Number of observations varies from one per day to several per hour.

Visibility means number of days with less than ½-mile visibility.

Note: "—" indicates slight trace or possibility blank indicates no statistics available

MARSEILLE

FRANCE

Month	Daily Average Max.	Daily Average Min.	Extreme Max.	Extreme Min.	No. of Days Over 90°	No. of Days Under 32°	No. of Days 0°	Rel. Humidity A.M. 7:30	Rel. Humidity P.M. 1:00	Inches Total Prec.	Inches Snow Only	No. of Days Total Prec. 0.04"	No. of Days Total Prec. 0.1"	No. of Days Snow Only 1.5"	Thun. Stms.	Max. Inches in 24 Hrs.	Wind Velocity 18 M.P.H.	Wind Velocity 30 M.P.H.	Visibility ½ Mile
Jan.	50	35	68	13	0	23	0	77	62	1.7	—	7	6	—	0.4	2.2	12	1.8	1.2
Feb.	53	36	71	2	0	—	0	75	58	1.3	—	7	4	—	0.6	2.3	13	2.5	1.2
March	59	41	75	14	0	—	0	74	56	1.7	—	5	5	—	0.4	2.0	18	2.7	0.4
April	64	46	83	28	0	—	0	71	56	1.7	—	5	5	—	0.7	1.9	18	4.2	0.1
May	71	52	91	32	—	—	0	69	53	1.8	0	6	6	0	1.4	2.9	15	2.0	0.1
June	79	58	99	42	—	0	0	70	55	0.9	0	3	3	0	1.8	1.8	16	1.0	0.0
July	84	63	102	46	7	0	0	68	51	0.4	0	2	1	0	1.1	4.1	13	1.0	0.1
Aug.	83	63	98	47	6	0	0	71	50	1.3	0	4	4	0	2.3	2.9	11	0.5	0.1
Sept.	77	58	94	34	—	0	0	78	53	2.4	0	6	6	0	2.1	8.3	12	0.5	0.2
Oct.	68	51	86	28	0	—	0	83	58	3.0	—	7	7	—	1.9	8.7	8	0.9	0.6
Nov.	58	43	73	22	0	—	0	82	61	2.7	—	8	7	—	0.7	5.9	13	1.3	1.1
Dec.	52	37	68	9	0	8	0	79	63	2.6	—	15	8	—	1.3	3.6	12	1.4	1.0
Year	67	49	102	2	13+	31+	0	75	56	22.0	0+	75	62	0+	15.0	8.7	13	1.7	6.0

TEMPERATURES—IN ° F

PRECIPITATION

Elevation 39 Feet

Latitude 43° 26' N   Longitude 5° 13' E

Total precipitation means the total of rain, snow, hail, etc. expressed as inches of rain.

Ten inches of snow equals approx. 1 inch of rain. Temperatures are as ° F.

Wind velocity is the percentage of observations greater than 18 and 30 M.P.H. Number of observations varies from one per day to several per hour.

Visibility means number of days with less than ½-mile visibility.

Note: "—" indicates slight trace or possibility blank indicates no statistics available

MESSINA (Sicily) ITALY

Month	Daily Average Max.	Daily Average Min.	Extreme Max.	Extreme Min.	Over 90°	Under 32°	0°	A.M. 6:00	P.M. 1:00	Total Prec.	Snow Only	0.04"	0.1"	Snow Only 1.5"	Thun. Stms.	Max. Inches in 24 Hrs.	18 M.P.H.	30 M.P.H.	½ MILE
Jan.	57	49	66	34	0	0	0	75	67	3.7	0	14	9	0		2			
Feb.	59	48	73	37	0	0	0	74	68	3.3	0	11	9	0		2			
March	62	50	80	34	0	0	0	75	63	2.9	0	9	7	0		2			
April	66	54	80	44	0	0	0	74	61	2.6	0	7	7	0		3			
May	73	59	91	45	—	0	0	74	60	1.5	0	6	5	0		3			
June	81	64	98	56	4	0	0	70	54	0.9	0	3	3	0		2			
July	86	72	95	64	9	0	0	71	54	0.5	0	2	1	0		2			
Aug.	86	74	99	61	9	0	0	71	56	1.0	0	3	3	0		2			
Sept.	82	69	99	57	5	0	0	76	59	2.0	0	6	6	0		2			
Oct.	74	62	87	52	0	0	0	79	66	3.9	0	11	9	0		4			
Nov.	69	56	75	41	0	0	0	76	66	4.8	0	12	9	0		4			
Dec.	60	51	69	38	0	0	0	76	63	4.3	0	12	10	0		2			
Year	71	59	99	34	27	0	0	74	61	31.0	0	96	78	0		4			

Elevation 167 Feet

Latitude 38° 12' N   Longitude 15° 33' E

Total precipitation means the total of rain, snow, hail, etc. expressed as inches of rain.

Ten inches of snow equals approx. 1 inch of rain.

Temperatures are as ° F.

Wind velocity is the percentage of observations greater than 18 and 30 M.P.H. Number of observations varies from one per day to several per hour.

Visibility means number of days with less than ½-mile visibility.

Note: "—" indicates slight trace or possibility
blank indicates no statistics available

MILAN

ITALY

Month	Daily Average Max.	Min.	Extreme Max.	Min.	Over 90°	Under 32°	0°	A.M. 5:30	P.M. 12:30	Total Prec.	Snow Only	0.1"	Snow Only 1.5"	Thun. Stms.	Max. Inches in 24 Hrs.	18 M.P.H.	30 M.P.H.	VISI-BIL-ITY ½ MILE
Jan.	40	29	64	5	0	22	0	90	82	2.4	—	7	—	—	2	—	0	25
Feb.	47	33	69	4	0	19	0	87	73	2.3	—	7	—	—	2	—	0	17
March	56	38	76	19	0	7	0	88	65	2.8	—	7	—	1	2	—	0	10
April	66	46	86	28	0	1	0	86	57	3.4	—	7	—	2	2	1.0	0	2
May	72	54	89	40	0	0	0	86	59	3.9	0	7	0	4	3	—	0	1
June	80	61	94	42	—	0	0	84	56	3.2	0	8	0	8	3	—	0	1
July	84	64	96	51	2	0	0	85	61	2.7	0	7	0	10	3	—	0	1
Aug.	82	63	96	49	1	0	0	89	58	3.1	0	8	0	5	3	—	0	2
Sept.	76	58	88	42	0	0	0	91	63	3.6	0	8	0	4	3	0.0	0	6
Oct.	64	49	80	31	0	—	0	94	73	4.7	0	9	0	1	3	—	0	16
Nov.	51	39	69	23	0	6	0	92	80	4.3	—	9	—	—	4	—	0	22
Dec.	42	33	60	8	0	13	0	94	89	3.0	—	9	—	—	2	—	0	27
Year	63	47	96	4	3+	68	0	89	68	39.0	0+	93	0+	35+	4	0.2	0	130

Column groups: TEMPERATURES—IN °F (Daily Average, Extreme, No. of Days); RELATIVE HUMIDITY; PRECIPITATION (Inches, No. of Days, Max. Inches in 24 Hrs.); WIND VELOCITY M.P.H.

Elevation 767 Feet

Latitude 45° 37' N   Longitude 8° 43' E

Total precipitation means the total of rain, snow, hail, etc. expressed as inches of rain.

Ten inches of snow equals approx. 1 inch of rain.

Temperatures are as ° F.

Wind velocity is the percentage of observations greater than 18 and 30 M.P.H. Number of observations varies from one per day to several per hour.

Visibility means number of days with less than ½-mile visibility.

Note: "—" indicates slight trace or possibility
blank indicates no statistics available

MITILINI (Lesbos Island)　　　GREECE

Month	Daily Average Max.	Daily Average Min.	Extreme Max.	Extreme Min.	No. of Days Over 90°	No. of Days Under 32°	No. of Days 0°	Relative Humidity A.M. 8:00	Relative Humidity P.M. 2:00	Inches Total Prec.	Inches Snow Only	No. of Days Total Prec. 0.04"	No. of Days Total Prec. 0.1"	No. of Days Snow Only 1.5"	No. of Days Thun. Stms.	Max. Inches in 24 Hrs.	Wind Velocity 18 M.P.H.	Wind Velocity 30 M.P.H.	Visibility ½ Mile
Jan.	54	42	66	25				87	77	4.5		12				3			
Feb.	53	40	68	22				85	75	4.3		9				2			
March	61	45	76	28				82	66	3.1		6				2			
April	68	52	86	33				78	61	2.3		4				3			
May	77	59	89	45				75	57	1.1		3				2			
June	85	67	106	55				69	50	0.8		1				1			
July	90	71	107	55				64	47	0.1		—				1			
Aug.	89	70	105	60				69	50	0.1		—				1			
Sept.	82	65	104	50				75	57	0.3		1				2			
Oct.	74	58	91	39				83	64	2.4		5				2			
Nov.	64	51	79	32				87	73	4.5		9				6			
Dec.	57	45	69	28				90	79	5.4		11				4			
Year	71	55	107	22				79	63	28.0		61				6			

Elevation 11 Feet

Latitude 39° 6' N　　　Longitude 26° 35' E

Total precipitation means the total of rain, snow, hail, etc. expressed as inches of rain.

Ten inches of snow equals approx. 1 inch of rain.

Temperatures are as ° F.

Wind velocity is the percentage of observations greater than 18 and 30 M.P.H. Number of observations varies from one per day to several per hour.

Visibility means number of days with less than ½-mile visibility.

Note: "—" indicates slight trace or possibility blank indicates no statistics available

MONTE CARLO

MONACO

Month	Daily Average Max.	Daily Average Min.	Extreme Max.	Extreme Min.	No. of Days Over 90°	No. of Days Under 32°	No. of Days 0°	Rel. Hum. A.M.	Rel. Hum. P.M.	Total Prec.	Snow Only	Total Prec. 0.04"	0.1"	Snow Only 1.5"	Thun. Stms.	Max. Inches in 24 Hrs.	Wind 18 M.P.H.	Wind 30 M.P.H.	Visibility ½ Mile
Jan.	54	46	72	31						2.4		6				2.8			
Feb.	55	47	73	28						2.3		5				2.1			
March	57	49	72	32						3.1		7				2.8			
April	61	53	78	42						2.2		6				2.3			
May	66	59	82	44						2.1		6				2.3			
June	73	65	93	54						1.4		3				2.2			
July	77	70	89	60						0.7		1				1.4			
Aug.	78	71	93	60						1.1		3				2.1			
Sept.	74	67	88	52						2.3		5				5.1			
Oct.	67	60	81	39						4.7		7				3.4			
Nov.	61	53	75	36						4.3		8				3.8			
Dec.	56	49	68	27						3.5		7				2.3			
Year	65	57	93	27						30.0		64				5.0			

Elevation 180 Feet

Latitude 43° 44′ N   Longitude 7° 25′ E

Total precipitation means the total of rain, snow, hail, etc. expressed as inches of rain.

Ten inches of snow equals approx. 1 inch of rain.

Temperatures are as ° F.

Wind velocity is the percentage of observations greater than 18 and 30 M.P.H. Number of observations varies from one per day to several per hour.

Visibility means number of days with less than ½-mile visibility.

Note: "—" indicates slight trace or possibility blank indicates no statistics available

MULLINGAR

IRELAND

| Month | TEMPERATURES—IN °F | | | | | | | RELATIVE HUMIDITY | | PRECIPITATION | | | | | | | WIND VELOCITY | | VISIBILITY |
| | Daily Average | | Extreme | | No. of Days | | | | | Inches | | No. of Days | | | | Max. Inches in 24 Hrs. | 18 M.P.H. | 30 M.P.H. | ½ MILE |
	Max.	Min.	Max.	Min.	Over 90°	Under 32°	0°	A.M. 9:30	P.M. 5:30	Total Prec.	Snow Only	Total Prec. 0.04"	0.1"	Snow Only 1.5"	Thun. Stms.				
Jan.	45	35	56	16	0	—	0	89	87	3.5	—	15	9	—		1.1			
Feb.	46	36	58	10	0	—	0	88	83	3.0	—	13	8	—		1.1			
March	51	37	59	16	0	—	0	85	76	2.3	—	10	6	—		1.1			
April	55	40	71	26	0	—	0	81	74	2.7	—	14	7	—		1.4			
May	60	43	76	30	0	—	0	76	71	2.4	0	10	7	0		1.0			
June	65	48	86	38	0	0	0	79	75	3.3	0	14	8	0		0.9			
July	67	51	83	42	0	0	0	81	77	3.6	0	15	8	0		1.5			
Aug.	67	51	83	40	0	0	0	83	77	3.5	0	13	8	0		1.4			
Sept.	62	48	74	35	0	0	0	86	82	4.3	0	16	9	0		1.6			
Oct.	56	44	71	28	0	—	0	89	87	3.4	—	13	8	—		1.1			
Nov.	50	39	63	24	0	—	0	91	89	3.5	—	15	8	—		1.1			
Dec.	45	35	58	16	0	—	0	91	89	4.2	—	17	10	—		1.2			
Year	56	42	86	10	0	0+	0	85	81	39.7	0+	165	96	0+		1.6			

Elevation 361 Feet
Latitude 53° 31' N   Longitude 7° 21' W

Total precipitation means the total of rain, snow, hail, etc. expressed as inches of rain.

Ten inches of snow equals approx. 1 inch of rain.
Temperatures are as ° F.

Wind velocity is the percentage of observations greater than 18 and 30 M.P.H. Number of observations varies from one per day to several per hour.

Visibility means number of days with less than ½-mile visibility.

Note: "—" indicates slight trace or possibility blank indicates no statistics available

MUNICH   (West) FED. REP. GERMANY

| Month | TEMPERATURES—IN °F | | | | | | | RELATIVE HUMIDITY | | PRECIPITATION | | | | | | | WIND VELOCITY | | VISI-BIL-ITY ½ MILE |
| | Daily Average | | Extreme | | No. of Days | | | | | Inches | | No. of Days | | | | Max. Inches in 24 Hrs. | | | |
	Max.	Min.	Max.	Min.	Over 90°	Under 32°	0°	A.M. 7:00	P.M. 2:00	Total Prec.	Snow Only	Total Prec. 0.04"	0.1"	Snow Only 1.5"	Thun. Stms.		18 M.P.H.	30 M.P.H.	
Jan.	36	22	62	−14	0	28	1	91	79	2.7	12.0	10	7	3.0	0	1	7	1	8
Feb.	36	21	64	−29	0	24	3	89	69	2.0	11.0	9	6	2.5	—	1	6	—	7
March	48	29	75	1	0	21	0	90	60	1.9	3.0	10	6	0.6	—	1	5	1	3
April	56	36	78	10	0	11	0	83	55	2.3	2.4	13	7	0.3	2	3	4	—	2
May	64	42	86	23	0	2	0	81	56	4.3	0.6	13	10	0.2	5	2	1	0	1
June	69	49	93	34	—	0	0	79	55	4.8	0.0	14	11	0.0	8	2	1	—	1
July	73	53	96	22	1	—	0	80	55	6.0	0.0	14	11	0.0	7	2	2	0	1
Aug.	72	51	95	32	1	—	0	87	58	3.8	0.0	13	9	0.0	5	3	1	0	1
Sept.	67	46	89	31	0	1	0	92	60	3.0	0.0	11	7	0.0	2	1	2	—	4
Oct.	56	37	78	21	0	10	0	93	67	1.7	1.9	10	5	0.5	0	2	1	0	7
Nov.	44	31	68	−2	0	19	—	94	80	2.3	3.4	9	6	0.7	0	1	4	—	8
Dec.	38	26	60	0	0	26	—	91	82	1.9	9.5	11	6	1.8	—	1	7	1	11
Year	55	37	96	−29	2	142	4	87	65	37.0	44.0	137	91	9.6	29	3	3	—	54

Elevation 1,807 Feet

Latitude 48° 4' N   Longitude 11° 38' E

Total precipitation means the total of rain, snow, hail, etc. expressed as inches of rain.

Ten inches of snow equals approx. 1 inch of rain.

Temperatures are as ° F.

Wind velocity is the percentage of observations greater than 18 and 30 M.P.H. Number of observations varies from one per day to several per hour.

Visibility means number of days with less than ½-mile visibility.

Note: "—" indicates slight trace or possibility blank indicates no statistics available

FRANCE

NANCY

Month	TEMPERATURES—IN °F							RELATIVE HUMIDITY		PRECIPITATION							WIND VELOCITY		VISIBILITY
	Daily Average		Extreme		No. of Days			A.M. 7:30	P.M. 1:30	Inches		No. of Days				Max. Inches in 24 Hrs.	18 M.P.H.	30 M.P.H.	½ MILE
	Max.	Min.	Max.	Min.	Over 90°	Under 32°	0°			Total Prec.	Snow Only	Total Prec. 0.004"	0.1"	Snow Only 1.5"	Thun. Stms.				
Jan.	40	29	57	−5	0.0	—	—	91	82	2.1	—	15	7	—	0.2	—			
Feb.	43	29	68	−13	0.0	—	—	90	71	1.9	—	14	6	—	0.2	0.8			
March	50	34	74	8	0.0	—	—	89	62	2.1	—	14	6	—	0.3	1.0			
April	58	39	85	20	0.0	5	0	84	58	2.1	—	15	6	—	1.0	1.1			
May	66	46	91	24	—	1	0	83	58	2.2	—	15	6	—	4.0	1.7			
June	71	51	97	35	0.3	0	0	83	58	2.4	0	14	6	0	6.0	1.1			
July	74	55	97	39	1.2	0	0	82	57	2.7	0	13	7	0	5.0	1.1			
Aug.	73	54	94	39	1.0	0	0	86	58	2.5	0	14	7	0	5.0	1.2			
Sept.	68	49	93	30	—	—	0	90	60	2.4	0	13	6	0	2.0	2.3			
Oct.	57	42	78	18	0.0	4	0	93	72	2.6	—	14	7	—	0.3	1.6			
Nov.	46	35	70	15	0.0	8	0	93	79	2.6	—	15	7	—	0.1	1.5			
Dec.	41	25	64	−6	0.0	24	—	92	83	2.7	—	17	8	—	0.0	1.1			
Year	57	41	97	−13	2.5	42	0+	88	67	28.0	0+	173	79	0+	24.0	2.3			

Elevation 774 Feet

Latitude 48° 42' N   Longitude 6° 14' E

Total precipitation means the total of rain, snow, hail, etc. expressed as inches of rain.

Ten inches of snow equals approx. 1 inch of rain.

Temperatures are as ° F.

Wind velocity is the percentage of observations greater than 18 and 30 M.P.H. Number of observations varies from one per day to several per hour.

Visibility means number of days with less than ½-mile visibility.

Note: "—" indicates slight trace or possibility blank indicates no statistics available

NANTES

FRANCE

Elevation 92 Feet
Latitude 47° 11' N **Longitude 1° 33' W**

| Month | TEMPERATURES—IN °F | | | | | | | RELATIVE HUMIDITY | | PRECIPITATION | | | | | | | Max. Inches in 24 Hrs. | WIND VELOCITY | | VISIBILITY ½ MILE |
| | Daily Average | | Extreme | | No. of Days | | | A.M. 6:00 | P.M. 1:00 | Inches | | No. of Days | | | | | | 18 M.P.H. | 30 M.P.H. | |
	Max.	Min.	Max.	Min.	Over 90°	Under 32°	0°			Total Prec.	Snow Only	Total Prec. 0.04"	0.1"	Snow Only 1.5"	Thun. Stms.					
Jan.	45	36	59	8	0	—	0	91	80	2.6	—	11	8	—	0.4	1.1			
Feb.	49	36	71	9	0	—	0	90	73	2.5	—	11	7	—	1.0	1.4			
March	54	38	72	23	0	—	0	91	67	2.6	—	12	7	—	1.0	1.3			
April	61	42	84	30	0	—	0	90	61	2.2	0	9	6	0	1.0	1.1			
May	67	46	92	31	—	—	0	89	61	2.2	0	10	6	0	2.0	1.0			
June	73	53	100	40	—	0	0	90	61	2.1	0	9	6	0	2.0	2.2			
July	77	56	104	44	—	0	0	91	59	2.1	0	8	6	0	2.0	1.4			
Aug.	77	55	97	45	—	0	0	93	59	1.8	0	7	5	0	2.0	1.8			
Sept.	72	51	92	38	—	0	0	94	62	1.9	0	7	6	0	2.0	1.3			
Oct.	62	45	81	26	0	—	0	95	73	3.7	—	13	8	—	1.0	2.2			
Nov.	52	39	70	22	0	—	0	94	79	3.5	—	13	8	—	1.0	1.3			
Dec.	46	36	65	12	0	—	0	92	82	3.8	—	14	10	—	1.0	1.8			
Year	61	44	104	8	0+	0+	0	92	68	31.0	0+	124	83	0+	16.0	2.2			

Total precipitation means the total of rain, snow, hail, etc. expressed as inches of rain.

Ten inches of snow equals approx. 1 inch of rain.

Temperatures are as ° F.

Wind velocity is the percentage of observations greater than 18 and 30 M.P.H. Number of observations varies from one per day to several per hour.

Visibility means number of days with less than ½-mile visibility.

Note: "—" indicates slight trace or possibility blank indicates no statistics available

NAPLES

ITALY

Month	TEMPERATURES—IN °F Daily Average Max.	Daily Average Min.	Extreme Max.	Extreme Min.	No. of Days Over 90°	No. of Days Under 32°	No. of Days 0°	RELATIVE HUMIDITY A.M. 6:00	RELATIVE HUMIDITY P.M. 1:00	PRECIPITATION Inches Total Prec.	Inches Snow Only	No. of Days Total Prec. 0.04"	No. of Days 0.1"	No. of Days Snow Only 1.5"	Thun. Stms.	Max. Inches in 24 Hrs.	WIND VELOCITY 18 M.P.H.	WIND VELOCITY 30 M.P.H.	VISIBILITY ½ MILE
Jan.	53	41	65	26	0	3	0	77	68	3.7	0	11	9	0	3	2	3	—	2
Feb.	54	41	70	24	0	2	0	78	66	2.8	0	11	8	0	4	2	4	—	2
March	59	44	76	27	0	1	0	77	62	2.9	0	7	7	0	2	1	3	0	1
April	67	50	81	34	0	0	0	79	61	2.9	0	7	7	0	2	2	1	0	2
May	74	56	91	42	—	0	0	85	63	2.0	0	6	6	0	2	2	1	0	2
June	81	62	95	51	3	0	0	75	58	1.5	0	4	4	0	2	3	—	0	1
July	86	67	101	53	8	0	0	73	53	0.7	0	2	2	0	2	3	—	0	1
Aug.	86	67	102	58	8	0	0	74	53	0.9	0	3	3	0	3	2	—	0	1
Sept.	81	63	103	49	2	0	0	78	59	2.9	0	7	7	0	4	4	1	0	1
Oct.	71	55	87	44	0	0	0	79	63	5.3	0	10	10	0	4	5	1	—	—
Nov.	62	49	77	33	0	0	0	81	68	4.5	0	11	9	0	2	3	3	—	1
Dec.	57	45	68	30	0	—	0	80	70	4.7	0	11	10	0	3	2	3	0	1
Year	69	53	103	24	21	6	0	78	62	35.0	0	92	82	0	33	5	2	—	15

Elevation 85 Feet

Latitude 40° 55' N   Longitude 14° 23' E

Total precipitation means the total of rain, snow, hail, etc. expressed as inches of rain.

Ten inches of snow equals approx. 1 inch of rain.

Temperatures are as ° F.

Wind velocity is the percentage of observations greater than 18 and 30 M.P.H. Number of observations varies from one per day to several per hour.

Visibility means number of days with less than ½-mile visibility.

Note: "—" indicates slight trace or possibility blank indicates no statistics available

NÁXOS (Island)

GREECE

| Month | TEMPERATURES—IN °F | | | | | | | RELATIVE HUMIDITY | | PRECIPITATION | | | | | | WIND VELOCITY | | VISIBILITY |
| | Daily Average | | Extreme | | No. of Days | | | A.M. 7:30 | P.M. 1:30 | Inches | | No. of Days | | | | 18 M.P.H. | 30 M.P.H. | ½ MILE |
	Max.	Min.	Max.	Min.	Over 90°	Under 32°	0°			Total Prec.	Snow Only	Total Prec. 0.1" / Trace	Snow Only 1.5"	Thun. Stms.	Max. Inches in 24 Hrs.			
Jan.	59	49	69	36				70	66	3.1		12			3			
Feb.	57	49	72	35				68	64	2.3		11			1			
March	62	52	77	39				68	63	1.4		8			1			
April	68	57	90	43				66	59	0.9		5			1			
May	73	62	89	51				70	66	0.7		5			2			
June	79	68	94	51				70	64	0.1		2			1			
July	80	71	95	62				71	67	0.0		—			—			
Aug.	81	72	92	59				73	67	0.0		—			—			
Sept.	78	69	92	58				72	67	0.2		2			—			
Oct.	73	63	89	44				72	67	1.1		5			1			
Nov.	67	58	82	39				73	67	2.1		8			1			
Dec.	62	52	74	40				71	67	3.0		13			2			
Year	70	60	95	35				70	65	15.0		72			3			

Elevation 29 Feet

Latitude 37° 6' N   **Longitude 25° 24' E**

Total precipitation means the total of rain, snow, hail, etc. expressed as inches of rain.

Ten inches of snow equals approx. 1 inch of rain.

Temperatures are as ° F.

Wind velocity is the percentage of observations greater than 18 and 30 M.P.H. Number of observations varies from one per day to several per hour.

Visibility means number of days with less than ½-mile visibility.

Note: "—" indicates slight trace or possibility blank indicates no statistics available

FRANCE

NICE

Month	Daily Average Max.	Daily Average Min.	Extreme Max.	Extreme Min.	No. of Days Over 90°	No. of Days Under 32°	No. of Days 0°	Rel. Humidity A.M. 7:30	Rel. Humidity P.M. 1:30	Inches Total Prec.	Inches Snow Only	No. of Days Total Prec. 0.1"	No. of Days Snow Only 1.5"	No. of Days Thun. Stms.	Max. Inches in 24 Hrs.	Wind Velocity 18 M.P.H.	Wind Velocity 30 M.P.H.	Visibility ½ Mile
Jan.	55	40	70	18	0.0	2	0	59	52	2.5	—	8	—	0.6				
Feb.	56	40	72	20	0.0	3	0	62	53	2.2	—	7	—	0.6				
March	59	44	72	23	0.0	—	0	64	58	2.7	—	7	—	1.3				
April	63	48	81	29	0.0	—	0	70	61	2.2	0	6	0	1.7				
May	69	54	87	36	0.0	0	0	72	63	2.4	0	7	0	3.0				
June	75	61	92	36	—	0	0	71	62	1.6	0	5	0	5.4				
July	80	65	100	49	2.1	0	0	70	64	1.0	0	2	0	3.8				
Aug.	80	64	97	50	2.0	0	0	70	61	1.0	0	3	0	3.6				
Sept.	77	61	92	40	—	0	0	73	63	2.4	0	6	0	4.0				
Oct.	69	54	82	34	0.0	0	0	72	61	5.8	0	10	0	3.1				
Nov.	62	46	77	28	0.0	—	0	70	62	4.4	—	9	—	1.7				
Dec.	56	41	68	18	0.0	1	0	66	59	3.1	—	9	—	0.8				
Year	67	53	100	18	4.0	6	0	68	60	31.0	0+	79	0+	30.0				

Elevation 13 Feet
Latitude 43° 39′ N   **Longitude 7° 12′ E**

Total precipitation means the total of rain, snow, hail, etc. expressed as inches of rain.

Ten inches of snow equals approx. 1 inch of rain.

Temperatures are as ° F.

Wind velocity is the percentage of observations greater than 18 and 30 M.P.H. Number of observations varies from one per day to several per hour.

Visibility means number of days with less than ½-mile visibility.

Note: "—" indicates slight trace or possibility
blank indicates no statistics available

NORTH FRONT (United Kingdom) GIBRALTAR

| Month | TEMPERATURES—IN °F | | | | | | | RELATIVE HUMIDITY | | PRECIPITATION | | | | | | | WIND VELOCITY | | VISIBILITY ½ MILE |
| | Daily Average | | Extreme | | No. of Days | | | | | Inches | | No. of Days | | | | Max. Inches in 24 Hrs | | | |
	Max.	Min.	Max.	Min.	Over 90°	Under 32°	0°	A.M. 5:30	P.M. 2:30	Total Prec.	Snow Only	Total Prec. 0.04"	0.1"	Snow Only 1.5"	Thun. Stms.		18 M.P.H.	30 M.P.H.	MILE
Jan.	60	50	74	36	0	0	0	80	70	6.1	0	11	11	0	2	5	13	—	—
Feb.	62	51	75	33	0	0	0	80	66	3.9	0	10	10	0	1	4	22	5.0	—
March	65	54	81	38	0	0	0	84	66	4.7	0	9	8	0	1	2	18	1.0	—
April	69	56	82	45	0	0	0	83	65	2.4	0	7	7	0	1	2	8	—	—
May	73	60	87	47	0	0	0	82	60	1.1	0	4	4	0	1	1	6	0.0	1
June	79	65	91	57	1	0	0	82	63	0.2	0	—	—	0	1	—	3	0.0	2
July	83	68	101	58	1	0	0	83	58	0.0	0	—	0	0	—	—	3	0.0	5
Aug.	84	70	99	57	—	0	0	82	58	0.1	0	—	—	0	1	—	3	0.0	5
Sept.	80	66	92	57	1	0	0	84	65	0.4	0	2	2	0	1	4	1	0.0	3
Oct.	74	63	92	50	—	0	0	84	69	2.5	0	7	7	0	1	2	11	—	—
Nov.	67	57	84	46	0	0	0	84	73	5.7	0	10	10	0	1	4	7	1.0	1
Dec.	62	53	75	36	0	0	0	83	71	4.9	0	11	11	0	1	3	18	1.0	1
Year	72	59	101	33	3	0	0	83	65	32.0	0	73	70	0	12	5	9	0.6	18

Elevation 8 Feet

Latitude 36° 9' N   Longitude 5° 21' W

Total precipitation means the total of rain, snow, hail, etc. expressed as inches of rain.

Ten inches of snow equals approx. 1 inch of rain.

Temperatures are as ° F.

Wind velocity is the percentage of observations greater than 18 and 30 M.P.H. Number of observations varies from one per day to several per hour.

Visibility means number of days with less than ½-mile visibility.

Note: "—" indicates slight trace or possibility blank indicates no statistics available

NUREMBERG   (West) FED. REP. GERMANY

Month	Daily Average Max.	Min.	Extreme Max.	Min.	No. of Days Over 90°	Under 32°	0°	RELATIVE HUMIDITY A.M. 6:30	P.M. 1:30	PRECIP. Inches Total Prec.	Snow Only	No. of Days Total Prec. 0.04"	0.1"	Snow Only 1.5"	Thun. Stms.	Max. Inches in 24 Hrs.	Wind Vel. 18 M.P.H.	30 M.P.H.	Visib. ½ Mile
Jan.	35	26	58	−18	0	20	—	91	80	1.5		9	5		—	1			
Feb.	39	27	64	−15	0	18	—	90	72	1.2		8	4		—	1			
March	47	32	73	2	0	19	0	87	62	1.3		9	4		1	1			
April	56	38	81	13	0	7	0	82	54	1.7		9	5		3	1			
May	65	46	92	25	—	2	0	79	53	2.2		10	6		7	1			
June	71	52	93	36	1	0	0	79	54	2.7		10	7	0	7	2			
July	74	55	97	34	—	0	0	79	55	3.1		11	8	0	7	1			
Aug.	72	54	99	40	1	0	0	84	57	3.1		11	8	0	7	2			
Sept.	66	48	93	27	—	—	0	88	62	2.1		9	6		3	1			
Oct.	55	41	82	18	0	4	0	90	70	2.1		8	6		—	1			
Nov.	44	34	70	4	0	11	0	90	77	1.9		8	6		—	2			
Dec.	37	29	64	−17	0	19	—	91	82	1.7		10	6		—	1			
Year	55	40	99	−18	2	101	0+	86	65	24.0		112	71		37	2			

Elevation 1,045 Feet

Latitude 49° 29' N   Longitude 11° 4' E

Total precipitation means the total of rain, snow, hail, etc. expressed as inches of rain.

Ten inches of snow equals approx. 1 inch of rain.

Temperatures are as °F.

Wind velocity is the percentage of observations greater than 18 and 30 M.P.H. Number of observations varies from one per day to several per hour.

Visibility means number of days with less than ½-mile visibility.

Note: "—" indicates slight trace or possibility
blank indicates no statistics available

ODENSE

DENMARK

Month	TEMPERATURES—IN °F							RELATIVE HUMIDITY		PRECIPITATION							WIND VELOCITY		VISIBILITY ½ MILE
	Daily Average		Extreme		No. of Days			A.M. 7:30	P.M. 1:30	Inches		No. of Days				Max. Inches in 24 Hrs.	18 M.P.H.	30 M.P.H.	
	Max.	Min.	Max.	Min.	Over 90°	Under 32°	0°			Total Prec.	Snow Only	Total Prec. 0.004"	0.1"	Snow Only 1.5"	Thun. Stms.				
Jan.	36	28	51	−10	0	17	—	90	88	1.8	—	17	6	—	—	1.0	27	3	5
Feb.	36	27	57	−11	0	19	—	90	84	1.5	—	13	5	—	—	0.9	18	—	6
March	42	30	65	−6	0	17	—	89	77	1.8	—	12	6	—	—	0.7	15	1	5
April	51	36	76	20	0	9	0	83	70	1.6	—	13	5	—	—	0.8	13	—	3
May	61	42	84	26	0	1	0	78	64	1.6	—	10	5	—	1	1.4	16	—	1
June	67	49	88	32	0	—	0	78	65	1.8	0	12	5	0	2	1.8	15	0	2
July	70	53	89	38	0	0	0	81	68	2.4	0	13	6	0	2	2.7	11	—	1
Aug.	69	53	90	37	—	0	0	86	69	3.0	0	15	7	0	2	1.9	10	0	3
Sept.	63	48	83	31	0	—	0	90	73	2.1	0	14	6	0	1	2.4	6	0	5
Oct.	54	41	68	21	0	3	0	92	79	2.5	—	16	7	—	—	1.0	10	—	7
Nov.	45	37	59	8	0	2	0	92	86	2.1	—	18	6	—	—	1.1	8	0	3
Dec.	39	31	54	1	0	10	0	91	89	2.4	—	17	7	—	—	0.7	26	0	2
Year	53	40	90	−11	0⁺	78	0⁺	87	76	24.0	—	170	71	—	10	2.7	15	1	42

Elevation 56 Feet

Latitude 55° 28' N   Longitude 10° 20' E

Total precipitation means the total of rain, snow, hail, etc. expressed as inches of rain.

Ten inches of snow equals approx. 1 inch of rain.

Temperatures are as ° F.

Wind velocity is the percentage of observations greater than 18 and 30 M.P.H. Number of observations varies from one per day to several per hour.

Visibility means number of days with less than ½-mile visibility.

Note: "—" indicates slight trace or possibility
blank indicates no statistics available

FRANCE

ORLEANS

Month	TEMPERATURES—IN °F Daily Average Max.	Daily Average Min.	Extreme Max.	Extreme Min.	No. of Days Over 90°	No. of Days Under 32°	No. of Days 0°	RELATIVE HUMIDITY A.M. 7:00	RELATIVE HUMIDITY P.M. 1:00	PRECIPITATION Inches Total Prec.	Inches Snow Only	No. of Days Total Prec. 0.004"	No. of Days 0.1"	No. of Days Snow Only 1.5"	Thun. Stms.	Max. Inches in 24 Hrs.	WIND VELOCITY 18 M.P.H.	WIND VELOCITY 30 M.P.H.	VISIBILITY ½ MILE
Jan.	42	32	60	0	0	—	0	92	81	2.0	0	17	6	0	0.2	0.7	8	0	
Feb.	45	33	67	6	0	—	0	90	70	1.7	0	14	6	0	0.3	1.1	7	0	
March	54	36	80	18	0	—	0	86	60	1.5	0	12	5	0	1.0	1.1	6	0	
April	60	40	86	24	0	—	0	85	60	1.7	0	14	5	0	1.0	1.3	7	0	
May	66	46	89	26	0	—	0	85	59	2.1	0	15	6	0	3.0	1.0	5	0	
June	72	52	97	34	—	0	0	83	58	1.9	0	11	5	0	3.0	1.1	3	0	
July	76	55	105	40	—	0	0	83	55	1.9	0	12	5	0	2.0	1.6	4	0	
Aug.	75	55	98	40	—	0	0	87	56	2.3	0	12	6	0	3.0	3.0	5	0	
Sept.	70	51	91	31	—	—	0	91	60	1.9	0	12	6	0	1.0	0.9	3	0	
Oct.	60	44	81	23	0	—	0	94	69	2.2	0	14	6	0	1.0	1.5	4	0	
Nov.	50	38	71	15	0	—	0	93	80	2.1	0	16	6	0	0.2	1.5	5	0	
Dec.	40	34	65	2	0	—	0	92	82	2.2	0	17	7	0	0.0	1.7	6	0	
Year	59	43	105	0	0+	0+	0	88	66	24.0	0	166	69	0	16.0	3.0	5	0	

Elevation 413 Feet

Latitude 47° 59' N   Longitude 1° 45' E

Total precipitation means the total of rain, snow, hail, etc. expressed as inches of rain.

Ten inches of snow equals approx. 1 inch of rain.

Temperatures are as °F.

Wind velocity is the percentage of observations greater than 18 and 30 M.P.H. Number of observations varies from one per day to several per hour.

Visibility means number of days with less than ½-mile visibility.

Note: "—" indicates slight trace or possibility blank indicates no statistics available

OSLO

NORWAY

Elevation 56 Feet
Latitude 59° 54' N   **Longitude 10° 37' E**

| Month | TEMPERATURES—IN °F | | | | | | | RELATIVE HUMIDITY | | PRECIPITATION | | | | | | | WIND VELOCITY | | VISIBILITY ½ MILE |
| | Daily Average | | Extreme | | No. of Days | | | A.M. 7:30 | P.M. 1:30 | Inches | | No. of Days | | | | Max. Inches in 24 Hrs. | 18 M.P.H. | 30 M.P.H. | |
	Max.	Min.	Max.	Min.	Over 90°	Under 32°	Under 0°			Total Prec.	Snow Only	Total Prec. 0.04"	0.1"	Snow Only 1.5"	Thun. Stms.				
Jan.	30	20	53	−21	0	29	1	85	82	1.7	—	8	6	—	0	1	2	—	8
Feb.	32	20	57	−18	0	25	1	84	75	1.3	—	7	4	—	0	1	1	0	6
March	40	25	63	−10	0	28	—	81	65	1.4	—	7	5	—	0	1	2	—	4
April	50	34	75	5	0	15	0	74	56	1.6	—	7	5	—	0	1	2	—	2
May	62	43	84	26	0	1	0	67	51	1.8	—	7	6	—	1	2	2	0	—
June	69	51	93	33	—	0	0	68	54	2.4	0	8	6	0	3	2	3	0	0
July	73	56	91	42	—	0	0	73	57	2.9	0	10	7	0	6	2	2	—	0
Aug.	69	53	88	37	0	0	0	79	61	3.8	0	11	9	0	4	2	1	—	0
Sept.	60	45	77	26	0	1	0	84	65	2.5	—	8	7	—	1	3	1	—	2
Oct.	49	37	70	12	0	6	0	86	72	2.9	—	10	7	—	—	2	2	0	4
Nov.	37	29	57	2	0	16	0	87	82	2.3	—	9	6	—	—	2	2	—	4
Dec.	31	24	54	−10	0	23	1	87	85	2.3	—	10	7	—	—	1	3	—	9
Year	50	36	93	−21	0+	143	3	80	67	27.0	—	102	74	—	16	3	2	—	39

Total precipitation means the total of rain, snow, hail, etc. expressed as inches of rain.

Ten inches of snow equals approx. 1 inch of rain.

Temperatures are as ° F.

Wind velocity is the percentage of observations greater than 18 and 30 M.P.H. Number of observations varies from one per day to several per hour.

Visibility means number of days with less than ½-mile visibility.

Note: "—" indicates slight trace or possibility blank indicates no statistics available

OSTEND

BELGIUM

Month	Temperatures—IN °F Daily Average Max.	Min.	Extreme Max.	Min.	No. of Days Over 90°	Under 32°	0°	Relative Humidity A.M. 7:00	P.M. —	Precipitation Inches Total Prec.	Snow Only	No. of Days Total Prec. 0.04"	0.1"	Snow Only 1.5"	Thun. Stms.	Max. Inches in 24 Hrs.	Wind Velocity 18 M.P.H.	30 M.P.H.	Visibility ½ Mile
Jan.	43	34	57	12	0.0	11	0	91		2.4	—	12	7	—	0.3	1.2			
Feb.	43	33	61	—2	0.0	11	—	90		1.8	—	9	6	—	0.3	1.4			
March	48	37	71	17	0.0	8	0	89		2.2	—	11	6	—	0.3	1.0			
April	52	40	82	26	0.0	—	0	84		2.1	—	10	6	—	0.3	1.1			
May	60	47	93	31	—	—	0	78		2.0	0	9	6	0	1.0	1.2			
June	65	52	91	39	—	0	0	79		2.5	0	8	7	0	1.0	1.7			
July	68	55	95	39	0.1	0	0	81		2.5	0	9	7	0	1.0	1.6			
Aug.	69	55	92	41	0.1	0	0	83		3.0	0	10	7	0	1.0	1.8			
Sept.	65	52	93	37	0.1	0	0	85		2.6	0	10	7	0	0.3	1.5			
Oct.	58	45	79	28	0.0	—	0	87		3.2	—	12	8	—	0.3	1.4			
Nov.	49	38	66	17	0.0	4	0	89		3.2	—	12	8	—	0.3	1.4			
Dec.	44	35	66	7	0.0	6	0	91		3.2	—	14	9	—	0.3	1.3			
Year	55	44	95	—2	0.3	40	0+	86		31.0	0+	126	84	0+	6.4	1.8			

Elevation 13 Feet

Latitude 51° 11' N   Longitude 2° 51' E

Total precipitation means the total of rain, snow, hail, etc. expressed as inches of rain.

Ten inches of snow equals approx. 1 inch of rain.

Temperatures are as ° F.

Wind velocity is the percentage of observations greater than 18 and 30 M.P.H. Number of observations varies from one per day to several per hour.

Visibility means number of days with less than ½-mile visibility.

Note: "—" indicates slight trace or possibility blank indicates no statistics available

PALERMO (Sicily) ITALY

| Month | TEMPERATURES—IN °F | | | | | | | RELATIVE HUMIDITY | | PRECIPITATION | | | | | | WIND VELOCITY | | VISI- BIL- ITY |
| | Daily Average | | Extreme | | No. of Days | | | | | Inches | | No. of Days | | | Max. | | | ½ MILE |
	Max.	Min.	Max.	Min.	Over 90°	Under 32°	0°	A.M. 6:00	P.M. 1:00	Total Prec.	Snow Only	Total Prec. 0.1"	Snow Only 1.5"	Thun. Stms.	Inches in 24 Hrs.	18 M.P.H.	30 M.P.H.	
Jan.	58	47	71	34	0	0	0	76	67	4.4	0	10	0	3	2	21	2	0
Feb.	60	47	78	32	0	—	0	72	63	3.8	0	10	0	4	2	13	2	1
March	62	49	91	31	—	—	0	72	60	2.6	0	7	0	1	2	9	1	0
April	67	53	86	42	0	0	0	70	60	2.1	0	6	0	0	2	5	1	0
May	83	59	98	48	6	0	0	71	58	1.4	0	5	0	1	2	5	1	0
June	82	66	104	52	5	0	0	68	54	0.6	0	2	0	0	2	2	0	0
July	86	71	108	58	9	0	0	64	52	0.3	0	1	0	0	1	2	—	0
Aug.	87	72	113	61	11	0	0	64	52	0.6	0	2	0	1	2	2	—	—
Sept.	83	69	106	51	6	0	0	66	53	1.8	0	5	0	1	2	3	—	0
Oct.	75	62	90	48	—	0	0	72	61	3.7	0	8	0	3	3	5	—	0
Nov.	67	55	84	41	0	0	0	74	64	4.2	0	9	0	1	4	8	1	0
Dec.	61	50	79	37	0	0	0	74	65	5.2	0	11	0	3	4	4	—	0
Year	73	58	113	31	37	0+	0	70	59	31.0	0	76	0	18	4	7	1	1

Elevation 345 Feet

Latitude 38° 6' N   Longitude 13° 18' E

Total precipitation means the total of rain, snow, hail, etc. expressed as inches of rain.

Ten inches of snow equals approx. 1 inch of rain.
Temperatures are as ° F.

Wind velocity is the percentage of observations greater than 18 and 30 M.P.H. Number of observations varies from one per day to several per hour.

Visibility means number of days with less than ½-mile visibility.

Note: "—" indicates slight trace or possibility blank indicates no statistics available

PALMA (Island Of Mallorca) (Balearic Islands) SPAIN

Month	TEMPERATURES—IN °F							RELATIVE HUMIDITY		PRECIPITATION								WIND VELOCITY		VISI-BILITY
	Daily Average		Extreme		No. of Days					Inches		No. of Days								½ MILE
	Max.	Min.	Max.	Min.	Over 90°	Under 32°	0°	A.M. 7:00	P.M. 1:00	Total Prec.	Snow Only	Total Prec. 0.004"	0.1"	Snow Only 1.5"	Thun. Stms.	Max. Inches in 24 Hrs.	18 M.P.H.	30 M.P.H.	
Jan.	57	42	75	28	0	—	0	83	71	1.4	—	8	5	—	3	2			
Feb.	59	43	77	28	0	—	0	84	69	1.6	—	8	5	—	2	1			
March	62	45	79	30	0	—	0	81	67	1.5	0	8	5	0	1	2			
April	66	49	84	33	0	0	0	77	64	1.3	0	5	4	0	1	2			
May	73	55	91	41	—	0	0	74	63	1.3	0	5	4	0	2	2			
June	80	61	100	48	—	0	0	69	60	1.0	0	3	3	0	2	2			
July	84	66	101	54	7	0	0	68	60	0.2	0	1	—	0	1	1			
Aug.	86	67	102	56	9	0	0	74	63	0.8	0	2	2	0	2	3			
Sept.	81	64	94	48	4	0	0	80	67	2.5	0	7	7	0	2	5			
Oct.	74	57	89	35	0	0	0	83	68	2.8	0	8	7	0	4	4			
Nov.	65	50	78	33	0	0	0	84	70	2.8	0	9	7	0	1	3			
Dec.	59	44	73	27	0	—	0	82	70	2.2	—	10	7	—	2	2			
Year	71	54	102	27	20	0+	0	78	66	19.0	—	74	57	—	23	5			

Elevation 13 Feet
Latitude 39° 33' N   **Longitude 2° 43' E**
Total precipitation means the total of rain, snow, hail, etc. expressed as inches of rain.
Ten inches of snow equals approx. 1 inch of rain.
Temperatures are as ° F.

Wind velocity is the percentage of observations greater than 18 and 30 M.P.H. Number of observations varies from one per day to several per hour.
Visibility means number of days with less than ½-mile visibility.
Note: "—" indicates slight trace or possibility
blank indicates no statistics available

PANTELLERIA (Mediterranean Island)　　ITALY

Month	TEMPERATURES—IN °F							RELATIVE HUMIDITY		PRECIPITATION							WIND VELOCITY		VISI-BIL-ITY
	Daily Average		Extreme		No. of Days					Inches		No. of Days							½ MILE
	Max.	Min.	Max.	Min.	Over 90°	Under 32°	0°	A.M. 6:00	P.M. 1:00	Total Prec.	Snow Only	Total Prec. 0.1"	Snow Only 1.5"	Thun. Stms.	Max. Inches in 24 Hrs.	18 M.P.H.	30 M.P.H.	
Jan.	54	47	66	32	0	—	0	82	79	1.9	0	6	0						
Feb.	55	47	74	37	0	0	0	80	78	1.2	0	4	0						
March	58	49	88	39	0	0	0	80	75	0.8	0	3	0						
April	62	53	85	42	0	0	0	80	72	1.0	0	3	0						
May	66	56	87	47	—	0	0	77	68	0.3	0	—	0						
June	78	64	106	50	—	0	0	73	64	0.1	0	0	0						
July	81	68	98	51	4	0	0	77	66	0.1	0	0	0						
Aug.	82	69	104	62	5	0	0	79	68	0.1	0	0	0						
Sept.	77	68	95	55	—	0	0	82	72	0.2	0	2	0						
Oct.	71	62	88	49	0	0	0	80	74	2.2	0	6	0						
Nov.	63	55	79	38	0	0	0	81	77	1.3	0	4	0						
Dec.	58	50	69	38	0	0	0	83	78	1.5	0	5	0						
Year	67	57	106	32	9	0+	0	80	73	11.0	0	33	0						

Elevation 833 Feet
Latitude 36° 48' N　　Longitude 11° 57' E

Total precipitation means the total of rain, snow, hail, etc. expressed as inches of rain.
Ten inches of snow equals approx. 1 inch of rain.
Temperatures are as ° F.

Wind velocity is the percentage of observations greater than 18 and 30 M.P.H. Number of observations varies from one per day to several per hour.
Visibility means number of days with less than ½-mile visibility.
Note: "—" indicates slight trace or possibility blank indicates no statistics available

PARIS

FRANCE

Month	Daily Average Max.	Daily Average Min.	Extreme Max.	Extreme Min.	No. of Days Over 90°	No. of Days Under 32°	No. of Days 0°	Rel. Hum. A.M. 7:00	Rel. Hum. P.M. 1:00	Inches Total Prec.	Inches Snow Only	No. of Days Total Prec. 0.004"	No. of Days 0.1"	No. of Days Snow Only 1.5"	No. of Days Thun. Stms.	Max. Inches in 24 Hrs.	Wind Velocity 18 M.P.H.	Wind Velocity 30 M.P.H.	Visibility ½ Mile
Jan.	43	33	58	9	0.0	19	0	90	77	2.2	0.0	15	7	0.0	0.0	0.9	7.0	0.2	7
Feb.	45	34	63	5	0.0	14	0	90	69	1.8	3.5	13	6	1.0	0.2	1.0	14.0	1.2	4
March	53	37	76	23	0.0	9	0	88	59	1.4	11.0	15	5	2.0	0.7	1.1	5.0	0.2	3
April	59	42	86	29	0.0	1	0	80	50	1.7	0.0	14	5	0.0	1.1	11	8.0	0.3	—
May	66	48	89	32	0.0	—	0	79	52	2.1	0.0	13	6	0.0	3.3	1.6	4.0	0.1	—
June	72	53	99	39	0.8	0	0	79	55	2.2	0.0	11	6	0.0	3.1	1.5	2.0	0.0	—
July	76	57	103	46	2.4	0	0	81	55	2.2	0.0	12	6	0.0	3.7	1.4	3.0	0.0	—
Aug.	75	56	98	45	1.1	0	0	84	54	2.6	0.0	12	7	0.0	3.2	1.3	4.0	0.0	—
Sept.	69	52	91	38	0.3	0	0	91	59	2.2	0.0	11	6	0.0	1.7	1.2	4.0	0.2	1
Oct.	60	46	82	25	0.0	1	0	94	68	2.0	0.0	14	6	0.0	0.2	2.0	5.0	0.3	4
Nov.	48	39	68	21	0.0	7	0	93	76	2.1	0.0	15	7	0.0	0.2	1.1	5.0	0.1	7
Dec.	42	34	61	9	0.0	9	0	92	80	2.0	0.0	17	6	0.0	0.0	1.3	6.0	0.2	7
Year	59	44	103	5	4.6	59	0	87	63	24.0	15.0	162	73	3.0	17.0	2.0	5.5	0.2	33

Elevation 292 Feet

Latitude 48° 43' N   **Longitude 2° 24' E**

Total precipitation means the total of rain, snow, hail, etc. expressed as inches of rain.

Ten inches of snow equals approx. 1 inch of rain. Temperatures are as ° F.

Wind velocity is the percentage of observations greater than 18 and 30 M.P.H. Number of observations varies from one per day to several per hour.

Visibility means number of days with less than ½-mile visibility.

Note: "—" indicates slight trace or possibility blank indicates no statistics available

PEMBROKE (United Kingdom) WALES

Month	Daily Average Max.	Daily Average Min.	Extreme Max.	Extreme Min.	No. of Days Over 90°	No. of Days Under 32°	No. of Days 0°	Rel. Humidity A.M.	Rel. Humidity P.M.	Total Prec. (in)	Snow Only (in)	No. of Days Total Prec. 0.1"	No. of Days Snow Only 1.5"	No. of Days Thun. Stms.	Max. Inches in 24 Hrs.	Wind Velocity 18 M.P.H.	Wind Velocity 30 M.P.H.	Visibility ½ Mile
Jan.	47	41	56	20	0	1.4	0	—	89	4.5	—	10	—	0.2		39	12	2.2
Feb.	46	40	56	23	0	3.1	0	—	89	2.9	—	8	—	0.0		33	8	3.3
March	48	40	64	25	0	2.4	0		86	2.3	—	6	—	0.0		25	3	2.5
April	51	43	68	30	0	0.2	0		86	1.8	0	6	0	0.0		16	2	2.6
May	56	47	75	32	0	—	0		85	2.3	0	6	0	0.0		10	1	4.1
June	60	52	80	40	0	0.0	0		86	1.9	0	5	0	0.5		13	2	5.1
July	63	55	82	46	0	0.0	0		87	2.9	0	7	0	0.1		14	2	4.7
Aug.	63	56	79	44	0	0.0	0		87	2.9	0	7	0	0.2		10	1	4.5
Sept.	61	54	73	38	0	0.0	0		86	3.0	0	7	0	0.2		14	3	2.3
Oct.	57	50	69	33	0	0.0	0		86	4.0	0	9	0	0.0		26	7	1.2
Nov.	51	45	61	30	0	—	0		87	4.6	0	9	0	0.0		27	9	1.7
Dec.	48	42	58	23	0	2.0	0		89	4.1	—	10	—	0.0		39	13	1.8
Year	54	47	82	20	0	9.0	0		87	37.0	0+	90	0+	1.2		22	5	36.0

Elevation 141 Feet

Latitude 51° 41' N   Longitude 5° 10' W

Total precipitation means the total of rain, snow, hail, etc. expressed as inches of rain.

Ten inches of snow equals approx. 1 inch of rain.

Temperatures are as ° F.

Wind velocity is the percentage of observations greater than 18 and 30 M.P.H. Number of observations varies from one per day to several per hour.

Visibility means number of days with less than ½-mile visibility.

Note: "—" indicates slight trace or possibility blank indicates no statistics available

PERPIGNAN

FRANCE

| Month | TEMPERATURES—IN °F | | | | | | | RELATIVE HUMIDITY | | PRECIPITATION | | | | | | | WIND VELOCITY | | VISIBILITY ½ MILE |
| | Daily Average | | Extreme | | No. of Days | | | | | Inches | | No. of Days | | | | Max. Inches in 24 Hrs. | 18 M.P.H. | 30 M.P.H. | |
	Max.	Min.	Max.	Min.	Over 90°	Under 32°	0°	A.M. 7:00	P.M. 1:00	Total Prec.	Snow Only	Total Prec. 0.004"	0.1"	Snow Only 1.5"	Thun. Stms.				
Jan.	53	38	76	12	0.0	4	0	73	55	2.3	—	10	7	—	0.3	1.1			
Feb.	56	39	80	19	0.0	5	0	72	51	1.3	—	10	4	—	0.3	1.1			
March	60	42	84	23	0.0	1	0	76	54	2.1	—	12	6	—	1.0	1.6			
April	65	47	90	30	—	—	0	68	51	1.9	0	10	6	0	1.0	1.5			
May	71	53	94	34	0.1	0	0	67	50	2.0	0	9	6	0	3.0	1.7			
June	78	59	106	36	1.4	0	0	65	49	1.4	0	5	4	0	4.0	1.5			
July	83	64	108	45	5.1	0	0	63	46	0.8	0	3	2	0	4.0	0.6			
Aug.	83	63	104	32	4.8	—	0	67	48	1.2	0	4	4	0	4.0	2.3			
Sept.	77	59	99	32	0.8	—	0	74	53	1.7	0	6	5	0	4.0	1.2			
Oct.	69	51	92	25	—	—	0	77	56	3.3	—	8	8	—	1.0	2.1			
Nov.	60	45	83	22	0.0	1	0	78	59	2.9	—	9	7	—	0.3	2.3			
Dec.	55	40	80	14	0.0	5	0	76	60	2.3	—	9	7	—	0.3	1.9			
Year	68	50	108	12	12.0	17	0	71	53	23.0	0+	95	66	0+	23.0	2.3			

Elevation 144 Feet

Latitude 42° 44' N   Longitude 2° 52' E

Total precipitation means the total of rain, snow, hail, etc. expressed as inches of rain.

Ten inches of snow equals approx. 1 inch of rain.

Temperatures are as ° F.

Wind velocity is the percentage of observations greater than 18 and 30 M.P.H. Number of observations varies from one per day to several per hour.

Visibility means number of days with less than ½-mile visibility.

Note: "—" indicates slight trace or possibility blank indicates no statistics available

PERTH   (United Kingdom) SCOTLAND

Month	Daily Average Max.	Daily Average Min.	Extreme Max.	Extreme Min.	No. of Days Over 90°	No. of Days Under 32°	No. of Days 0°	Rel. Humidity A.M.	Rel. Humidity P.M.	Inches Total Prec.	Inches Snow Only	No. of Days Total Prec. 0.01"	No. of Days 0.1"	No. of Days Snow Only 1.5"	No. of Days Thun. Stms.	Max. Inches in 24 Hrs.	Wind 18 M.P.H.	Wind 30 M.P.H.	Visibility ½ Mile
Jan.	43	32	57	0	0	12	0	—	85	2.5	—	18	8	—		1.0			
Feb.	44	33	58	3	0	11	0	—	82	2.3	—	14	7	—		0.9			
March	48	35	68	4	0	11	0	—	80	2.5	—	14	7	—		1.1			
April	53	38	73	22	0	5	0	—	77	1.8	—	13	6	—		1.1			
May	59	42	78	25	0	2	0	—	76	2.2	—	14	6	—		1.2			
June	65	47	89	31	0	—	0	—	74	2.0	0	13	6	0		1.4			
July	68	51	88	38	0	0	0	—	77	2.9	0	16	7	0		1.8			
Aug.	66	50	84	31	0	—	0	—	79	3.4	0	15	8	0		1.6			
Sept.	62	46	78	25	0	—	0	—	80	2.2	—	15	6	—		1.5			
Oct.	55	41	77	20	0	3	0	—	82	3.0	—	17	7	—		1.5			
Nov.	47	35	64	14	0	8	0	—	84	2.9	—	16	7	—		1.5			
Dec.	44	33	59	7	0	9	0	—	85	3.2	—	17	9	—		1.2			
Year	55	40	89	0	0	59	0	—	80	31.0	0+	182	84	0+		1.8			

Elevation 397 Feet

Latitude 56° 26' N   Longitude 3° 22' W

Total precipitation means the total of rain, snow, hail, etc. expressed as inches of rain.

Ten inches of snow equals approx. 1 inch of rain.

Temperatures are as ° F.

Wind velocity is the percentage of observations greater than 18 and 30 M.P.H. Number of observations varies from one per day to several per hour.

Visibility means number of days with less than ½-mile visibility.

Note: "—" indicates slight trace or possibility blank indicates no statistics available

PERUGIA — ITALY

| Month | TEMPERATURES—IN °F | | | | | | | RELATIVE HUMIDITY | | PRECIPITATION | | | | | | | WIND VELOCITY | | VISIBILITY |
| | Daily Average | | Extreme | | No. of Days | | | | | Inches | | No. of Days | | | | Max. Inches in 24 Hrs. | | | ½ MILE |
	Max.	Min.	Max.	Min.	Over 90°	Under 32°	0°	A.M. 6:00	P.M. 1:00	Total Prec.	Snow Only	Total Prec. 0.04"	0.1"	Snow Only 1.5"	Thun. Stms.		18 M.P.H.	30 M.P.H.	
Jan.	45	36	62	17	0	—	0	82	72	2.0	—	8	6	—					
Feb.	48	37	68	19	0	—	0	79	67	2.1	—	9	7	—					
March	54	41	70	22	0	—	0	75	58	3.2	—	7	7	—					
April	62	47	82	33	0	0	0	76	57	3.6	0	7	7	0					
May	69	53	85	36	0	0	0	80	55	3.2	0	9	7	0					
June	77	59	93	46	—	0	0	76	49	3.4	0	8	8	0					
July	83	64	100	49	6	0	0	76	45	1.4	0	5	4	0					
Aug.	83	64	98	52	6	0	0	72	46	1.9	0	5	5	0					
Sept.	76	60	95	48	—	0	0	80	54	2.8	0	7	7	0					
Oct.	65	52	82	39	0	0	0	82	65	5.2	0	10	10	0					
Nov.	54	44	68	29	0	—	0	84	73	4.3	0	9	9	0					
Dec.	47	38	65	24	0	—	0	83	69	2.7	—	9	8	—					
Year	64	50	100	17	12+	0+	0	79	59	36.0	0+	93	85	0+					

Elevation 1,677 Feet

Latitude 43° 7' N   Longitude 12° 23' E

Total precipitation means the total of rain, snow, hail, etc. expressed as inches of rain.

Ten inches of snow equals approx. 1 inch of rain.

Temperatures are as ° F.

Wind velocity is the percentage of observations greater than 18 and 30 M.P.H. Number of observations varies from one per day to several per hour.

Visibility means number of days with less than ½-mile visibility.

Note: "—" indicates slight trace or possibility blank indicates no statistics available

PIC DU MIDI

FRANCE

Month	TEMPERATURES—IN °F Daily Average Max.	Daily Average Min.	Extreme Max.	Extreme Min.	No. of Days Over 90°	No. of Days Under 32°	No. of Days 0°	RELATIVE HUMIDITY A.M.	RELATIVE HUMIDITY P.M.	PRECIPITATION Inches Total Prec.	Inches Snow Only	No. of Days Total Prec. 0.1"	No. of Days Snow Only 1.5"	Thun. Stms.	Max. Inches in 24 Hrs.	WIND VELOCITY 18 M.P.H.	WIND VELOCITY 30 M.P.H.	VISIBILITY ½ MILE
Jan.	24	14	50	−23	0	29	2.9	—	70	7				0				
Feb.	23	13	50	−27	0	26	1.5		70	18				0				
March	28	17	55	−8	0	30	2.2		72	6				0				
April	31	19	52	−5	0	27	—		75	3				1				
May	37	25	57	0	0	30	0.0		79	6				2				
June	45	33	72	10	0	17	0.0		82	2				5				
July	51	39	68	16	0	7	0.0		72	—				5				
Aug.	50	38	70	18	0	3	0.0		73	—				5				
Sept.	45	34	62	7	0	19	0.0		75	7				4				
Oct.	36	27	62	−4	0	25	—		74	7				1				
Nov.	30	20	49	−7	0	24	—		69	5				0				
Dec.	24	15	45	−13	0	30	4.0		70	14				0				
Year	35	25	72	−27	0	267	11.0		73	75				23				

Elevation 9,378 Feet

Latitude 42° 56' N   Longitude 0° 8' E

Total precipitation means the total of rain, snow, hail, etc. expressed as inches of rain.

Ten inches of snow equals approx. 1 inch of rain.

Temperatures are as ° F.

Wind velocity is the percentage of observations greater than 18 and 30 M.P.H. Number of observations varies from one per day to several per hour.

Visibility means number of days with less than ½-mile visibility.

Note: "—" indicates slight trace or possibility blank indicates no statistics available

PLYMOUTH   (United Kingdom) ENGLAND

| | TEMPERATURES—IN °F | | | | | | | RELATIVE HUMIDITY | | PRECIPITATION | | | | | | | WIND VELOCITY | | VISI-BILITY |
| | Daily Average | | Extreme | | No. of Days | | | | | Inches | | No. of Days | | | | Max. Inches in 24 Hrs. | | | ½ MILE |
Month	Max.	Min.	Max.	Min.	Over 90°	Under 32°	0°	A.M. 6:30	P.M. 1:00	Total Prec.	Snow Only	Total Prec. 0.01"	0.1"	Snow Only 1.5"	Thun. Stms.		18 M.P.H.	30 M.P.H.	
Jan.	47	40	57	16	0	2.5	0	89	83	4.3	—	20	10	—	0.3	1.4			
Feb.	47	39	59	17	0	3.1	0	88	78	3.1	—	15	9	—	0.3	1.4			
March	50	40	67	23	0	3.6	0	88	74	2.7	—	14	7	—	0.3	1.5			
April	54	43	72	31	0	0.2	0	86	72	2.2	0	14	6	0	0.3	1.0			
May	59	47	79	31	0	—	0	86	72	2.4	0	14	7	0	1.0	1.5			
June	64	52	82	35	0	0.0	0	84	74	2.0	0	11	6	0	0.3	1.4			
July	66	55	86	43	0	0.0	0	88	77	2.6	0	14	7	0	1.0	1.5			
Aug.	66	55	88	43	0	0.0	0	90	77	2.9	0	14	7	0	1.0	1.7			
Sept.	64	53	78	38	0	0.0	0	90	75	2.9	0	14	7	0	0.3	1.6			
Oct.	58	49	75	31	0	0.1	0	90	78	3.8	0	17	8	0	0.3	2.0			
Nov.	52	43	63	25	0	0.9	0	89	79	4.7	—	18	9	—	0.3	2.5			
Dec.	48	41	58	23	0	3.4	0	89	83	4.5	—	18	10	-.-	0.3	1.7			
Year	56	46	88	16	0	14.0	0	88	77	38.0	0+	183	93	0+	5.7	2.5			

Elevation 87 Feet

Latitude 50° 22' N   Longitude 4° 7' W

Total precipitation means the total of rain, snow, hail, etc. expressed as inches of rain.

Ten inches of snow equals approx. 1 inch of rain.

Temperatures are as °F.

Wind velocity is the percentage of observations greater than 18 and 30 M.P.H. Number of observations varies from one per day to several per hour.

Visibility means number of days with less than ½-mile visibility.

Note: "—" indicates slight trace or possibility blank indicates no statistics available

POITIERS   FRANCE

| Month | TEMPERATURES—IN °F | | | | | | | RELATIVE HUMIDITY | | PRECIPITATION | | | | | | | WIND VELOCITY | | VISIBILITY ½ MILE |
| | Daily Average | | Extreme | | No. of Days | | | A.M. 7:00 | P.M. 1:00 | Inches | | No. of Days | | | | Max. Inches in 24 Hrs. | 18 M.P.H. | 30 M.P.H. | |
	Max.	Min.	Max.	Min.	Over 90°	Under 32°	0°			Total Prec.	Snow Only	Total Prec. 0.004"	0.1"	Snow Only 1.5"	Thun. Stms.				
Jan.	45	33	60	8	0	—	0	92	78	1.7	—	17	6	—	0.1	2.9			
Feb.	49	34	67	18	0	—	0	91	70	1.9	—	14	6	—	0.4	2.0			
March	55	36	72	21	0	—	0	90	63	2.2	—	14	6	—	1.0	3.0			
April	61	41	78	25	0	—	0	87	62	1.9	—	16	6	—	1.0	1.3			
May	67	47	92	29	—	—	0	86	61	2.0	0	14	6	0	3.0	2.7			
June	73	52	89	38	0	0	0	84	59	2.0	0	10	6	0	4.0	0.9			
July	76	56	102	44	—	0	0	83	54	1.8	0	11	5	0	4.0	1.9			
Aug.	77	55	93	43	—	0	0	87	55	1.7	0	12	5	0	3.0	1.4			
Sept.	72	51	95	35	—	0	0	90	60	1.8	0	10	5	0	1.0	0.7			
Oct.	62	45	83	26	0	—	0	94	68	2.9	—	15	7	—	1.0	1.1			
Nov.	51	39	67	19	0	—	0	93	80	2.8	—	16	7	—	1.0	2.0			
Dec.	47	35	62	10	0	—	0	93	83	2.4	—	16	7	—	0.2	1.2			
Year	61	44	102	8	0+	0+	0	89	66	25.0	0+	165	72	0+	20.0	3.0			

Elevation 416 Feet
Latitude 46° 35' N   Longitude 0° 18' E

Total precipitation means the total of rain, snow, hail, etc. expressed as inches of rain.

Ten inches of snow equals approx. 1 inch of rain.
Temperatures are as ° F.

Wind velocity is the percentage of observations greater than 18 and 30 M.P.H. Number of observations varies from one per day to several per hour.

Visibility means number of days with less than ½-mile visibility.

Note: "—" indicates slight trace or possibility blank indicates no statistics available

PORTO   PORTUGAL

Month	Daily Average Max.	Daily Average Min.	Extreme Max.	Extreme Min.	Over 90°	Under 32°	0°	R.H. A.M. 8:30	R.H. P.M. 2:30	Total Prec.	Snow Only	Total Prec. 0.04"	0.1"	Snow Only 1.5"	Thun. Stms.	Max. Inches in 24 Hrs.	Wind 18 M.P.H.	30 M.P.H.	Vis. ½ Mile
Jan.	56	40	71	25	0	3	0	86	70	6.0	—	13	11	—	1	3.7	11	2.0	
Feb.	58	41	78	23	0	3	0	84	66	4.6	—	10	10	—	2	2.3	10	1.0	
March	61	45	83	29	0	1	0	81	67	5.5	0	13	8	0	2	1.9	10	1.0	
April	64	48	89	33	0	0	0	76	65	4.1	0	11	8	0	3	1.6	10	1.0	
May	66	51	92	38	—	0	0	76	67	3.3	0	9	7	0	3	1.9	6	—	
June	73	56	98	41	—	0	0	75	65	1.7	0	5	5	0	2	2.6	6	1.0	
July	76	58	104	46	1	0	0	74	63	0.9	0	3	3	0	1	1.7	5	—	
Aug.	77	58	103	46	2	0	0	74	61	0.7	0	3	2	0	1	1.1	7	—	
Sept.	75	56	99	41	—	0	0	79	64	2.1	0	6	6	0	2	2.8	5	—	
Oct.	69	51	92	35	—	0	0	83	67	4.3	0	10	9	0	2	2.4	8	—	
Nov.	61	46	78	30	0	—	0	87	71	5.9	0	13	10	0	2	4.6	11	3.0	
Dec.	56	41	73	25	0	2	0	86	71	6.5	—	13	11	—	2	2.6	11	3.0	
Year	66	49	104	23	4	9	0	80	66	46.0	—	109	89	—	23	4.6	8	1.1	

Elevation 259 Feet
Latitude 41° 14' N   Longitude 8° 41' W
Total precipitation means the total of rain, snow, hail, etc. expressed as inches of rain.
Ten inches of snow equals approx. 1 inch of rain.
Temperatures are as ° F.

Wind velocity is the percentage of observations greater than 18 and 30 M.P.H. Number of observations varies from one per day to several per hour.
Visibility means number of days with less than ½-mile visibility.
Note: "—" indicates slight trace or possibility blank indicates no statistics available

PRESWICK (United Kingdom) SCOTLAND

Month	Daily Average Max.	Min.	Extreme Max.	Min.	Over 90°	Under 32°	0°	RH A.M.	P.M.	Total Prec.	Snow Only	Days Total Prec. 0.1"	Snow Only 1.5"	Thun. Stms.	Max. Inches in 24 Hrs.	Wind 18 M.P.H.	30 M.P.H.	Visibility ½ Mile
Jan.	43	34	56	11	0.0	12	0	—	85	3.5	—	10	—	0.2		20	1.5	1.2
Feb.	44	34	55	12	0.0	11	0	—	83	2.1	—	7	—	0.2		18	1.0	1.0
March	49	36	70	8	0.0	9	0	—	80	2.2	—	6	—	0.3		15	0.3	1.4
April	53	40	71	26	0.0	3	0	—	77	2.0	—	7	—	0.1		17	0.2	0.0
May	60	44	80	30	0.0	1	0	—	74	2.1	0	7	0	0.7		12	0.1	0.1
June	63	49	84	35	0.0	0	0	—	78	2.6	0	8	0	1.0		8	0.1	0.7
July	65	53	90	40	0.1	0	0	—	80	3.1	0	9	0	1.1		7	0.0	0.4
Aug.	65	52	84	38	0.0	0	0	—	81	3.2	0	9	0	0.8		6	0.1	0.7
Sept.	61	50	79	31	0.0	—	0	—	82	3.9	0	11	0	1.0		13	0.7	0.3
Oct.	56	45	67	26	0.0	2	0	—	82	2.2	—	9	—	0.6		31	0.5	0.4
Nov.	50	40	62	19	0.0	5	0	—	85	3.4	—	10	—	0.5		15	0.7	1.0
Dec.	46	37	58	8	0.0	9	0	—	85	3.6	—	12	—	0.5		23	2.5	0.6
Year	55	43	90	8	0.1	52	0	—	81	35.0	0+	105	0+	7.0		14	0.6	7.7

Elevation 64 Feet

Latitude 55° 30' N   Longitude 4° 35' W

Total precipitation means the total of rain, snow, hail, etc. expressed as inches of rain.

Ten inches of snow equals approx. 1 inch of rain.

Temperatures are as °F.

Wind velocity is the percentage of observations greater than 18 and 30 M.P.H. Number of observations varies from one per day to several per hour.

Visibility means number of days with less than ½-mile visibility.

Note: "—" indicates slight trace or possibility blank indicates no statistics available

RÉYKJAVIK

ICELAND

| Month | TEMPERATURES—IN °F | | | | | | | RELATIVE HUMIDITY | | PRECIPITATION | | | | | | | WIND VELOCITY | | VISIBILITY |
| | Daily Average | | Extreme | | No. of Days | | | A.M. 7:30 | P.M. 1:30 | Inches | | No. of Days | | | | Max. Inches in 24 Hrs. | 18 M.P.H. | 30 M.P.H. | ½ MILE |
	Max.	Min.	Max.	Min.	Over 90°	Under 32°	0°			Total Prec.	Snow Only	Total Prec. 0.004"	0.1"	Snow Only 1.5"	Thun. Stms.				
Jan.	36	28	50	5				80	80	4.0		20				1.7			
Feb.	37	28	50	4				82	78	3.1		18				1.4			
March	39	30	58	6				79	77	3.0		20				1.3			
April	43	33	59	13				76	68	2.1		18				0.9			
May	50	39	64	19				70	66	1.6		16				0.7			
June	55	44	69	32				77	72	1.7		16				1.2			
July	58	48	74	39				75	72	2.0		16				1.1			
Aug.	57	47	71	34				76	72	2.6		18				1.3			
Sept.	51	42	68	27				79	72	3.1		20				1.3			
Oct.	44	36	58	14				82	77	3.4		20				1.5			
Nov.	39	32	53	10				83	81	3.6		19				1.6			
Dec.	38	30	53	4				81	80	3.7		21				2.2			
Year	46	37	74	4				78	75	34.0		222				2.2			

Elevation 92 Feet
Latitude 64° 9' N   Longitude 21° 56' W
Total precipitation means the total of rain, snow, hail, etc. expressed as inches of rain.
Ten inches of snow equals approx. 1 inch of rain.
Temperatures are as ° F.

Wind velocity is the percentage of observations greater than 18 and 30 M.P.H. Number of observations varies from one per day to several per hour.
Visibility means number of days with less than ½-mile visibility.
Note: "—" indicates slight trace or possibility
blank indicates no statistics available

RHODES

GREECE

| Month | TEMPERATURES—IN °F | | | | | | | RELATIVE HUMIDITY | | PRECIPITATION | | | | | | | WIND VELOCITY | | VISI-BIL-ITY |
| | Daily Average | | Extreme | | No. of Days | | | | | Inches | | No. of Days | | | | Max. Inches in 24 Hrs. | | | ½ MILE |
	Max.	Min.	Max.	Min.	Over 90°	Under 32°	0°	A.M.	P.M.	Total Prec.	Snow Only	Total Prec. 0.04"	0.1"	Snow Only 1.5"	Thun. Stms.		18 M.P.H.	30 M.P.H.	
Jan.	59	51	66	30				—	68	5.7		13				3			
Feb.	59	52	67	35					64	3.9		10				2			
March	63	55	75	41					69	2.6		7				2			
April	67	59	82	48					71	1.7		4				3			
May	73	65	94	54					71	0.5		2				2			
June	78	70	95	60					70	0.3		—				2			
July	83	74	97	70					71	0.0		0				0			
Aug.	83	76	89	73					69	—		—				—			
Sept.	81	72	95	63					65	0.4		1				1			
Oct.	76	68	85	56					67	1.7		4				3			
Nov.	68	52	77	45					67	5.2		7				4			
Dec.	62	55	70	38					67	6.7		13				4			
Year	71	63	97	30					68	29.0		62				4			

Elevation 289 Feet

Latitude 36° 26' N   Longitude 28° 15' E

Total precipitation means the total of rain, snow, hail, etc. expressed as inches of rain.

Ten inches of snow equals approx. 1 inch of rain.

Temperatures are as ° F.

Wind velocity is the percentage of observations greater than 18 and 30 M.P.H. Number of observations varies from one per day to several per hour.

Visibility means number of days with less than ½-mile visibility.

Note: "—" indicates slight trace or posiblility blank indicates no statistics available

ROME

ITALY

| Month | TEMPERATURES—IN °F Daily Average | | Extreme | | No. of Days | | | RELATIVE HUMIDITY | | PRECIPITATION Inches | | No. of Days | | | | Max. Inches in 24 Hrs. | WIND VELOCITY | | VISIBILITY ½ MILE |
	Max.	Min.	Max.	Min.	Over 90°	Under 32°	0°	A.M. 6:00	P.M. 1:00	Total Prec.	Snow Only	Total Prec. 0.04"	0.1"	Snow Only 1.5"	Thun. Stms.		18 M.P.H.	30 M.P.H.	
Jan.	52	37	65	23	0	9	0	85	68	3.2	0	9	9	0	1		7	—	2
Feb.	54	38	67	20	0	6	0	86	64	2.7	0	11	8	0	1		11	1	2
March	61	43	77	23	0	2	0	83	56	2.9	0	7	7	0	1		9	1	2
April	67	48	79	28	0	—	0	83	54	2.6	0	7	7	0	2		5	1	3
May	75	55	92	42	—	0	0	77	54	2.2	0	6	6	0	2		4	—	1
June	82	61	99	49	4	0	0	74	48	1.6	0	5	5	0	1		3	—	1
July	88	65	103	54	14	0	0	70	42	0.7	0	2	2	0	3		2	0	1
Aug.	87	65	105	55	9	0	0	73	43	1.0	0	3	3	0	2		3	—	—
Sept.	82	61	103	46	3	0	0	82	50	2.6	0	7	7	0	3		4	—	1
Oct.	72	53	87	36	0	0	0	86	59	5.0	0	9	9	0	4		5	1	2
Nov.	62	45	74	28	0	—	0	87	66	4.4	0	9	9	0	3		7	1	1
Dec.	56	41	67	29	0	2	0	86	70	3.7	0	10	10	0	1		7	—	2
Year	70	51	105	20	30	19	0	81	56	33.0	0	85	82	0	24		6	—	18

Elevation 430 Feet
Latitude 41° 48' N   Longitude 12° 36' E
Total precipitation means the total of rain, snow, hail, etc. expressed as inches of rain.
Ten inches of snow equals approx. 1 inch of rain.
Temperatures are as ° F.

Wind velocity is the percentage of observations greater than 18 and 30 M.P.H. Number of observations varies from one per day to several per hour.
Visibility means number of days with less than ½-mile visibility.
Note: "—" indicates slight trace or possibility blank indicates no statistics available

ROROS

NORWAY

| Month | TEMPERATURES—IN °F | | | | | | | RELATIVE HUMIDITY | | PRECIPITATION | | | | | | WIND VELOCITY | | VISI-BIL-ITY ½ MILE |
| | Daily Average | | Extreme | | No. of Days | | | | | Inches | | No. of Days | | | Max. Inches in 24 Hrs. | | | |
	Max.	Min.	Max.	Min.	Over 90°	Under 32°	0°	A.M. 8:00	P.M. 2:00	Total Prec.	Snow Only	Total Prec. 0.1"	Snow Only 1.5"	Thun. Stms.		18 M.P.H.	30 M.P.H.	
Jan.	19	3	43	−59	0	31	—	87	85	1.3		4		0	1			
Feb.	19	4	46	−44	0	28	—	87	81	1.1		4	·	0	1			
March	30	9	48	−42	0	31	—	87	74	1.1		4		0	1			
April	38	21	61	−26	0	27	—	84	70	0.8		3		0	1			
May	48	31	75	−2	0	16	—	76	61	1.1		4		—	3			
June	56	39	86	23	0	4	0	75	61	2.0		6		1	2			
July	62	44	86	28	0	1	0	78	62	2.8		7		3	2			
Aug.	58	42	82	23	0	2	0	84	64	2.9		7		1	2			
Sept.	50	35	73	14	0	10	0	87	69	1.7		5		—	2			
Oct.	39	27	59	−19	0	24	—	88	78	1.3		4		0	2			
Nov.	27	16	51	−28	0	29	—	89	86	1.1		4		0	1			
Dec.	21	8	43	−46	0	31	—	89	88	1.2		4		0	1			
Year	39	23	86	−59	0	231	0+	84	73	18.0		56		5	3			

Elevation 2,060 Feet

Latitude 62° 34' N   Longitude 11° 23' E

Total precipitation means the total of rain, snow, hail, etc. expressed as inches of rain.

Ten inches of snow equals approx. 1 inch of rain.

Temperatures are as ° F.

Wind velocity is the percentage of observations greater than 18 and 30 M.P.H. Number of observations varies from one per day to several per hour.

Visibility means number of days with less than ½-mile visibility.

Note: "—" indicates slight trace or possibility blank indicates no statistics available

ROTTERDAM

THE NETHERLANDS

| Month | TEMPERATURES—IN °F | | | | | | | RELATIVE HUMIDITY | | PRECIPITATION | | | | | | | WIND VELOCITY | | VISI-BILITY ½ MILE |
| | Daily Average | | Extreme | | No. of Days | | | | | Inches | | Total Prec. | | No. of Days | | Max. Inches in 24 Hrs. | | | |
	Max.	Min.	Max.	Min.	Over 90°	Under 32°	0°	A.M. 8:00	P.M. 2:00	Total Prec.	Snow Only	0.004"	0.1"	Snow Only 1.5"	Thun. Stms.		18 M.P.H.	30 M.P.H.	
Jan.	41	33	57	2	0	12	0	88	81	2.3	—	19	7	—	1	1.2	21	0.8	8
Feb.	42	33	63	2	0	12	0	87	76	1.6	—	15	5	—	1	0.8	3	0.0	11
March	48	37	71	9	0	9	0	84	66	1.4	—	13	5	—	1	1.4	4	0.0	8
April	54	42	79	28	0	3	0	80	62	1.9	—	15	6	—	2	1.4	13	0.0	6
May	62	49	90	30	—	1	0	77	60	2.0	0	13	6	0	4	1.7	8	0.0	0
June	66	54	100	39	—	0	0	76	62	2.1	0	12	6	0	5	1.8	4	0.0	2
July	70	58	94	43	—	0	0	79	64	2.6	0	14	7	0	5	1.6	9	0.4	8
Aug.	70	58	95	45	—	0	0	81	64	2.9	0	15	7	0	5	1.5	—	0.0	2
Sept.	65	54	92	41	—	0	0	83	66	3.0	0	16	7	0	3	2.0	3	0.0	4
Oct.	57	47	81	27	0	—	0	86	71	3.0	—	17	7	—	2	1.5	11	0.9	2
Nov.	48	40	69	15	0	6	0	87	79	3.3	—	19	8	—	1	1.6	4	0.0	10
Dec.	42	35	57	6	0	7	0	88	84	2.5	—	18	8	—	1	1.2	16	0.0	5
Year	55	45	100	2	0+	50	0	83	70	29.0	0+	186	78	0+	31	2.0	8	0.2	66

Elevation 15 Feet
Latitude 51° 57' N   Longitude 4° 26' E
Total precipitation means the total of rain, snow, hail, etc. expressed as inches of rain.
Ten inches of snow equals approx. 1 inch of rain.
Temperatures are as ° F.

Wind velocity is the percentage of observations greater than 18 and 30 M.P.H. Number of observations varies from one per day to several per hour.
Visibility means number of days with less than ½-mile visibility.
Note: "—" indicates slight trace or possibility blank indicates no statistics available

ST. HUBERT

BELGIUM

Month	TEMPERATURES—IN °F							RELATIVE HUMIDITY		PRECIPITATION						WIND VELOCITY		VISIBILITY ½ MILE
	Daily Average		Extreme		No. of Days			A.M.	P.M.	Inches		No. of Days			Max. Inches in 24 Hrs.	18 M.P.H.	30 M.P.H.	
	Max.	Min.	Max.	Min.	Over 90°	Under 32°	Under 0°			Total Prec.	Snow Only	Total Prec. 0.1"	Snow Only 1.5"	Thun. Stms.				
Jan.	35	26	57	0	0	24	0.0	—	92	3.7	—	9	—	0.3				
Feb.	37	27	57	—6	0	21	0.8	—	88	3.6	—	9	—	0.3				
March	46	31	68	9	0	17	0.0	—	79	2.6	—	7	—	1.0				
April	53	36	79	23	0	12	0.0	—	75	3.0	—	7	—	2.0				
May	59	42	81	27	0	2	0.0	—	75	3.1	—	7	—	5.0				
June	63	47	90	33	0	0	0.0	—	80	4.1	0	9	0	5.0				
July	67	51	90	37	—	0	0.0	—	79	3.6	0	8	0	5.0				
Aug.	67	51	87	39	0	0	0.0	—	81	3.8	0	9	0	4.0				
Sept.	62	48	85	31	0	—	0.0	—	85	3.5	0	8	0	3.0				
Oct.	53	40	72	23	0	4	0.0	—	87	2.5	—	7	—	1.0				
Nov.	43	34	62	18	0	11	0.0	—	92	4.2	—	9	—	1.0				
Dec.	37	29	56	—1	0	20	0.1	—	94	4.5	—	10	—	0.3				
Year	52	39	90	—6	0+	111	0.9	—	84	42.0	0+	99	0+	28.0				

Elevation 1,831 Feet

Latitude 50° 2' N   Longitude 5° 24' E

Total precipitation means the total of rain, snow, hail, etc. expressed as inches of rain.

Ten inches of snow equals approx. 1 inch of rain.

Temperatures are as ° F.

Wind velocity is the percentage of observations greater than 18 and 30 M.P.H. Number of observations varies from one per day to several per hour.

Visibility means number of days with less than ½-mile visibility.

Note: "—" indicates slight trace or possibility blank indicates no statistics available

SALZBURG

AUSTRIA

Month	TEMPERATURES—IN °F Daily Average Max.	Daily Average Min.	Extreme Max.	Extreme Min.	No. of Days Over 90°	No. of Days Under 32°	No. of Days 0°	RELATIVE HUMIDITY A.M. 7:00	RELATIVE HUMIDITY P.M. 2:00	PRECIPITATION Inches Total Prec.	Inches Snow Only	No. of Days Total Prec. 0.04"	No. of Days Total Prec. 0.1"	No. of Days Snow Only 1.5"	No. of Days Thun. Stms.	Max. Inches in 24 Hrs.	WIND VELOCITY 18 M.P.H.	WIND VELOCITY 30 M.P.H.	VISIBILITY ½ MILE
Jan.	36	24	64	−24	0	27	3	83	74	2.8	—	12	8	—	—	2.3	2.0	—	5
Feb.	38	25	66	−24	0	26	2	82	66	2.5	—	10	7	—	—	2.0	1.0	—	4
March	49	32	76	0	0	20	—	80	57	2.6	—	10	7	—	—	1.3	2.0	0	2
April	56	39	86	18	0	7	0	81	57	3.7	0	14	7	0	1	1.5	1.0	0	1
May	66	47	91	26	—	1	0	80	55	5.1	0	14	8	0	4	2.8	—	0	—
June	71	51	95	36	—	0	0	81	56	6.8	0	16	10	0	5	4.3	1.0	0	—
July	74	55	97	41	1	0	0	82	56	7.9	0	15	12	0	6	5.3	1.0	—	—
Aug.	73	55	97	36	1	0	0	85	58	6.9	0	14	10	0	4	2.6	—	0	1
Sept.	68	50	91	27	—	—	0	86	60	5.1	—	12	10	—	2	2.3	—	0	3
Oct.	57	41	82	17	—	5	0	87	65	3.5	—	11	8	—	—	2.0	—	0	6
Nov.	45	33	74	2	—	16	0	87	73	2.9	—	11	7	—	—	2.2	1.0	0	7
Dec.	36	26	61	−18	—	24	2	86	77	2.8	—	12	8	—	—	2.0	1.0	—	11
Year	56	40	97	−24	2	126	7	83	63	53.0	—	151	102	—	24	5.3	0.8	—	40

Elevation 1,410 Feet

Latitude 43° 47' N   Longitude 13° 0' E

Total precipitation means the total of rain, snow, hail, etc. expressed as inches of rain.

Ten inches of snow equals approx. 1 inch of rain.

Temperatures are as ° F.

Wind velocity is the percentage of observations greater than 18 and 30 M.P.H. Number of observations varies from one per day to several per hour.

Visibility means number of days with less than ½-mile visibility.

Note: "—" indicates slight trace or possibility
blank indicates no statistics available

SAN SEBASTIAN

SPAIN

Month	TEMPERATURES—IN °F Daily Average Max.	Min.	Extreme Max.	Min.	No. of Days Over 90°	Under 32°	0°	RELATIVE HUMIDITY A.M.	P.M.	PRECIPITATION Inches Total Prec.	Snow Only	No. of Days Total Prec. 0.1"	Snow Only 1.5"	Thun. Stms.	Max. Inches in 24 Hrs.	WIND VELOCITY 18 M.P.H.	30 M.P.H.	VISI-BIL-ITY ½ MILE
Jan.	53	40	75	15	0	4	0	—	74	5	—	10	—	2		10	2	
Feb.	55	42	77	21	0	3	0	—	72	4	—	10	—	2		10	2	
March	57	43	85	24	0	1	0	—	70	5	—	8	—	2		15	2	
April	59	46	88	30	0	1	0	—	71	5	0	8	0	3		10	1	
May	66	52	98	36	—	—	0	—	71	4	0	8	0	4		11	—	
June	69	56	100	44	—	0	0	—	75	3	0	8	0	4		9	—	
July	73	59	102	46	1	0	0	—	74	3	0	7	0	4		6	1	
Aug.	75	60	105	48	1	0	0	—	75	2	0	7	0	4		5	0	
Sept.	73	57	100	42	1	0	0	—	73	5	0	9	0	5		6	—	
Oct.	67	52	91	30	—	—	0	—	72	6	0	10	0	2		16	5	
Nov.	59	46	82	21	0	—	0	—	74	6	—	10	—	2		15	2	
Dec.	56	44	75	19	0	6	0	—	75	6	—	11	—	3		15	3	
Year	64	50	105	15	3	15	0		73	53	—	104	—	37		11	2	

Elevation 2 Feet

Latitude 43° 22' N   Longitude 1° 48' W

Total precipitation means the total of rain, snow, hail, etc. expressed as inches of rain.

Ten inches of snow equals approx. 1 inch of rain.

Temperatures are as ° F.

Wind velocity is the percentage of observations greater than 18 and 30 M.P.H. Number of observations varies from one per day to several per hour.

Visibility means number of days with less than ½-mile visibility.

Note: "___" indicates slight trace or possibility blank indicates no statistics available

SANTIS

SWITZERLAND

| Month | TEMPERATURES—IN °F | | | | | | | RELATIVE HUMIDITY | | PRECIPITATION | | | | | | | WIND VELOCITY | | VISIBILITY |
| | Daily Average | | Extreme | | No. of Days | | | | | Inches | | No. of Days | | | | Max. | | | ½ MILE |
	Max.	Min.	Max.	Min.	Over 90°	Under 32°	0°	A.M. 7:00	P.M. 1:00	Total Prec.	Snow Only	Total Prec. 0.01"	0.1"	Snow Only 1.5"	Thun. Stms.	Inches in 24 Hrs.	18 M.P.H.	30 M.P.H.	
Jan.	20	12	38	−15	0	—	—	78	76	8.8	—	15			0	4.5			
Feb.	20	12	40	−23	0	—	—	78	76	7.3	—	14			0	3.5			
March	24	16	40	−11	0	—	—	80	79	7.4	—	16			—	3.2			
April	29	20	53	−4	0	—	—	84	83	9.3	—	17			1	3.8			
May	37	28	64	7	0	—	0	82	83	8.4	—	18			3	6.8			
June	42	33	63	15	0	—	0	82	85	11.0	—	19			3	2.8			
July	47	37	65	22	0	—	0	82	84	12.0	—	18			4	4.3			
Aug.	47	38	66	22	0	—	0	81	82	11.0	—	17			3	4.4			
Sept.	42	34	60	9	0	—	0	80	80	8.6	—	15			1	3.8			
Oct.	35	27	55	2	0	—	0	78	77	7.4	—	14			—	3.2			
Nov.	27	20	51	−6	0	—	—	78	75	7.7	—	13			0	6.6			
Dec.	21	14	42	−22	0	—	—	79	76	9.0	—	16			0	4.1			
Year	33	24	66	−23	0	0+	0+	80	80	108.0	—	192			16	6.8			

Elevation 8,203 Feet
Latitude 47° 15' N   Longitude 9° 20' E
Total precipitation means the total of rain, snow, hail, etc. expressed as inches of rain.
Ten inches of snow equals approx. 1 inch of rain.
Temperatures are as ° F.

Wind velocity is the percentage of observations greater than 18 and 30 M.P.H. Number of observations varies from one per day to several per hour.
Visibility means number of days with less than ½-mile visibility.
Note: "—" indicates slight trace or possibility blank indicates no statistics available

SÄRNA

SWEDEN

| Month | TEMPERATURES—IN °F | | | | | | | RELATIVE HUMIDITY | | PRECIPITATION | | | | | | | WIND VELOCITY | | VISIBILITY ½ MILE |
| | Daily Average | | Extreme | | No. of Days | | | A.M. 8:00 | P.M. 2:00 | Inches | | No. of Days | | | | Max. Inches in 24 Hrs. | 18 M.P.H. | 30 M.P.H. | |
	Max.	Min.	Max.	Min.	Over 90°	Under 32°	0°			Total Prec.	Snow Only	Total Prec. 0.04"	0.1"	Snow Only 1.5"	Thun. Stms.				
Jan.	19	4	45	−51	0	31	12	90	85	1.2	—	8	4	—	0	1.4			
Feb.	24	5	48	−40	0	28	12	89	79	0.7	—	5	2	—	0	0.5			
March	33	11	57	−35	0	30	9	85	67	0.9	—	5	3	—	0	0.8			
April	42	23	64	−22	0	27	—	81	61	1.0	—	7	3	—	0	0.8			
May	56	32	81	12	0	18	0	69	48	2.0	—	6	6	—	1	1.2			
June	63	41	91	21	—	4	0	68	53	2.5	—	11	7	—	1	1.5			
July	69	46	91	30	—	—	0	76	56	3.3	0	13	8	0	3	2.2			
Aug.	65	44	86	25	0	2	0	85	62	3.6	—	12	8	—	1	1.9			
Sept.	54	36	79	18	0	9	0	90	67	1.9	—	10	6	—	0	1.4			
Oct.	42	28	61	−13	0	17	—	91	76	2.0	—	9	6	—	0	1.3			
Nov.	30	19	52	−18	0	26	2	93	87	1.4	—	9	4	—	0	0.8			
Dec.	24	11	45	−44	0	30	9	92	89	1.4	—	9	5	—	0	1.2			
Year	43	25	91	−51	0+	223	44	84	69	22.0	—	104	62	—	6	2.2			

Elevation 1,503 Feet

Latitude 61° 41' N   Longitude 13° 7' E

Total precipitation means the total of rain, snow, hail, etc. expressed as inches of rain.

Ten inches of snow equals approx. 1 inch of rain.

Temperatures are as ° F.

Wind velocity is the percentage of observations greater than 18 and 30 M.P.H. Number of observations varies from one per day to several per hour.

Visibility means number of days with less than ½-mile visibility.

Note: "—" indicates slight trace or possibility blank indicates no statistics available

SASSARI　　(Sardinia)　ITALY

| Month | TEMPERATURES—IN °F | | | | | | | RELATIVE HUMIDITY | | PRECIPITATION | | | | | | | | WIND VELOCITY | | VISIBILITY ½ MILE |
| | Daily Average | | Extreme | | No. of Days | | | | | Inches | | No. of Days | | | | | | | | |
	Max.	Min.	Max.	Min.	Over 90°	Under 32°	0°	A.M. 5:30	P.M. 12:30	Total Prec.	Snow Only	Total Prec. 0.04"	0.1"	Snow Only 1.5"	Thun. Stms.	Max. Inches in 24 Hrs.	18 M.P.H.	30 M.P.H.	
Jan.	54	44	64	32	0	—	0	85	75	2.4	0	10	7	0					
Feb.	53	43	70	24	0	—	0	82	71	2.0	—	9	6	—					
March	60	46	81	32	0	—	0	80	63	2.2	0	7	6	0					
April	65	50	86	38	0	0	0	78	61	2.2	0	6	6	0					
May	70	56	91	43	—	0	0	77	60	1.7	0	7	5	0					
June	79	62	94	53	—	0	0	72	50	1.0	0	3	3	0					
July	83	66	100	53	6	0	0	69	47	0.4	0	1	1	0					
Aug.	84	67	103	54	7	0	0	70	48	0.4	0	1	1	0					
Sept.	79	64	99	54	—	0	0	74	53	1.5	0	5	5	0					
Oct.	71	58	83	42	0	0	0	80	60	3.3	0	9	8	0					
Nov.	62	51	76	40	0	0	0	79	66	3.7	0	8	8	0					
Dec.	56	46	72	36	0	0	0	82	70	2.9	0	10	8	0					
Year	68	54	103	24	13+	0+	0	77	60	24.0	0+	76	64	0+					

Elevation 623 Feet
Latitude 40° 43' N　　　Longitude 8° 33' E
Total precipitation means the total of rain, snow, hail, etc. expressed as inches of rain.

Ten inches of snow equals approx. 1 inch of rain.
Temperatures are as ° F.

Wind velocity is the percentage of observations greater than 18 and 30 M.P.H. Number of observations varies from one per day to several per hour.
Visibility means number of days with less than ½-mile visibility.
Note: "—" indicates slight trace or possibility blank indicates no statistics available

SCILLY ISLANDS

(United Kingdom) ENGLAND

| Month | TEMPERATURES—IN °F | | | | | | RELATIVE HUMIDITY | | PRECIPITATION | | | | | | | WIND VELOCITY | | VISI-BIL-ITY ½ MILE |
| | Daily Average | | Extreme | | No. of Days | | A.M. 6:30 | P.M. 1:00 | Inches | | No. of Days | | | | | | | |
	Max.	Min.	Max.	Min.	Over 90°	Under 32°			Total Prec.	Snow Only	Total Prec. 0.01"	0.1"	Snow Only 1.5"	Thun. Stms.	Max. Inches in 24 Hrs.	18 M.P.H.	30 M.P.H.	
Jan.	49	44	57	25		0	89	86	3.6	0	22		0	0.4	0.9			
Feb.	49	43	56	26		0	89	85	2.7	0	17		0	0.4	2.4			
March	51	43	61	29		0	90	84	2.4	0	16		0	0.3	1.2			
April	53	45	67	33		0	89	82	2.2	0	15		0	0.2	2.1			
May	57	48	74	37		0	90	83	2.2	0	15		0	0.4	1.3			
June	62	53	78	43		0	91	81	1.7	0	13		0	0.5	1.8			
July	65	56	82	48		0	91	81	2.2	0	17		0	0.7	1.3			
Aug.	66	56	82	48		0	93	82	2.4	0	16		0	0.8	1.3			
Sept.	63	55	76	44		0	92	83	2.4	0	15		0	0.3	1.3			
Oct.	58	51	68	36		0	89	83	3.5	0	18		0	0.6	2.3			
Nov.	53	47	61	35		0	87	83	3.7	0	20		0	0.3	1.6			
Dec.	50	45	57	28		0	88	84	3.5	0	22		0	0.3	1.3			
Year	56	49	82	25		0	90	83	33.0	0	206		0	5.0	2.4			

Elevation 163 Feet

Latitude 49° 56' N   Longitude 6° 18' W

Total precipitation means the total of rain, snow, hail, etc. expressed as inches of rain.

Ten inches of snow equals approx. 1 inch of rain.

Temperatures are as ° F.

Wind velocity is the percentage of observations greater than 18 and 30 M.P.H. Number of observations varies from one per day to several per hour.

Visibility means number of days with less than ½-mile visibility.

Note: "—" indicates slight trace or possibility blank indicates no statistics available

SHANNON

IRELAND

| Month | TEMPERATURES—IN °F | | | | | | | RELATIVE HUMIDITY | | PRECIPITATION | | | | | | | WIND VELOCITY | | VISI-BIL-ITY ½ MILE |
| | Daily Average | | Extreme | | No. of Days | | | | | Inches | | No. of Days | | | | Max. Inches in 24 Hrs. | | | |
	Max.	Min.	Max.	Min.	Over 90°	Under 32°	0°	A.M. 6:30	P.M. 5:30	Total Prec.	Snow Only	Total Prec. 0.04"	0.1"	Snow Only 1.5"	Thun. Stms.		18 M.P.H.	30 M.P.H.	
Jan.	46	36	58	12	0	—	0	89	88	3.78	—	15	9.5	—	0.6	1.0	55	1.9	
Feb.	47	37	59	17	0	—	0	89	85	2.52	—	12	7.5	—	0.2	1.2	52	2.3	
March	52	39	65	18	0	—	0	91	80	2.2	—	10	6.2	—	0.4	1.2	47	0.9	
April	55	41	69	24	0	—	0	89	74	2.0	—	11	6.0	—	0.6	1.0	60	3.5	
May	61	44	77	29	0	—	0	87	70	2.5	0	11	6.7	0	0.0	1.3	45	0.3	
June	66	50	83	34	0	0	0	86	71	2.3	0	11	6.2	0	0.2	1.3	39	0.0	
July	66	53	87	40	0	0	0	89	77	3.0	0	15	7.4	0	0.8	2.1	45	0.0	
Aug.	68	53	84	37	0	0	0	91	77	3.3	0	14	7.8	0	1.2	1.1	46	0.2	
Sept.	64	51	75	31	0	—	0	91	82	3.5	0	14	8.0	0	0.4	1.8	50	2.2	
Oct.	58	45	70	26	0	—	0	41	86	3.4	—	14	8.0	—	0.2	1.0	51	2.4	
Nov.	52	40	65	19	0	—	0	92	89	3.5	—	17	8.0	—	0.0	1.4	47	1.6	
Dec.	48	37	61	19	0	—	0	91	89	4.7	—	17	10.0	—	0.0	1.5	57	3.2	
Year	57	44	87	12	0	0+	0	90	81	37.0	0+	161	92.0	0+	4.6	2.1	50	1.5	

Elevation 47 Feet

Latitude 52° 41′ N   Longitude 8° 54′ W

Total precipitation means the total of rain, snow, hail, etc. expressed as inches of rain.

Ten inches of snow equals approx. 1 inch of rain.

Temperatures are as ° F.

Wind velocity is the percentage of observations greater than 18 and 30 M.P.H. Number of observations varies from one per day to several per hour.

Visibility means number of days with less than ½-mile visibility.

Note: "—" indicates slight trace or possibility blank indicates no statistics available

SIRACUSA (Sicily) ITALY

Month	TEMPERATURES—IN °F Daily Average Max.	Daily Average Min.	Extreme Max.	Extreme Min.	No. of Days Over 90°	Under 32°	0°	RELATIVE HUMIDITY A.M. 8:00	P.M. 7:00	PRECIPITATION Inches Total Prec.	Snow Only	No. of Days Total Prec. 0.1"	Snow Only 1.5"	Thun. Stms.	Max. Inches in 24 Hrs.	WIND VELOCITY 18 M.P.H.	30 M.P.H.	VISIBILITY ½ MILE
Jan.	58	44	69	32	0	—	0	85	80	3.6	0	9	0		5			
Feb.	60	45	75	32	0	—	0	83	79	2.5	0	8	0		2			
March	63	47	79	34	0	0	0	81	79	1.7	0	5	0		3			
April	68	51	94	37	—	0	0	82	77	1.5	0	5	0		2			
May	74	57	91	45	—	0	0	81	77	0.7	0	2	0		2			
June	82	63	98	52	5	0	0	73	71	0.2	0	—	0		1			
July	88	68	104	57	13	0	0	65	66	0.7	0	2	0		1			
Aug.	89	69	105	58	14	0	0	71	73	0.4	0	1	0		1			
Sept.	83	66	104	53	6	0	0	79	75	2.2	0	6	0		3			
Oct.	76	60	95	47	—	0	0	81	77	3.8	0	8	0		3			
Nov.	67	53	80	35	0	0	0	79	76	5.6	0	10	0		4			
Dec.	62	47	77	32	0	—	0	77	77	3.7	0	9	0		2			
Year	73	56	105	32	38	0+	0	77	76	27.0	0	65	0		5			

Elevation 49 Feet

Latitude 37° 4' N   Longitude 15° 17' E

Total precipitation means the total of rain, snow, hail, etc. expressed as inches of rain.

Ten inches of snow equals approx. 1 inch of rain.

Temperatures are as ° F.

Wind velocity is the percentage of observations greater than 18 and 30 M.P.H. Number of observations varies from one per day to several per hour.

Visibility means number of days with less than ½-mile visibility.

Note: "—" indicates slight trace or possibility blank indicates no statistics available

SKAGEN

DENMARK

Month	Daily Average Max.	Daily Average Min.	Extreme Max.	Extreme Min.	No. of Days Over 90°	No. of Days Under 32°	No. of Days 0°	Rel. Hum. A.M.	Rel. Hum. P.M.	Precip. Inches Total Prec.	Precip. Inches Snow Only	No. of Days Total Prec. 0.04"	No. of Days Total Prec. 0.1"	No. of Days Snow Only 1.5"	Thun. Stms.	Max. Inches in 24 Hrs.	Wind 18 M.P.H.	Wind 30 M.P.H.	Visibility ½ Mile
Jan.	36	30	50	2	0	19	0	88	87	1.5	—	14	5	—	—	1.5	32	7	6
Feb.	35	29	52	3	0	21	0	88	86	1.1	—	11	4	—	—	0.8	29	6	6
March	38	31	59	7	0	19	0	87	83	1.3	—	12	4	—	0	0.6	24	3	8
April	46	36	68	18	0	8	0	86	83	1.5	—	11	5	—	—	1.4	18	1	5
May	56	44	83	30	0	—	0	80	72	1.5	0	11	5	0	1	1.1	14	2	3
June	65	51	90	37	—	0	0	79	72	1.6	0	9	5	0	1	1.5	16	1	1
July	68	55	90	37	—	0	0	83	73	2.2	0	11	6	0	1	2.1	15	2	1
Aug.	66	55	87	41	0	0	0	84	76	3.1	0	14	8	0	2	2.9	25	3	1
Sept.	60	51	76	36	0	0	0	84	76	2.1	0	12	6	0	—	1.2	23	3	1
Oct.	51	44	71	23	0	—	0	84	79	2.6	—	15	7	—	—	1.7	26	4	2
Nov.	44	37	55	19	0	3	0	85	83	2.1	—	15	6	—	—	1.1	32	5	2
Dec.	39	33	54	6	0	13	0	87	86	2.0	—	16	6	—	—	1.1	37	5	4
Year	50	41	90	2	0+	84	0	85	80	23.0	—	151	66	—	7	2.9	24	4	41

Elevation 23 Feet

Latitude 57° 46' N   Longitude 10° 39' E

Total precipitation means the total of rain, snow, hail, etc. expressed as inches of rain.

Ten inches of snow equals approx. 1 inch of rain.

Temperatures are as ° F.

Wind velocity is the percentage of observations greater than 18 and 30 M.P.H. Number of observations varies from one per day to several per hour.

Visibility means number of days with less than ½-mile visibility.

Note: "—" indicates slight trace or possibility blank indicates no statistics available

SODANKYLA

FINLAND

Month	TEMPERATURES—IN °F							RELATIVE HUMIDITY		PRECIPITATION							WIND VELOCITY		VISI-BILITY ½ MILE
	Daily Average		Extreme		No. of Days					Inches		No. of Days				Max. Inches in 24 Hrs.			
	Max.	Min.	Max.	Min.	Over 90°	Under 32°	0°	A.M. 7:00	P.M. 3:00	Total Prec.	Snow Only	Total Prec. 0.04"	0.1"	Snow Only 1.5"	Thun. Stms.		18 M.P.H.	30 M.P.H.	
Jan.	17	−1	43	−50	0	31	13	89	88	1.1		11	4		1	0.4	1		3
Feb.	16	−3	43	−56	0	28	13	87	85	0.9		9	3		1	0.5	1		2
March	26	3	48	−41	0	31	12	88	71	0.8		6	3		2	0.3	2		2
April	37	17	63	−33	0	28	5	85	63	1.1		9	4		1	0.8	1		2
May	49	31	78	−6	0	16	—	72	57	1.3		8	4		1	1.3	1		2
June	61	42	87	23	0	3	0	70	54	2.2		10	6		1	1.3	1		1
July	69	47	89	26	0	1	0	76	59	2.6		11	7		—	1.1	—		3
Aug.	63	44	83	22	0	2	0	89	67	2.6		12	7		—	1.8	—		6
Sept.	51	35	75	8	0	10	0	93	71	1.5		9	5		—	0.9	—		4
Oct.	36	24	57	−22	0	21	1	93	85	1.8		13	5		1	0.9	1		6
Nov.	27	14	46	−44	0	28	6	92	92	1.5		12	5		1	0.7	1		4
Dec.	21	5	42	−46	0	31	12	90	90	1.2		10	4		—	0.5	—		2
Year	39	22	89	−56	0	231	62	85	73	19.0		120	55		9	1.8	1		37

Elevation 580 Feet

Latitude 67° 22' N   Longitude 26° 39' E

Total precipitation means the total of rain, snow, hail, etc. expressed as inches of rain.

Ten inches of snow equals approx. 1 inch of rain.

Temperatures are as ° F.

Wind velocity is the percentage of observations greater than 18 and 30 M.P.H. Number of observations varies from one per day to several per hour.

Visibility means number of days with less than ½-mile visibility.

Note: "—" indicates slight trace or possibility blank indicates no statistics available

SONNBLICK

AUSTRIA

Month	TEMPERATURES—IN °F Daily Average Max.	Min.	Extreme Max.	Min.	No. of Days Over 90°	Under 32°	0°	RELATIVE HUMIDITY A.M. 7:00	P.M. 2:00	PRECIPITATION Inches Total Prec.	Snow Only	No. of Days Total Prec. 0.04"	0.1"	Snow Only 1.5"	Thun. Stms.	Max. Inches in 24 Hrs.	WIND VELOCITY 18 M.P.H.	30 M.P.H.	VISIBILITY ½ MILE
Jan.	13	4	34	−35	0	31	10	77	77	4.3	—	17	10	—	0	1.9			
Feb.	13	4	38	−35	0	28	12	78	78	4.4	—	17	10	—	0	2.6			
March	16	7	38	−30	0	31	8	82	84	5.2	—	20	8	—	0	2.0			
April	21	12	39	−16	0	30	2	86	88	6.0	—	21	9	—	—	2.2			
May	28	21	49	−4	0	30	—	89	90	5.9	—	21	9	—	2	4.3			
June	34	26	55	4	0	26	0	89	91	4.9	—	21	9	—	4	2.6			
July	38	30	57	12	0	18	0	88	91	5.4	—	21	10	—	7	1.7			
Aug.	38	30	54	14	0	20	0	86	91	4.8	—	19	9	—	6	1.9			
Sept.	34	27	50	3	0	22	0	83	88	4.1	—	16	9	—	2	1.5			
Oct.	27	21	47	−5	0	29	—	81	84	4.6	—	16	9	—	—	2.1			
Nov.	20	13	42	−19	0	30	3	79	80	4.2	—	16	9	—	—	1.9			
Dec.	15	7	34	−27	0	31	6	79	79	4.7	—	17	10	—	0	2.4			
Year	25	17	57	−35	0	326	41	83	85	59.0	—	222	112	—	22	4.3			

Elevation 10,190 Feet

Latitude 47° 3' N   Longitude 12° 57' E

Total precipitation means the total of rain, snow, hail, etc. expressed as inches of rain.

Ten inches of snow equals approx. 1 inch of rain.

Temperatures are as ° F.

Wind velocity is the percentage of observations greater than 18 and 30 M.P.H. Number of observations varies from one per day to several per hour.

Visibility means number of days with less than ½-mile visibility.

Note: "—" indicates slight trace or possibility blank indicates no statistics available

STOCKHOLM

SWEDEN

Month	TEMPERATURES—IN °F Daily Average Max.	Min.	Extreme Max.	Min.	No. of Days Over 90°	Under 32°	0°	RELATIVE HUMIDITY A.M. 8:00	P.M. 2:00	PRECIPITATION Inches Total Prec.	Snow Only	No. of Days Total Prec. 0.04"	0.1"	Snow Only 1.5"	Thun. Stms.	Max. Inches in 24 Hrs.	WIND VELOCITY 18 M.P.H.	30 M.P.H.	VISIBILITY ½ MILE
Jan.	31	23	51	−26	0.0	24	1	85	82	1.5	—	8	5	—	0	0.7			
Feb.	31	22	54	−22	0.0	24	2	82	75	1.1	—	7	4	—	0	0.9			
March	37	26	59	−14	0.0	28	1	80	67	1.1	—	7	4	—	0	0.8			
April	45	32	77	−8	0.0	16	—	77	62	1.5	—	6	5	—	—	1.5			
May	57	41	84	19	0.0	2	0	67	54	1.6	—	8	5	—	1	1.8			
June	65	49	91	32	—	—	0	66	55	1.9	0	7	5	0	2	1.3			
July	70	55	97	40	0.3	0	0	71	59	2.8	0	9	7	0	4	2.7			
Aug.	66	53	91	36	0.1	0	0	79	64	3.1	0	10	8	0	2	2.7			
Sept.	58	46	84	23	0.0	1	0	85	68	2.1	—	8	6	—	1	1.7			
Oct.	48	39	68	16	0.0	7	0	88	76	3.1	—	9	6	—	—	1.4			
Nov.	38	31	57	0	0.0	13	—	88	84	1.9	—	9	6	—	0	1.6			
Dec.	33	26	52	−11	0.0	20	1	87	86	1.9	—	9	6	—	0	1.1			
Year	48	37	97	−26	0.4	134	4	80	69	23.0	—	97	66	—	11	2.7			

Elevation 46 Feet

Latitude 59° 21' N   Longitude 17° 56' E

Total precipitation means the total of rain, snow, hail, etc. expressed as inches of rain.

Ten inches of snow equals approx. 1 inch of rain.

Temperatures are as ° F.

Wind velocity is the percentage of observations greater than 18 and 30 M.P.H. Number of observations varies from one per day to several per hour.

Visibility means number of days with less than ½-mile visibility.

Note: "—" indicates slight trace or possibility blank indicates no statistics available

STORNOWAY (Lewis Isle, Outer Hebrides) (United Kingdom) SCOTLAND

| Month | TEMPERATURES—IN °F | | | | | | | RELATIVE HUMIDITY | | PRECIPITATION | | | | | | | WIND VELOCITY | | VISIBILITY |
| | Daily Average | | Extreme | | No. of Days | | | | | Inches | | No. of Days | | | | Max. Inches in 24 Hrs. | | | ½ MILE |
	Max.	Min.	Max.	Min.	Over 90°	Under 32°	0°	A.M. 6:30	Noon	Total Prec.	Snow Only	Total Prec. 0.01"	0.1"	Snow Only 1.5"	Thun. Stms.		18 M.P.H.	30 M.P.H.	
Jan.	44	37	55	11	0	3.9	0	89	86	4.9	—	26	11	—	0.5	1.8	39	15.0	0.1
Feb.	44	37	55	12	0	4.3	0	90	83	4.2	—	22	10	—	0.4	3.3	32	10.0	0.1
March	47	37	59	13	0	5.8	0	89	80	3.9	—	23	7	—	0.2	0.9	29	7.0	0.4
April	49	39	67	16	0	1.9	0	88	77	3.9	—	21	7	—	0.0	1.0	28	6.0	0.5
May	54	43	75	26	0	0.3	0	89	76	2.4	—	19	7	—	0.3	0.5	21	3.0	0.8
June	58	47	78	31	0	—	0	86	77	2.2	0	18	6	0	0.2	1.8	19	2.0	0.7
July	61	51	77	32	0	—	0	89	81	2.9	0	20	7	0	0.5	1.6	15	1.0	0.5
Aug.	61	51	75	33	0	0.0	0	91	81	3.8	0	22	9	0	0.2	1.4	17	1.0	0.8
Sept.	57	48	77	32	0	—	0	90	79	3.7	0	22	8	0	0.1	1.9	20	4.0	0.7
Oct.	53	44	70	26	0	0.7	0	89	81	4.9	—	25	9	—	0.1	1.4	30	9.0	0.2
Nov.	48	40	58	20	0	1.2	0	88	84	5.5	—	25	10	—	0.4	1.4	30	10.0	0.1
Dec.	45	38	56	12	0	2.5	0	88	86	5.9	—	25	10	—	0.5	1.5	35	12.0	0.2
Year	52	43	78	11	0	21.0	0	89	81	48.0	0+	268	101	0+	3.4	3.3	26	6.8	5.0

Elevation 30 Feet

Latitude 58° 12' N   Longitude 6° 19' W

Total precipitation means the total of rain, snow, hail, etc. expressed as inches of rain.

Ten inches of snow equals approx. 1 inch of rain.

Temperatures are as ° F.

Wind velocity is the percentage of observations greater than 18 and 30 M.P.H. Number of observations varies from one per day to several per hour.

Visibility means number of days with less than ½-mile visibility.

Note: "—" indicates slight trace or possibility blank indicates no statistics available

STUTTGART — (West) FED. REP. GERMANY

Month	Temperatures—in °F							Relative Humidity		Precipitation							Wind Velocity		Visibility
	Daily Average		Extreme		No. of Days			A.M. 6:30	P.M. 1:30	Inches		No. of Days				Max. Inches in 24 Hrs.	18 M.P.H.	30 M.P.H.	½ Mile
	Max.	Min.	Max.	Min.	Over 90°	Under 32°	0°			Total Prec.	Snow Only	Total Prec. 0.04"	0.1"	Snow Only 1.5"	Thun. Stms.				
Jan.	38	28	63	−13	0	23	—	84	77	1.5	6.8	9	5	1.4	—	1	5	—	6
Feb.	42	29	69	−14	0	18	1	83	70	1.3	7.2	8	4	1.1	—	1	4	0	6
March	50	34	75	11	0	15	0	82	62	1.7	1.3	9	5	0.2	1	1	3	0	3
April	57	40	84	22	0	5	0	77	57	2.2	1.0	11	6	0.2	2	1	2	0	2
May	66	48	94	28	—	1	0	77	56	2.7	0.0	11	7	0.0	5	2	2	0	2
June	72	54	91	38	—	0	0	74	57	3.3	0.0	11	8	0.0	6	2	1	0	1
July	75	57	102	39	1	0	0	75	57	3.1	0.0	11	8	0.0	5	2	1	0	1
Aug.	74	51	98	39	—	0	0	80	57	2.6	0.0	11	7	0.0	4	2	1	0	2
Sept.	67	51	91	34	0	0	0	84	63	2.5	0.0	10	7	0.0	2	2	1	0	7
Oct.	56	43	84	22	0	4	0	85	69	1.9	0.0	9	6	0.0	—	1	1	0	9
Nov.	46	35	71	10	0	10	0	86	76	1.8	1.2	9	5	0.3	—	2	2	0	8
Dec.	40	31	65	0	0	18	0	85	80	1.9	2.2	11	6	0.5	0	1	4	—	8
Year	57	42	102	−14	1	94	1	81	65	27.0	20.0	120	74	3.7	25	2	2	—	55

Elevation 1,300 Feet

Latitude 48° 41' N   Longitude 9° 12' E

Total precipitation means the total of rain, snow, hail, etc. expressed as inches of rain.

Ten inches of snow equals approx. 1 inch of rain.

Temperatures are as ° F.

Wind velocity is the percentage of observations greater than 18 and 30 M.P.H. Number of observations varies from one per day to several per hour.

Visibility means number of days with less than ½-mile visibility.

Note: "—" indicates slight trace or possibility blank indicates no statistics available

TAMPERE

FINLAND

Month	Temperatures—in °F Daily Average Max.	Min.	Extreme Max.	Min.	No. of Days Over 90°	Under 32°	0°	Relative Humidity A.M. 6:30	P.M. 2:30	Precipitation Inches Total Prec.	Snow Only	No. of Days Total Prec. 0.04"	0.1"	Snow Only 1.5"	Thun. Stms.	Max. Inches in 24 Hrs.	Wind Velocity 18 M.P.H.	30 M.P.H.	Visibility ½ Mile
Jan.	24	13	45	−32	0	—	—	88	87	1.5	—	10	5	—	—	0.8			
Feb.	24	12	48	−23	0	—	—	86	82	1.3	—	9	4	—	0	1.1			
March	33	17	52	−23	0	—	—	85	69	1.1	—	7	4	—	0	0.9			
April	44	28	68	−6	0	—	—	81	62	1.2	—	8	4	—	—	0.9			
May	58	38	82	19	0	—	0	75	55	1.7	—	8	5	—	1	1.0			
June	66	47	88	25	—	—	0	73	55	2.4	—	9	7	—	3	1.0			
July	73	55	91	39	—	0	0	78	59	2.3	0	11	6	0	5	1.6			
Aug.	68	52	89	32	0	—	0	86	64	3.0	0	11	7	0	3	1.9			
Sept.	57	44	80	22	0	—	0	90	70	2.5	—	10	7	—	1	1.0			
Oct.	45	35	61	7	0	—	0	89	78	2.4	—	13	7	—	—	0.9			
Nov.	36	29	52	−7	0	—	—	89	87	2.0	—	12	6	—	—	1.1			
Dec.	29	20	46	−23	0	—	—	89	88	1.8	—	9	6	—	—	0.9			
Year	46	33	91	−32	0+	150+	0+	84	71	23.0	—	117	67	—	15	1.9			

Elevation 302 Feet
Latitude 61° 27' N   Longitude 23° 44' E

Total precipitation means the total of rain, snow, hail, etc. expressed as inches of rain.

Ten inches of snow equals approx. 1 inch of rain.

Temperatures are as ° F.

Wind velocity is the percentage of observations greater than 18 and 30 M.P.H. Number of observations varies from one per day to several per hour.

Visibility means number of days with less than ½-mile visibility.

Note: "—" indicates slight trace or possibility blank indicates no statistics available

TEIGARHORN

ICELAND

Month	Temperatures—in °F — Daily Average Max.	Min.	Extreme Max.	Min.	No. of Days Over 90°	Under 32°	0°	Relative Humidity A.M. 8:00	P.M. 2:00	Precipitation Inches Total Prec.	Snow Only	No. of Days Total Prec. 0.004"	0.1"	Snow Only 1.5"	Thun. Stms.	Max. Inches in 24 Hrs.	Wind Velocity 18 M.P.H.	30 M.P.H.	Visibility ½ Mile
Jan.	37	29	55	5				79	79	6.0		16				2.4			
Feb.	37	28	57	2				78	79	4.5		13				4.3			
March	39	29	58	2				78	79	3.6		13				2.8			
April	43	31	64	13				78	80	3.3		10				1.9			
May	50	37	74	17				81	80	3.1		9				2.0			
June	56	42	87	31				82	79	3.1		8				2.8			
July	53	45	80	34				85	81	3.6		11				2.7			
Aug.	57	45	74	33				83	80	3.9		12				3.2			
Sept.	52	41	70	26				83	83	5.1		12				2.8			
Oct.	45	35	67	18				82	81	4.9		13				2.0			
Nov.	40	31	64	11				79	80	4.8		13				2.9			
Dec.	38	29	55	5				80	79	5.7		17				2.9			
Year	46	35	87	2				81	80	52.0		147				4.3			

Elevation 59 Feet

Latitude 64° 41' N   Longitude 14° 22' W

Total precipitation means the total of rain, snow, hail, etc. expressed as inches of rain.

Ten inches of snow equals approx. 1 inch of rain.

Temperatures are as ° F.

Wind velocity is the percentage of observations greater than 18 and 30 M.P.H. Number of observations varies from one per day to several per hour.

Visibility means number of days with less than ½-mile visibility.

Note: "—" indicates slight trace or possibility
blank indicates no statistics available

THESSALONIKI

GREECE

| Month | TEMPERATURES—IN °F | | | | | | | RELATIVE HUMIDITY | | PRECIPITATION | | | | | | WIND VELOCITY | | VISIBILITY ½ MILE |
| | Daily Average | | Extreme | | No. of Days | | | | | Inches | | No. of Days | | | | | | |
	Max.	Min.	Max.	Min.	Over 90°	Under 32°	0°	A.M. 7:30	P.M. 1:30	Total Prec.	Snow Only	Total Prec. 0.1"	Snow Only 1.5"	Thun. Stms.	Max. Inches in 24 Hrs.	18 M.P.H.	30 M.P.H.	
Jan.	49	36	67	7	0	12	0	83	72	1.3	—	4	—	—	1			
Feb.	52	38	75	10	0	9	0	77	62	1.3	—	4	—	—	1			
March	58	44	86	21	0	5	0	71	59	1.5	—	5	—	—	2			
April	67	50	88	33	0	0	0	70	62	1.7	0	5	0	1	3			
May	77	59	96	40	2	0	0	67	59	2.1	0	6	0	4	4			
June	85	66	101	47	11	0	0	65	54	1.4	0	4	0	6	2			
July	91	71	107	45	21	0	0	59	49	1.0	0	3	0	3	2			
Aug.	90	70	109	56	20	0	0	62	51	1.0	0	3	0	2	1			
Sept.	81	64	99	46	6	0	0	66	56	1.0	0	4	0	2	2			
Oct.	72	57	97	33	—	0	0	78	66	2.3	0	6	0	1	2			
Nov.	60	47	81	23	0	2	0	78	66	2.5	—	7	—	1	3			
Dec.	52	41	71	19	0	7	0	81	70	1.9	—	6	—	—	1			
Year	70	54	109	7	60	35	0	71	60	19.0	—	57	—	21	4			

Elevation 20 Feet
Latitude 40° 31' N   Longitude 22° 58' E
Total precipitation means the total of rain, snow, hail, etc. expressed as inches of rain.
Ten inches of snow equals approx. 1 inch of rain.
Temperatures are as ° F.

Wind velocity is the percentage of observations greater than 18 and 30 M.P.H. Number of observations varies from one per day to several per hour.
Visibility means number of days with less than ½-mile visibility.
Note: "—" indicates slight trace or possibility blank indicates no statistics available

TIREE (Inner Hebrides)   (United Kingdom) SCOTLAND

| Month | TEMPERATURES—IN °F | | | | | | | RELATIVE HUMIDITY | | PRECIPITATION | | | | | | | WIND VELOCITY | | VISIBILITY |
| | Daily Average | | Extreme | | No. of Days | | | | | Inches | | No. of Days | | | | Max. Inches in 24 Hrs | | | ½ MILE |
	Max.	Min.	Max.	Min.	Over 90°	Under 32°	0°	A.M. 6:30	Noon	Total Prec.	Snow Only	Total Prec. 0.01"	0.1"	Snow Only 1.5"	Thun. Stms		18 M.P.H.	30 M.P.H.	
Jan.	45	38	53	20	0	2.1	0	86	83	4.8	—	23	10	—	0.4	1.5			
Feb.	45	38	53	20	0	3.2	0	86	81	3.1	—	19	9	—	0.0	1.3			
March	47	39	59	21	0	3.4	0	88	79	2.7	—	18	7	—	0.2	1.0			
April	51	41	64	25	0	0.6	0	85	77	2.6	—	18	7	—	0.1	1.3			
May	55	45	72	31	0	—	0	84	76	2.1	0	15	6	0	0.0	1.3			
June	59	49	78	36	0	0.0	0	85	77	2.8	0	17	7	0	0.3	1.3			
July	61	52	78	43	0	0.0	0	90	82	3.5	0	20	8	0	0.3	1.9			
Aug.	61	53	76	41	0	0.0	0	89	82	3.6	0	19	8	0	0.5	1.5			
Sept.	59	51	69	34	0	0.0	0	87	80	4.5	0	20	9	0	0.4	2.9			
Oct.	54	47	64	33	0	0.0	0	85	80	5.6	0	24	10	0	0.2	2.3			
Nov.	49	43	58	27	0	0.9	0	85	81	4.9	—	23	9	—	0.3	1.5			
Dec.	47	40	57	23	0	2.7	0	85	83	4.7	—	23	10	—	0.5	1.4			
Year	53	45	78	20	0	13.0	0	86	80	44.9	0+	239	100	0+	3.2	2.9			

Elevation 38 Feet

Latitude 56° 31' N   Longitude 6° 52' W

Total precipitation means the total of rain, snow, hail, etc. expressed as inches of rain.

Ten inches of snow equals approx. 1 inch of rain.

Temperatures are as ° F.

Wind velocity is the percentage of observations greater than 18 and 30 M.P.H. Number of observations varies from one per day to several per hour.

Visibility means number of days with less than ½-mile visibility.

Note: "—" indicates slight trace or possibility blank indicates no statistics available

TOULOUSE

FRANCE

Elevation 499 Feet
Latitude 43° 38' N. Longitude 1° 22' E

| Month | TEMPERATURES—IN °F | | | | | | | RELATIVE HUMIDITY | | PRECIPITATION | | | | | | | WIND VELOCITY | | VISIBILITY ½ MILE |
| | Daily Average | | Extreme | | No. of Days | | | A.M. 7:00 | P.M. 1:00 | Inches | | No. of Days | | | Thun. Stms. | Max. Inches in 24 Hrs. | 18 M.P.H. | 30 M.P.H. | |
	Max.	Min.	Max.	Min.	Over 90°	Under 32°	0°			Total Prec.	Snow Only	Total Prec. 0.004"	0.1"	Snow Only 1.5"					
Jan.	47	33	74	1	0	13	0	90	74	1.9	—	14	6	—	0.2	0.9	13	0.2	
Feb.	50	34	77	12	0	13	0	89	66	1.8	—	11	6	—	1.0	1.1	15	0.4	
March	58	39	83	20	0	6	0	87	62	2.1	—	11	6	—	1.0	1.3	18	1.0	
April	62	43	86	24	0	1	0	84	59	2.0	—	15	6	—	1.0	1.5	17	0.5	
May	69	49	93	33	—	0	0	83	58	2.0	0	14	7	0	3.0	1.4	12	0.1	
June	76	55	104	42	—	0	0	81	56	2.4	0	9	6	0	4.0	2.0	10	0.0	
July	80	59	105	45	—	0	0	79	50	1.7	0	8	5	0	4.0	1.2	8	0.1	
Aug.	80	59	111	44	—	0	0	81	48	2.1	0	8	6	0	5.0	3.9	9	0.0	
Sept.	75	55	99	32	—	—	0	88	56	2.5	0	9	7	0	2.0	5.5	8	0.0	
Oct.	65	47	96	26	—	2	0	91	64	1.8	—	10	5	—	1.0	1.2	9	0.0	
Nov.	55	40	81	12	0	4	0	90	73	2.0	—	13	6	—	—	2.3	11	0.2	
Dec.	48	36	80	3	0	10	0	90	78	2.6	—	16	8	—	—	1.5	15	0.2	
Year	64	46	111	1	0⁺	50	0	86	62	26.0	0⁺	138	74	0⁺	23.0	5.5	12	0.2	

Total precipitation means the total of rain, snow, hail, etc. expressed as inches of rain.

Ten inches of snow equals approx. 1 inch of rain.

Temperatures are as °F.

Wind velocity is the percentage of observations greater than 18 and 30 M.P.H. Number of observations varies from one per day to several per hour.

Visibility means number of days with less than ½-mile visibility.

Note: "—" indicates slight trace or possibility blank indicates no statistics available

TRIER   (West) FED. REP. GERMANY

| Month | Temperatures—in °F | | | | | | | Relative Humidity | | Precipitation | | | | | | Wind Velocity | | Visibility |
| | Daily Average | | Extreme | | No. of Days | | | | | Inches | | No. of Days | | | Max. Inches in 24 Hrs. | | | ½ MILE |
	Max.	Min.	Max.	Min.	Over 90°	Under 32°	0°	A.M.	P.M.	Total Prec.	Snow Only	Total Prec. 0.1"	Snow Only 1.5"	Thun. Stms.		18 M.P.H.	30 M.P.H.	
Jan.	39	28	58	−3	0	19	—	—	85	2.0		6		0		24	3	7
Feb.	43	30	65	−6	0	15	—		81	1.7		6		—		26	6	5
March	50	33	72	7	0	12	0		76	1.9		6		1		21	5	4
April	58	38	86	21	0	4	0		70	1.9		6		1		23	5	1
May	67	45	94	28	—	1	0		69	2.2		6		4		14	2	2
June	73	50	94	35	—	0	0		70	2.6	0	7	0	5		10	1	4
July	75	53	99	40	2	0	0		71	2.9	0	7	0	5		12	1	2
Aug.	74	52	96	38	1	0	0		75	2.8	0	7	0	4		12	2	5
Sept.	68	48	95	31	1	—	0		79	2.2	0	6	0	1		13	—	6
Oct.	57	41	83	21	0	3	0		84	2.8		7		—		13	2	10
Nov.	46	35	69	8	0	8	0		85	2.3		6		—		20	3	6
Dec.	40	31	64	1	0	15	0		86	2.7		8		—		20	3	10
Year	58	40	99	−6	4	77	0+		78	28.0		78		21		17	3	62

Elevation 443 Feet

Latitude 49° 43' N   **Longitude 6° 36' E**

Total precipitation means the total of rain, snow, hail, etc. expressed as inches of rain.

Ten inches of snow equals approx. 1 inch of rain.

Temperatures are as ° F.

Wind velocity is the percentage of observations greater than 18 and 30 M.P.H. Number of observations varies from one per day to several per hour.

Visibility means number of days with less than ½-mile visibility.

Note: "—" indicates slight trace or possibility blank indicates no statistics available

TROMSO — NORWAY

Month	Temperatures—in °F Daily Average Max.	Min.	Extreme Max.	Min.	No. of Days Over 90°	Under 32°	0°	Relative Humidity A.M. 8:30	P.M. 2:30	Precipitation Inches Total Prec.	Snow Only	No. of Days Total Prec. 0.04"	0.1"	Snow Only 1.5"	Thun. Stms.	Max. Inches in 24 Hrs.	Wind Velocity 18 M.P.H.	30 M.P.H.	Visibility ½ Mile
Jan.	30	22	45	1	0	27	0	76	77	4.1	—	14	10	—	—	2	14	2	5
Feb.	29	21	47	−1	0	26	—	77	75	3.8	—	14	10	—	0	2	10	2	2
March	32	21	47	3	0	28	0	76	71	3.3	—	13	7	—	0	3	15	1	3
April	37	27	57	6	0	22	0	73	67	2.4	—	11	7	—	—	2	10	1	3
May	44	34	69	16	0	10	0	72	66	2.1	—	11	6	—	0	1	4	—	1
June	53	42	82	24	0	—	0	73	66	2.1	—	10	6	—	—	1	3	0	1
July	59	48	83	34	0	0	0	78	70	2.3	0	11	6	0	1	2	2	—	1
Aug.	57	46	80	30	0	—	0	81	72	2.9	0	12	7	0	—	2	3	0	1
Sept.	49	40	72	24	0	2	0	83	75	4.7	—	16	9	—	—	2	2	0	1
Oct.	40	33	60	8	0	15	0	80	77	4.5	—	16	9	—	0	3	9	1	2
Nov.	34	27	52	3	0	23	0	79	79	4.0	—	14	9	—	0	2	9	1	2
Dec.	31	24	47	2	0	28	0	77	77	3.9	—	12	10	—	—	3	8	1	3
Year	41	32	83	−1	0	180	0+	77	73	40.0	—	154	95	—	3	3	7	1	25

Elevation 31 Feet

Latitude 69° 41' N   Longitude 18° 56' E

Total precipitation means the total of rain, snow, hail, etc. expressed as inches of rain.

Ten inches of snow equals approx. 1 inch of rain.

Temperatures are as ° F.

Wind velocity is the percentage of observations greater than 18 and 30 M.P.H. Number of observations varies from one per day to several per hour.

Visibility means number of days with less than ½-mile visibility.

Note: "—" indicates slight trace or possibility blank indicates no statistics available

TRONDHEIM

NORWAY

Month	TEMPERATURES—IN °F							RELATIVE HUMIDITY		PRECIPITATION							WIND VELOCITY		VISIBILITY ½ MILE
	Daily Average		Extreme		No. of Days			A.M. 7:30	P.M. 1:30	Inches		No. of Days				Max. Inches in 24 Hrs.	18 M.P.H.	30 M.P.H.	
	Max.	Min.	Max.	Min.	Over 90°	Under 32°	0°			Total Prec.	Snow Only	Total Prec. 0.04"	0.1"	Snow Only 1.5"	Thun. Stms.				
Jan.	29	20	48	−8	0	27	2	75	73	2.1	—	13	8	—		2	7	1	1
Feb.	31	21	54	−11	0	25	3	75	71	1.6	—	11	6	—	0	2	8	—	1
March	38	25	52	3	0	23	0	76	67	2.8	—	11	9	—	0	3	9	2	1
April	45	33	64	18	0	16	0	75	63	2.1	—	10	7	—	0	2	9	2	1
May	54	39	75	27	0	6	0	74	61	1.8	—	10	6	—	2	1	6	—	—
June	61	47	84	32	0	—	0	76	64	2.7	0	11	8	0	2	1	7	—	0
July	64	51	84	37	0	0	0	79	65	3.0	0	11	9	0	1	2	7	—	—
Aug.	64	50	82	37	0	0	0	82	67	3.5	0	12	8	0	1	2	4	—	—
Sept.	57	43	73	25	0	2	0	85	72	3.5	—	15	10	—	0	2	4	—	1
Oct.	48	37	61	18	0	8	0	83	75	2.5	—	14	8	—	0	3	4	—	1
Nov.	38	31	54	3	0	17	0	79	76	2.9	—	12	9	—	0	2	9	1	—
Dec.	34	26	52	−9	0	21	1	77	76	5.5	—	12	12	—	0	3	8	1	1
Year	47	35	84	−11	0	146	6	78	69	34.0	—	142	99	—	6	3	7	1	6

Elevation 61 Feet

Latitude 63° 27' N   Longitude 10° 55' E

Total precipitation means the total of rain, snow, hail, etc. expressed as inches of rain.

Ten inches of snow equals approx. 1 inch of rain.

Temperatures are as °F.

Wind velocity is the percentage of observations greater than 18 and 30 M.P.H. Number of observations varies from one per day to several per hour.

Visibility means number of days with less than ½-mile visibility.

Note: "—" indicates slight trace or possibility blank indicates no statistics available

TURIN — ITALY

Month	Temp Daily Avg Max	Temp Daily Avg Min	Temp Extreme Max	Temp Extreme Min	Days Over 90°	Days Under 32°	Days 0°	Rel. Hum. A.M.	Rel. Hum. P.M.	Prec. Inches Total	Prec. Inches Snow Only	No. Days Total Prec.	No. Days 0.1"	No. Days Snow Only 1.5"	Thun. Stms.	Max. Inches in 24 Hrs.	Wind 18 M.P.H.	Wind 30 M.P.H.	Visibility ½ Mile
Jan.	39	26	60	−1	0	—	—	—	88	1.2	—	5	4	—		3			
Feb.	47	30	72	11	0	—	0	—	76	0.9	—	4	3	—		1			
March	57	39	76	20	0	—	0	—	73	1.9	—	6	6	—		2			
April	65	46	86	30	0	—	0	—	71	3.3	0	9	7	0		1			
May	69	53	84	37	0	0	0	—	79	6.3	0	10	10	0		3			
June	80	59	95	44	—	0	0	—	68	1.6	0	7	5	0		2			
July	85	64	98	53	8	0	0	—	65	1.4	0	5	4	0		2			
Aug.	82	63	98	52	5	0	0	—	68	2.0	0	6	6	0		1			
Sept.	76	57	95	45	—	0	0	—	75	2.2	0	6	6	0		1			
Oct.	64	49	81	30	0	—	0	—	84	2.1	0	6	6	0		2			
Nov.	52	37	68	26	0	—	0	—	85	1.0	—	9	4	—		2			
Dec.	41	28	57	7	0	—	0	—	86	1.3	—	6	4	—		2			
Year	63	46	98	−1	13+	0+	0+	—	77	25.0	0+	79	65	0+		3			

Elevation 986 Feet

Latitude 45° 11' N   Longitude 7° 39' E

Total precipitation means the total of rain, snow, hail, etc. expressed as inches of rain.

Ten inches of snow equals approx. 1 inch of rain.

Temperatures are as °F.

Wind velocity is the percentage of observations greater than 18 and 30 M.P.H. Number of observations varies from one per day to several per hour.

Visibility means number of days with less than ½-mile visibility.

Note: "—" indicates slight trace or possibility blank indicates no statistics available

TWENTHE (Winterswijk—Enschede) THE NETHERLANDS

| Month | TEMPERATURES—IN °F | | | | | | | RELATIVE HUMIDITY | | PRECIPITATION | | | | | | | | WIND VELOCITY | | VISIBILITY |
| | Daily Average | | Extreme | | No. of Days | | | | | Inches | | No. of Days | | | Thun. Stms. | Max. Inches in 24 Hrs. | | | ½ MILE |
	Max.	Min.	Max.	Min.	Over 90°	Under 32°	0°	A.M. 8:00	P.M. 2:00	Total Prec.	Snow Only	Total Prec. 0.004"	0.1"	Snow Only 1.5"			18 M.P.H.	30 M.P.H.	
Jan.	39	32	55	9	0	17	0	89	81	2.7	—	20	9	—	0.2	1.0	13	1.1	6
Feb.	40	31	64	—6	0	16	—	88	75	2.2	—	18	7	—	0.2	2.3	9	0.9	6
March	48	35	70	14	0	13	0	83	63	1.8	—	16	6	—	0.3	0.9	9	1.4	5
April	56	39	81	25	0	6	0	77	59	1.6	—	17	5	—	1.0	2.1	8	0.3	3
May	63	45	86	28	0	—	0	73	57	2.2	—	14	7	—	3.8	1.9	4	0.2	2
June	69	51	93	36	—	0	0	74	59	2.3	0	14	7	0	4.4	1.4	3	0.0	2
July	71	55	97	41	—	0	0	78	62	4.2	0	16	9	0	6.5	2.5	4	0.1	3
Aug.	71	55	91	37	—	0	0	81	62	3.1	0	16	9	0	6.6	1.4	3	0.0	3
Sept.	66	51	91	36	—	0	0	85	66	2.9	0	17	8	0	2.8	1.4	5	0.1	4
Oct.	57	45	79	27	0	1	0	86	72	2.2	—	21	6	—	0.8	1.5	5	0.2	7
Nov.	47	38	66	21	0	6	0	90	80	2.6	—	21	8	—	0.5	1.5	6	0.2	6
Dec.	42	35	61	10	0	11	0	90	84	2.8	—	21	9	—	0.6	1.2	11	1.2	7
Year	56	43	97	—6	0+	70	0+	83	68	31.0	0+	211	90	0+	28.0	2.5	7	0.5	54

Elevation 113 Feet
Latitude 52° 16' N   Longitude 6° 53' E

Total precipitation means the total of rain, snow, hail, etc. expressed as inches of rain.

Ten inches of snow equals approx. 1 inch of rain.

Temperatures are as ° F.

Wind velocity is the percentage of observations greater than 18 and 30 M.P.H. Number of observations varies from one per day to several per hour.

Visibility means number of days with less than ½-mile visibility.

Note: "—" indicates slight trace or possibility blank indicates no statistics available

TYNEMOUTH (United Kingdom) ENGLAND

Month	Daily Average Max.	Daily Average Min.	Extreme Max.	Extreme Min.	Over 90°	Under 32°	0°	A.M. 7:00	P.M. 1:00	Total Prec.	Snow Only	0.01"	0.1"	Snow Only 1.5"	Thun. Stms.	Max. Inches in 24 Hrs	18 M.P.H.	30 M.P.H.	½ MILE
Jan.	43	37	58	16	0	6	0	88	82	2.4	—	17	7	—	0.0	1.3	16	2.0	4
Feb.	44	37	59	11	0	4	0	87	79	1.7	—	15	6	—	0.0	2.5	16	1.0	2
March	47	38	70	18	0	4	0	87	74	1.6	—	14	5	—	0.0	0.7	11	1.0	2
April	50	41	71	28	0	1	0	85	75	1.7	—	15	5	—	0.0	0.9	12	—	3
May	54	45	81	30	0	—	0	85	75	2.1	0	14	6	0	0.0	3.0	7	—	4
June	60	50	86	39	0	0	0	83	75	1.8	0	13	5	0	0.1	1.4	5	0.0	3
July	64	54	80	42	0	0	0	84	74	3.1	0	15	8	0	0.2	2.5	4	—	2
Aug.	64	54	81	42	0	0	0	86	75	3.1	0	15	8	0	0.5	2.0	4	0.0	4
Sept.	61	51	81	33	0	0	0	89	76	2.3	0	14	6	0	0.0	1.5	8	—	3
Oct.	55	46	74	30	0	—	0	87	75	2.5	0	16	7	0	0.1	1.9	11	1.0	2
Nov.	48	41	67	26	0	2	0	87	81	2.6	—	17	7	—	0.0	1.2	11	—	4
Dec.	44	38	58	22	0	4	0	87	83	2.3	—	17	7	—	0.0	1.1	15	1.0	7
Year	53	44	86	11	0	21	0	87	77	27.0	0+	182	77	0+	0.9	3.0	10	0.6	40

Elevation 108 Feet

Latitude 55° 1' N   Longitude 1° 25' W

Total precipitation means the total of rain, snow, hail, etc. expressed as inches of rain.

Ten inches of snow equals approx. 1 inch of rain.

Temperatures are as ° F.

Wind velocity is the percentage of observations greater than 18 and 30 M.P.H. Number of observations varies from one per day to several per hour.

Visibility means number of days with less than ½-mile visibility.

Note: "—" indicates slight trace or possibility blank indicates no statistics available

UMEÅ

SWEDEN

Month	Daily Average Max.	Daily Average Min.	Extreme Max.	Extreme Min.	Over 90°	Under 32°	0°	R.H. A.M. 8:30	R.H. P.M. 2:30	Total Prec. (in)	Snow Only (in)	Total Prec. 0.04"	0.1"	Snow Only 1.5"	Thun. Stms.	Max. Inches in 24 Hrs	Wind 18 M.P.H.	Wind 30 M.P.H.	Visibility ½ Mile
Jan.	25	11	46	−36	0	30	7	87	86	1.4	—	8	5	—	0	0.6			
Feb.	25	11	48	−32	0	27	10	85	82	1.1	—	6	4	—	0	0.7			
March	31	16	57	−26	0	29	4	85	75	1.2	—	6	4	—	0	1.0			
April	40	25	71	−14	0	27	3	78	69	1.3	—	6	4	—	—	1.5			
May	50	34	76	16	0	17	0	65	57	1.5	—	7	5	—	—	1.3			
June	60	44	92	24	—	2	0	64	58	1.8	—	7	5	—	1	1.6			
July	67	50	88	30	0	—	0	68	61	1.9	0	7	5	0	3	1.8			
Aug.	62	47	82	27	0	1	0	78	67	3.0	—	10	7	—	1	1.5			
Sept.	54	39	75	20	0	4	0	83	70	2.3	—	8	6	—	1	1.5			
Oct.	42	31	66	−2	0	17	—	87	79	2.5	—	10	7	—	0	1.9			
Nov.	32	21	51	−20	0	22	1	88	86	2.3	—	9	6	—	0	2.2			
Dec.	27	15	48	−29	0	29	7	86	88	1.9	—	10	6	—	0	1.3			
Year	43	29	92	−36	0+	205	31	80	73	22.0	—	91	65	—	7	2.2			

Elevation 22 Feet
Latitude 63° 47' N   Longitude 20° 17' E
Total precipitation means the total of rain, snow, hail, etc. expressed as inches of rain.
Ten inches of snow equals approx. 1 inch of rain.
Temperatures are as ° F.

Wind velocity is the percentage of observations greater than 18 and 30 M.P.H. Number of observations varies from one per day to several per hour.
Visibility means number of days with less than ½-mile visibility.
Note: "—" indicates slight trace or possibility blank indicates no statistics available

VÄÄSA — FINLAND

Month	Daily Average Max.	Daily Average Min.	Extreme Max.	Extreme Min.	Over 90°	Under 32°	Under 0°	R.H. A.M. 6:30	R.H. P.M. 2:30	Total Prec. (in)	Snow Only (in)	No. Days Total Prec. 0.04"	No. Days 0.1"	No. Days Snow Only 1.5"	Thun. Stms.	Max. Inches in 24 Hrs.	Wind 18 M.P.H.	Wind 30 M.P.H.	Visibility ½ Mile
Jan.	26	15	46	−29	0	30	6	90	89	1.3	—	8	4	—	0	0.7	6	—	3
Feb.	25	14	45	−35	0	27	6	89	85	1.0	—	6	3	—	0	0.7	6	—	2
March	31	18	53	−29	0	30	6	87	76	1.0	—	6	4	—	0	0.5	5	—	2
April	41	28	74	−6	0	24	—	85	68	1.3	—	6	4	—	—	0.8	6	—	2
May	53	38	83	19	0	11	0	79	62	1.5	—	7	5	—	1	0.9	3	0	1
June	62	48	89	27	0	1	0	78	64	2.2	—	7	6	—	2	1.3	3	0	—
July	69	55	89	32	0	—	0	82	65	2.2	0	8	6	0	3	1.9	2	0	2
Aug.	66	52	87	30	0	1	0	89	71	2.7	0	10	7	0	2	1.3	4	—	3
Sept.	56	44	78	21	0	4	0	90	75	2.3	—	10	6	—	1	2.0	3	—	3
Oct.	44	35	61	3	0	13	0	91	83	2.5	—	11	7	—	—	1.1	2	0	2
Nov.	35	28	50	−7	0	21	1	91	88	2.0	—	11	6	—	0	0.7	3	0	2
Dec.	30	21	47	−25	0	27	3	89	89	1.5	—	8	5	—	0	0.5	4	—	2
Year	45	33	89	−35	0	189	21	87	76	22.0	—	98	63	—	10	2.0	4	—	24

Elevation 13 Feet
Latitude 63° 2′ N   Longitude 21° 45′ E

Total precipitation means the total of rain, snow, hail, etc. expressed as inches of rain.

Ten inches of snow equals approx. 1 inch of rain.

Temperatures are as ° F.

Wind velocity is the percentage of observations greater than 18 and 30 M.P.H. Number of observations varies from one per day to several per hour.

Visibility means number of days with less than ½-mile visibility.

Note: "—" indicates slight trace or possibility blank indicates no statistics available

VALENCIA

SPAIN

Elevation 213 Feet
Latitude 39° 29' N   **Longitude 0° 28' W**

| Month | TEMPERATURES—IN °F | | | | | | | RELATIVE HUMIDITY | | PRECIPITATION | | | | | | | | WIND VELOCITY | | VISI-BIL-ITY ½ MILE |
| | Daily Average | | Extreme | | No. of Days | | | A.M. 7:00 | P.M. 1:00 | Inches | | No. of Days | | | Thun. Stms. | Max. Inches in 24 Hrs. | | 18 M.P.H. | 30 M.P.H. | |
	Max.	Min.	Max.	Min.	Over 90°	Under 32°	0°			Total Prec.	Snow Only	Total Prec. 0.004"	0.1"	Snow Only 1.5"						
Jan.	58	41	82	20	0	—	0	75	60	0.9	—	4	3	—	0	2.6		7	2.0	—
Feb.	60	43	82	19	0	2	0	72	56	1.6	—	5	5	—	—	2.6		8	—	1
March	63	47	87	29	0	—	0	72	52	1.1	0	5	4	0	—	0.6		7	1.0	1
April	67	51	95	38	—	0	0	72	57	1.2	0	5	4	0	—	1.3		6	1.0	—
May	73	56	95	41	—	0	0	71	57	1.3	0	5	4	0	2	1.7		3	—	1
June	78	63	97	46	1	0	0	70	58	1.1	0	4	3	0	2	5.2		4	—	—
July	83	68	107	51	2	0	0	71	58	0.4	0	2	1	0	1	1.3		—	0.0	0
Aug.	83	69	102	55	3	0	0	74	61	0.3	0	2	—	0	2	2.1		1	—	0
Sept.	80	64	97	45	1	0	0	78	62	2.8	0	7	7	0	1	4.9		1	0.0	1
Oct.	73	57	92	38	—	0	0	78	60	1.8	0	5	5	0	3	5.0		2	0.0	—
Nov.	65	48	85	31	0	—	0	75	60	2.8	0	7	7	0	—	5.0		4	—	—
Dec.	51	42	77	28	0	—	0	74	60	1.0	—	4	3	—	—	3.3		5	—	—
Year	70	54	107	19	8	2	0	73	58	16.0	—	55	49	—	13	5.2		4	0.5	5

Total precipitation means the total of rain, snow, hail, etc. expressed as inches of rain.

Ten inches of snow equals approx. 1 inch of rain.

Temperatures are as ° F.

Wind velocity is the percentage of observations greater than 18 and 30 M.P.H. Number of observations varies from one per day to several per hour.

Visibility means number of days with less than ½-mile visibility.

Note: "—" indicates slight trace or possibility blank indicates no statistics available

VALENTIA — IRELAND

Month	TEMPERATURES—IN °F Daily Average Max.	Min.	Extreme Max.	Min.	No. of Days Over 90°	Under 32°	0°	RELATIVE HUMIDITY A.M. 6:30	Noon	PRECIPITATION Inches Total Prec.	Snow Only	No. of Days Total Prec. 0.01"	0.1"	Snow Only 1.5"	Thun. Stms.	Max. Inches in 24 Hrs.	WIND VELOCITY 18 M.P.H.	30 M.P.H.	VISIBILITY ½ MILE
Jan.	49	41	57	20	0	—	0	84	81	6.0	—	25	11	—	0.8	1.9			
Feb.	49	40	58	22	0	—	0	82	77	5.0	—	19	10	—	0.6	2.4			
March	50	41	66	23	0	—	0	81	73	4.3	—	19	8	—	0.4	1.2			
April	53	43	70	29	0	—	0	81	72	3.7	0	18	7	0	0.3	1.2			
May	58	47	75	34	0	0	0	83	74	3.2	0	18	7	0	0.4	1.3			
June	62	51	81	35	0	0	0	85	77	3.4	0	16	8	0	0.6	1.5			
July	64	54	85	40	0	0	0	87	79	4.0	0	21	9	0	0.5	1.8			
Aug.	64	54	81	40	0	0	0	89	80	4.9	0	22	9	0	0.5	1.7			
Sept.	62	52	79	35	0	0	0	87	79	4.5	0	19	9	0	0.3	1.7			
Oct.	56	47	71	28	0	—	0	85	79	5.6	—	21	10	—	0.4	3.1			
Nov.	52	43	63	24	0	—	0	84	80	5.6	—	23	10	—	0.4	2.2			
Dec.	50	41	57	23	0	—	0	83	82	6.5	—	24	11	—	0.6	1.7			
Year	56	46	85	20	0	0+	0	84	78	57.0	0+	245	109	0+	5.8	3.1			

Elevation 46 Feet

Latitude 51° 56' N   Longitude 10° 15' W

Total precipitation means the total of rain, snow, hail, etc. expressed as inches of rain.

Ten inches of snow equals approx. 1 inch of rain.

Temperatures are as ° F.

Wind velocity is the percentage of observations greater than 18 and 30 M.P.H. Number of observations varies from one per day to several per hour.

Visibility means number of days with less than ½-mile visibility.

Note: "—" indicates slight trace or possibility blank indicates no statistics available

VARDO

NORWAY

Month	TEMPERATURES—IN °F							RELATIVE HUMIDITY		PRECIPITATION							WIND VELOCITY		VISI-BIL-ITY
	Daily Average		Extreme		No. of Days			A.M. 9:00	P.M. 3:00	Inches		No. of Days				Max. Inches in 24 Hrs.	18 M.P.H.	30 M.P.H.	½ MILE
	Max.	Min.	Max.	Min.	Over 90°	Under 32°	0°			Total Prec.	Snow Only	Total Prec. 0.04"	0.1"	Snow Only 1.5"	Thun. Stms.				
Jan.	27	19	42	−9	0	30	—	84	84	2.5	—	13	8	—	0	2			
Feb.	26	18	43	−11	0	28	—	84	84	2.5	—	13	8	—	—	2			
March	29	20	47	−3	0	29	—	83	82	2.3	—	12	6	—	0	2			
April	34	26	56	6	0	25	0	82	80	1.5	—	10	5	—	0	1			
May	40	32	69	14	0	16	0	81	79	1.3	—	9	4	—	0	1			
June	47	38	78	25	0	2	0	83	81	1.3	—	8	4	—	1	1			
July	53	44	80	30	0	—	0	86	84	1.5	0	8	4	0	1	2			
Aug.	53	44	77	30	0	—	0	85	83	1.7	0	9	5	0	—	1			
Sept.	47	40	68	23	0	1	0	85	82	1.9	—	12	6	—	—	1			
Oct.	38	32	56	8	0	16	0	85	83	2.5	—	14	7	—	0	1			
Nov.	33	26	49	5	0	26	0	85	85	2.1	—	13	6	—	0	1			
Dec.	30	22	44	−3	0	30	—	84	84	2.4	—	13	7	—	0	1			
Year	38	30	80	−11	0	204	0⁺	84	83	24.0	—	134	70	—	3	2			

Elevation 43 Feet

Latitude 70° 22' N   Longitude 31° 6' E

Total precipitation means the total of rain, snow, hail, etc. expressed as inches of rain.

Ten inches of snow equals approx. 1 inch of rain.

Temperatures are as ° F.

Wind velocity is the percentage of observations greater than 18 and 30 M.P.H. Number of observations varies from one per day to several per hour.

Visibility means number of days with less than ½-mile visibility.

Note: "—" indicates slight trace or possibility blank indicates no statistics available

VENICE

ITALY

Month	TEMPERATURES—IN °F Daily Average Max.	Min.	Extreme Max.	Min.	No. of Days Over 90°	Under 32°	0°	RELATIVE HUMIDITY A.M. 6:00	P.M. 1:00	PRECIPITATION Inches Total Prec.	Snow Only	No. of Days Total Prec. 0.04"	0.1"	Snow Only 1.5"	Thun. Stms.	Max. Inches in 24 Hrs.	WIND VELOCITY 18 M.P.H.	30 M.P.H.	VISI-BILITY ½ MILE
Jan.	43	33	58	18	0	14	0	86	•76	1.6	—	6	5	—	0	3	8	1	9
Feb.	46	35	65	12	0	13	0	80	76	1.6	—	5	5	—	—	2	14	4	7
March	54	41	70	27	0	4	0	86	68	2.0	—	6	6	—	—	2	14	3	7
April	63	49	80	35	0	0	0	86	67	2.4	0	7	7	0	3	3	10	1	3
May	71	57	90	44	—	0	0	85	69	2.9	0	8	7	0	8	4	5	—	2
June	78	64	89	48	0	0	0	83	65	3.0	0	8	7	0	9	2	3	—	1
July	82	67	96	52	2	0	0	82	64	2.3	0	8	6	0	10	2	2	—	—
Aug.	82	67	93	55	—	0	0	84	63	2.5	0	7	7	0	8	3	4	—	—
Sept.	78	62	91	48	—	0	0	87	64	2.8	0	7	7	0	4	5	4	—	3
Oct.	65	52	77	30	0	—	0	88	68	3.6	0	8	8	0	2	3	8	1	4
Nov.	54	43	67	28	0	1	0	88	75	2.7	—	7	7	—	—	3	5	—	7
Dec.	46	37	62	24	0	5	0	88	79	1.9	—	7	6	—	0	3	5	1	13
Year	64	51	96	12	2+	37	0	85	70	29.0	0+	84	78	0+	44+	5	7	1	56

Elevation 56 Feet
Latitude 45° 26' N   Longitude 12° 23' E
Total precipitation means the total of rain, snow, hail, etc. expressed as inches of rain.

Ten inches of snow equals approx. 1 inch of rain.
Temperatures are as °F.

Wind velocity is the percentage of observations greater than 18 and 30 M.P.H. Number of observations varies from one per day to several per hour.
Visibility means number of days with less than ½-mile visibility.
Note: "—" indicates slight trace or possibility
blank indicates no statistics available

VESTERVIG

DENMARK

Month	Daily Average Max.	Daily Average Min.	Extreme Max.	Extreme Min.	No. of Days Over 90°	No. of Days Under 32°	No. of Days 0°	Relative Humidity A.M. 7:30	Relative Humidity P.M. 1:30	Inches Total Prec.	Inches Snow Only	No. of Days Total Prec. 0.04"	No. of Days 0.1"	No. of Days Snow Only 1.5"	Thun. Stms.	Max. Inches in 24 Hrs.	Wind Velocity 18 M.P.H.	Wind Velocity 30 M.P.H.	Visibility ½ Mile
Jan.	36	29	50	−9	0	—	—	91	89	2.2	—	14	7	—	—	1.2			
Feb.	36	29	50	−3	0	—	—	91	87	1.7	—	9	6	—	—	2.4			
March	40	31	64	0	0	—	0	89	82	1.7	—	7	5	—	—	1.5			
April	47	36	71	20	0	—	0	85	74	1.6	—	10	5	—	1	1.0			
May	57	43	82	25	0	—	0	79	67	1.6	—	8	5	—	2	1.1			
June	63	50	86	36	0	0	0	80	71	1.7	0	8	5	0	2	1.7			
July	66	55	92	42	—	0	0	83	74	2.4	0	9	6	0	2	1.9			
Aug.	65	54	87	42	0	0	0	85	75	3.3	0	10	8	0	3	2.2			
Sept.	60	50	78	30	0	—	0	87	75	2.8	0	11	7	0	2	1.7			
Oct.	52	43	71	24	0	—	0	88	80	3.4	—	16	8	—	1	2.4			
Nov.	44	37	56	11	0	—	0	89	86	2.9	—	14	7	—	1	1.5			
Dec.	39	33	55	5	0	—	0	90	89	2.7	—	11	8	—	1	1.3			
Year	50	41	92	−9	0+	0+	0+	86	79	28.0	—	127	77	—	16	2.4			

TEMPERATURES—IN ° F

PRECIPITATION

Elevation 82 Feet
Latitude 56° 46' N   **Longitude 8° 19' E**

Total precipitation means the total of rain, snow, hail, etc. expressed as inches of rain.

Ten inches of snow equals approx. 1 inch of rain.
Temperatures are as ° F.

Wind velocity is the percentage of observations greater than 18 and 30 M.P.H. Number of observations varies from one per day to several per hour.

Visibility means number of days with less than ½-mile visibility.

Note: "—" indicates slight trace or possibility
blank indicates no statistics available

VESTMANNAEYJAR (Heymaey, Westman Islands)　　ICELAND

Month	TEMPERATURES—IN °F Daily Average Max.	Min.	Extreme Max.	Min.	No. of Days Over 90°	Under 32°	0°	RELATIVE HUMIDITY A.M. 7:30	P.M. 1:30	PRECIPITATION Inches Total Prec.	Snow Only	No. of Days Total Prec. 0.004"	0.1"	Snow Only 1.5"	Thun. Stms.	Max. Inches in 24 Hrs.	WIND VELOCITY 18 M.P.H.	30 M.P.H.	VISIBILITY ½ MILE
Jan.	39	31	50	6				83	81	6.2		21				2.4			
Feb.	39	32	48	5				81	80	4.8		19				4.2			
March	41	33	51	5				80	78	4.6		20				3.6			
April	43	35	55	14				80	77	3.7		18				2.1			
May	48	39	65	19				81	77	3.3		16				2.2			
June	52	44	66	32				81	77	3.0		16				1.6			
July	56	47	70	38				82	78	3.1		17				1.5			
Aug.	55	47	67	31				82	77	3.8		18				1.7			
Sept.	50	43	59	28				84	79	5.5		19				2.1			
Oct.	44	38	53	17				83	81	6.1		20				2.9			
Nov.	41	35	52	11				82	81	5.4		20				3.9			
Dec.	40	33	51	9				79	78	5.8		23				2.7			
Year	46	38	70	5				81	79	55.0		227				4.2			

Elevation 400 Feet

Latitude 63° 24' N　　Longitude 20° 17' W

Total precipitation means the total of rain, snow, hail, etc. expressed as inches of rain.

Ten inches of snow equals approx. 1 inch of rain.

Temperatures are as ° F.

Wind velocity is the percentage of observations greater than 18 and 30 M.P.H. Number of observations varies from one per day to several per hour.

Visibility means number of days with less than ½-mile visibility.

Note: "—" indicates slight trace or possibility blank indicates no statistics available

VIENNA   AUSTRIA

Month	TEMPERATURES—IN °F							RELATIVE HUMIDITY		PRECIPITATION							WIND VELOCITY		VISIBILITY ½ MILE
	Daily Average		Extreme		No. of Days			A.M. 5:00	P.M. 3:00	Inches		No. of Days				Max. Inches in 24 Hrs.	18 M.P.H.	30 M.P.H.	
	Max.	Min.	Max.	Min.	Over 90°	Under 32°	0°			Total Prec.	Snow Only	Total Prec. 0.04"	0.1"	Snow Only 1.5"	Thun. Stms.				
Jan.	34	26	62	−8	0	23	—	82	74	1.5	—	8	5	—	0	0.9	14	2	5
Feb.	38	28	67	−14	0	22	1	81	68	1.4	—	7	5	—	—	2.4	13	3	3
March	47	34	77	3	0	18	0	79	57	1.8	—	7	6	—	—	1.8	7	1	4
April	57	41	83	18	0	3	0	78	51	2.0	—	9	6	—	1	1.9	8	—	2
May	66	50	92	27	—	—	0	82	53	2.8	—	9	7	—	5	3.0	7	1	1
June	71	56	97	39	—	0	0	80	54	2.7	0	9	7	0	6	2.6	7	—	1
July	75	59	98	45	1	0	0	81	54	3.0	0	10	7	0	7	2.9	7	0	—
Aug.	73	58	97	42	2	0	0	82	55	2.7	0	7	7	0	6	1.7	5	—	0
Sept.	66	52	90	31	—	—	0	84	58	2.0	0	8	6	0	2	3.4	6	1	1
Oct.	55	44	82	16	0	2	0	86	67	2.0	—	8	6	—	—	2.7	6	1	4
Nov.	44	36	71	6	0	10	0	85	75	1.9	—	8	5	—	—	1.9	11	2	5
Dec.	37	30	66	−4	0	15	—	84	78	1.8	—	9	6	—	—	2.0	12	3	7
Year	55	43	98	−14	3	93	2	82	62	26.0	—	99	72	—	29	3.4	9	1	33

Elevation 666 Feet
Latitude 48° 15' N   Longitude 16° 22' E

Total precipitation means the total of rain, snow, hail, etc. expressed as inches of rain.

Ten inches of snow equals approx. 1 inch of rain.
Temperatures are as ° F.

Wind velocity is the percentage of observations greater than 18 and 30 M.P.H. Number of observations varies from one per day to several per hour.

Visibility means number of days with less than ½-mile visibility.

Note: "—" indicates slight trace or possibility blank indicates no statistics available

VISBY (Gotland Island) SWEDEN

Month	TEMPERATURES—IN °F Daily Average Max.	Min.	Extreme Max.	Min.	No. of Days Over 90°	Under 32°	0°	RELATIVE HUMIDITY A.M. 8:00	P.M. 2:00	PRECIPITATION Inches Total Prec.	Snow Only	No. of Days Total Prec. 0.04"	0.1"	Snow Only 1.5"	Thun. Stms.	Max. Inches in 24 Hrs.	WIND VELOCITY 18 M.P.H.	30 M.P.H.	VISIBILITY ½ MILE
Jan.	35	28	48	1	0	26	0	85	83	1.7	—	10	6	—	0	1.1			
Feb.	34	27	50	−9	0	25	—	82	79	1.1	—	8	4	—	0	0.9			
March	37	29	61	3	0	28	0	82	75	1.2	—	8	4	—	0	0.9			
April	44	33	73	5	0	20	0	79	71	1.4	—	6	5	—	0	1.0			
May	54	41	81	27	0	4	0	72	65	1.1	—	6	4	—	1	0.9			
June	62	48	88	30	0	—	0	72	66	1.4	0	6	4	0	1	1.5			
July	67	55	86	39	0	0	0	76	71	2.0	0	7	6	0	2	3.5			
Aug.	65	54	87	39	0	0	0	78	71	2.7	0	9	7	0	2	2.3			
Sept.	58	48	82	30	0	1	0	82	73	1.7	0	8	5	0	2	2.1			
Oct.	50	41	70	19	0	4	0	83	77	1.9	—	10	6	—	0	1.6			
Nov.	42	35	59	10	0	10	0	84	81	2.1	—	11	6	—	0	1.1			
Dec.	37	30	52	5	0	18	0	84	84	2.0	—	11	6	—	0	1.1			
Year	49	39	88	−9	0	135	0+	80	75	20.0	—	100	61	—	8	3.5			

Elevation 141 Feet
Latitude 57° 39' N   Longitude 18° 20' E
Total precipitation means the total of rain, snow, hail, etc. expressed as inches of rain.
Ten inches of snow equals approx. 1 inch of rain.
Temperatures are as ° F.

Wind velocity is the percentage of observations greater than 18 and 30 M.P.H. Number of observations varies from one per day to several per hour.
Visibility means number of days with less than ½-mile visibility.
Note: "—" indicates slight trace or possibility blank indicates no statistics available

WATERFORD

IRELAND

Month	Daily Average Max.	Daily Average Min.	Extreme Max.	Extreme Min.	No. of Days Over 90°	No. of Days Under 32°	No. of Days 0°	Rel. Humidity A.M. 8:00	Rel. Humidity P.M.	Inches Total Prec.	Inches Snow Only	No. of Days Total Prec. 0.04"	No. of Days 0.1"	No. of Days Snow Only 1.5"	No. of Days Thun. Stms.	Max. Inches in 24 Hrs.	Wind Velocity 18 M.P.H.	Wind Velocity 30 M.P.H.	Visibility ½ Mile
Jan.	47	37	57	16	0	—	0	91		3.7	—	16	9	—		2.2			
Feb.	47	37	59	19	0	—	0	90		3.2	—	11	9	—		2.3			
March	50	38	69	21	0	—	0	89		2.7	—	12	7	—		1.5			
April	54	41	70	26	0	—	0	83		2.5	—	10	7	—		1.3			
May	59	45	77	31	0	—	0	83		2.3	0	11	6	0		1.5			
June	64	50	81	37	0	0	0	82		2.6	0	10	7	0		1.5			
July	67	53	86	41	0	0	0	84		3.2	0	12	8	0		2.2			
Aug.	67	53	82	40	0	0	0	87		3.8	0	12	9	0		2.5			
Sept.	63	50	76	34	0	0	0	89		2.7	0	10	7	0		2.3			
Oct.	57	45	72	28	0	—	0	91		3.9	—	13	9	—		1.8			
Nov.	51	40	63	21	0	—	0	92		3.7	—	14	8	—		2.2			
Dec.	48	38	59	17	0	—	0	91		4.6	—	15	10	—		1.7			
Year	56	44	86	16	0	0+	0	88		39.0	0+	146	96	0+		2.5			

Elevation 136 Feet
Latitude 52° 16' N   Longitude 7° 7' W

Total precipitation means the total of rain, snow, hail, etc. expressed as inches of rain.

Ten inches of snow equals approx. 1 inch of rain.
Temperatures are as ° F.

Wind velocity is the percentage of observations greater than 18 and 30 M.P.H. Number of observations varies from one per day to several per hour.

Visibility means number of days with less than ½-mile visibility.

Note: "—" indicates slight trace or possibility blank indicates no statistics available

WICK (United Kingdom) SCOTLAND

Month	Daily Average Max.	Daily Average Min.	Extreme Max.	Extreme Min.	Over 90°	Under 32°	0°	R.H. A.M. 7:00	R.H. P.M. 1:00	Prec. Total	Prec. Snow Only	Days Total Prec. 0.01"	Days Total Prec. 0.1"	Days Snow Only 1.5"	Thun. Stms.	Max. in 24 Hrs	Wind 18 M.P.H.	Wind 30 M.P.H.	Visibility ½ MILE
Jan.	42	35	56	8	0	7	0	90	88	2.9	—	21	8	—	0.1	1.1	32	12	0.3
Feb.	43	35	58	9	0	7	0	91	87	2.0	—	19	6	—	0.0	0.9	31	7	0.3
March	45	36	62	14	0	7	0	90	83	1.8	—	20	6	—	0.1	0.8	32	7	0.5
April	48	38	67	19	0	3	0	89	81	2.0	—	19	6	—	0.0	2.3	25	6	0.8
May	51	42	69	25	0	1	0	89	83	1.8	—	16	6	—	0.5	1.1	12	2	2.2
June	56	46	80	30	0	—	0	88	81	2.0	0	17	6	0	0.5	1.7	8	—	3.1
July	59	50	80	34	0	0	0	92	85	2.6	0	18	7	0	0.7	1.0	6	0	3.5
Aug.	59	50	78	31	0	—	0	92	86	2.6	0	19	7	0	0.4	1.2	7	1	3.9
Sept.	57	47	75	29	0	1	0	92	83	2.9	0	20	7	0	0.1	1.5	15	1	2.5
Oct.	52	43	70	21	0	3	0	90	82	3.1	—	21	8	—	0.3	1.1	24	6	0.0
Nov.	47	39	62	12	0	5	0	89	85	3.1	—	22	8	—	0.0	0.9	27	8	0.0
Dec.	44	37	58	10	0	5	0	89	87	3.0	—	22	8	—	0.0	1.1	32	13	0.1
Year	50	42	80	8	0	34	0	90	84	30.0	0+	234	83	0+	2.7	2.3	21	5	17.0

Elevation 127 Feet
Latitude 58° 27' N   Longitude 3° 5' W
Total precipitation means the total of rain, snow, hail, etc. expressed as inches of rain.
Ten inches of snow equals approx. 1 inch of rain.
Temperatures are as °F.

Wind velocity is the percentage of observations greater than 18 and 30 M.P.H. Number of observations varies from one per day to several per hour.
Visibility means number of days with less than ½-mile visibility.
Note: "—" indicates slight trace or possibility blank indicates no statistics available

WIESBADEN   (West)   FED. REP. GERMANY

| Month | TEMPERATURES—IN °F | | | | | | | RELATIVE HUMIDITY | | PRECIPITATION | | | | | | WIND VELOCITY | | VISI-BILITY |
| | Daily Average | | Extreme | | No. of Days | | | | | Inches | | No. of Days | | | Max. Inches in 24 Hrs. | | | ½ MILE |
	Max.	Min.	Max.	Min.	Over 90°	Under 32°	0°	A.M.	P.M.	Total Prec.	Snow Only	Total Prec. 0.1"	Snow Only 1.5"	Thun. Stms.		18 M.P.H.	30 M.P.H.	
Jan.	37	28	56	3	0	20	0	—	82	1.4	4	5	8	—		4	—	8
Feb.	39	28	63	−3	0	17	—	—	79	1.2	4	4	6	—		4	—	6
March	50	35	71	9	0	11	0	—	72	1.1	1	3	4	—		4	0	4
April	58	42	84	28	0	1	0	—	67	1.0	—	3	1	1		3	0	1
May	65	48	89	33	0	0	0	—	65	1.5	0	5	1	3		3	—	1
June	72	54	92	38	1	0	0	—	68	2.0	0	6	1	5		1	0	1
July	74	57	99	45	—	0	0	—	70	1.9	0	6	1	6		1	0	1
Aug.	72	56	93	42	—	0	0	—	72	2.5	0	7	1	4		1	0	1
Sept.	68	52	93	36	—	0	0	—	74	1.5	0	5	4	1		1	—	4
Oct.	57	44	77	27	0	1	0	—	81	1.6	0	5	11	—		—	0	11
Nov.	46	37	62	15	0	6	0	—	84	1.2	1	4	9	0		2	0	9
Dec.	40	32	60	10	0	15	0	—	85	1.6	3	6	9	—		2	—	9
Year	57	43	99	−3	1	70	0+	—	75	18.5	13	59	56	20		2	—	56

Elevation 460 Feet

Latitude 50° 2' N   **Longitude 8° 19' E**

Total precipitation means the total of rain, snow, hail, etc. expressed as inches of rain.

Ten inches of snow equals approx. 1 inch of rain.

Temperatures are as ° F.

Wind velocity is the percentage of observations greater than 18 and 30 M.P.H. Number of observations varies from one per day to several per hour.

Visibility means number of days with less than ½-mile visibility.

Note: "—" indicates slight trace or possibility blank indicates no statistics available

YORK (United Kingdom) ENGLAND

Month	TEMPERATURES—IN °F Daily Average Max.	Min.	Extreme Max.	Min.	No. of Days Over 90°	Under 32°	0°	RELATIVE HUMIDITY A.M. 9:00	P.M. 9:00	PRECIPITATION Inches Total Prec.	Snow Only	No. of Days Total Prec. 0.01"	0.1"	Snow Only 1.5"	Thun. Stms.	Max. Inches in 24 Hrs.	WIND VELOCITY 18 M.P.H.	30 M.P.H.	VISI-BIL-ITY ½ MILE
Jan.	43	34	59	7				88	87	2.4		18				1.0			
Feb.	45	35	61	15				85	85	1.7		15				1.6			
March	49	36	73	9				80	82	1.3		13				1.1			
April	54	40	75	26				73	79	1.7		14				0.8			
May	61	44	85	30				71	77	2.0		14				2.8			
June	67	50	90	37				70	76	1.9		13				1.3			
July	70	54	89	37				74	77	2.4		14				2.1			
Aug.	69	53	89	39				77	80	2.5		14				2.3			
Sept.	64	49	84	31				80	83	2.1		13				1.9			
Oct.	56	44	75	23				84	85	2.2		16				1.3			
Nov.	48	38	66	18				87	87	2.5		18				1.0			
Dec.	44	36	58	19				88	88	2.0		17				0.9			
Year	56	43	90	7				80	82	25.0		179				2.8			

Elevation 57 Feet
Latitude 53° 57' N   Longitude 1° 5' W
Total precipitation means the total of rain, snow, hail, etc. expressed as inches of rain.
Ten inches of snow equals approx. 1 inch of rain.
Temperatures are as ° F.

Wind velocity is the percentage of observations greater than 18 and 30 M.P.H. Number of observations varies from one per day to several per hour.
Visibility means number of days with less than ½-mile visibility.
Note: "—" indicates slight trace or possibility blank indicates no statistics available

ZARAGOZA

SPAIN

Month	Daily Average Max.	Daily Average Min.	Extreme Max.	Extreme Min.	No. of Days Over 90°	No. of Days Under 32°	No. of Days 0°	Relative Humidity A.M. 7:00	Relative Humidity P.M. 1:00	Precipitation Inches Total Prec.	Precipitation Inches Snow Only	No. of Days Total Prec. 0.004"	No. of Days Total Prec. 0.1"	No. of Days Snow Only 1.5"	Thun. Stms.	Max. Inches in 24 Hrs	Wind Velocity 18 M.P.H.	Wind Velocity 30 M.P.H.	Visibility ½ Mile
Jan.	49	35	69	5	0	9	0	87	71	0.6	—	4	2	—	—	1			
Feb.	54	37	72	16	0	5	0	84	62	0.8	1	5	3	—	0	1			
March	60	41	79	23	0	1	0	80	55	0.9	0	7	3	0	—	1			
April	66	46	89	30	0	—	0	77	52	1.1	0	7	4	0	1	2			
May	73	52	96	36	2	0	0	74	51	1.6	0	8	5	0	4	2			
June	81	58	105	46	6	0	0	73	48	1.1	0	5	3	0	5	3			
July	87	62	104	46	12	0	0	71	46	1.8	0	3	2	0	5	5			
Aug.	86	62	104	49	10	0	0	75	47	0.6	0	3	2	0	4	2			
Sept.	79	58	96	40	4	0	0	80	54	1.1	0	5	4	0	5	3			
Oct.	69	50	88	30	0	—	0	83	59	1.4	0	6	5	0	1	2			
Nov.	57	42	75	22	0	2	0	85	64	1.2	0	6	4	0	—	4			
Dec.	50	37	68	16	0	7	0	87	71	0.9	1	7	3	—	—	2			
Year	68	48	105	5	32	24	0	80	57	12.0	2	66	38	—	25	5			

Elevation 844 Feet

Latitude 41° 40' N   Longitude 1° 2' W

Total precipitation means the total of rain, snow, hail, etc. expressed as inches of rain.

Ten inches of snow equals approx. 1 inch of rain.

Temperatures are as ° F.

Wind velocity is the percentage of observations greater than 18 and 30 M.P.H. Number of observations varies from one per day to several per hour.

Visibility means number of days with less than ½-mile visibility.

Note: "—" indicates slight trace or possibility blank indicates no statistics available

ZERMATT — SWITZERLAND

Month	TEMPERATURES—IN °F Daily Average Max.	Min.	Extreme Max.	Min.	No. of Days Over 90°	Under 32°	0°	RELATIVE HUMIDITY A.M. 7:00	P.M. 1:00	PRECIPITATION Inches Total Prec.	Snow Only	No. of Days Total Prec. 0.01"	0.1"	Snow Only 1.5"	Thun. Stms.	Max. Inches in 24 Hrs.	WIND VELOCITY 18 M.P.H.	30 M.P.H.	VISIBILITY ½ MILE
Jan.	29	17	50	−10	0	—	—	67	48	1.6	—	7	5	—	0	1.4			
Feb.	34	19	51	−10	0	—	—	66	44	1.7	—	7	6	—	0	1.9			
March	40	24	58	−3	0	—	—	68	42	2.1	—	8	6	—	0	2.5			
April	47	32	68	14	0	—	0	68	41	2.1	—	9	6	—	0	1.7			
May	55	41	78	19	0	—	0	68	41	2.7	—	11	7	—	—	1.6			
June	63	48	85	31	0	0	0	68	40	2.3	0	11	6	0	—	1.7			
July	66	51	84	34	0	0	0	71	41	2.2	0	10	6	0	1	2.0			
Aug.	65	47	84	34	0	0	0	77	43	2.7	0	11	7	0	1	1.5			
Sept.	60	42	77	26	0	—	0	79	44	2.3	—	9	6	—	—	3.2			
Oct.	51	35	73	13	0	—	0	77	45	2.5	—	9	7	—	—	3.1			
Nov.	40	27	63	5	0	—	0	71	47	1.9	—	8	6	—	0	2.2			
Dec.	31	20	51	−6	0	—	—	68	49	1.9	—	8	6	—	0	1.9			
Year	48	34	85	−10	0	0+	0+	71	44	26.0	—	108	73	—	2+	3.2			

Elevation 5,282 Feet

Latitude 46° 1′ N   Longitude 7° 45′ E

Total precipitation means the total of rain, snow, hail, etc. expressed as inches of rain.

Ten inches of snow equals approx. 1 inch of rain.

Temperatures are as ° F.

Wind velocity is the percentage of observations greater than 18 and 30 M.P.H. Number of observations varies from one per day to several per hour.

Visibility means number of days with less than ½-mile visibility.

Note: "—" indicates slight trace or possibility blank indicates no statistics available